Molecular Technologies for Detection of Chemical and Biological Agents

NATO Science for Peace and Security Series

This Series presents the results of scientific meetings supported under the NATO Programme: Science for Peace and Security (SPS).

The NATO SPS Programme supports meetings in the following Key Priority areas: (1) Defence Against Terrorism; (2) Countering other Threats to Security and (3) NATO, Partner and Mediterranean Dialogue Country Priorities. The types of meetings supported are generally "Advanced Study Institutes" and "Advanced Research Workshops". The NATO SPS Series collects together the results of these meetings. The meetings are co-organized by scientists from NATO countries and scientists from NATO's "Partner" or "Mediterranean Dialogue" countries. The observations and recommendations made at the meetings, as well as the contents of the volumes in the Series, reflect those of participants and contributors only; they should not necessarily be regarded as reflecting NATO views or policy.

Advanced Study Institutes (ASI) are high-level tutorial courses to convey the latest developments in a subject to an advanced-level audience.

Advanced Research Workshops (ARW) are expert meetings where an intense but informal exchange of views at the frontiers of a subject aims at identifying directions for future action.

Following a transformation of the programme in 2006, the Series has been re-named and re-organised. Recent volumes on topics not related to security, which result from meetings supported under the programme earlier, may be found in the NATO Science Series.

The Series is published by IOS Press, Amsterdam, and Springer, Dordrecht, in conjunction with the NATO Emerging Security Challenges Division.

Sub-Series

A.	Chemistry and Biology	Springer
B.	Physics and Biophysics	Springer
C.	Environmental Security	Springer
D.	Information and Communication Security	IOS Press
E.	Human and Societal Dynamics	IOS Press

http://www.nato.int/science
http://www.springer.com
http://www.iospress.nl

Series A: Chemistry and Biology

Molecular Technologies for Detection of Chemical and Biological Agents

edited by

Joseph H. Banoub
Fisheries and Oceans Canada, Science Branch, Special Projects,
Chemistry Department, Memorial University,
St John's, NL, Canada

and

Richard M. Caprioli
Department of Biochemistry, Mass Spectrometry Research Center,
Vanderbilt University School of Medicine,
Nashville, TN, USA

Springer

Published in Cooperation with NATO Emerging Security Challenges Division

Proceedings of the NATO Advanced Study Institute on Molecular
Technologies for Detection of Chemical and Biological Agents
Calabria, Italy
9–16 April, 2016

Library of Congress Control Number: 2017944715

ISBN 978-94-024-1114-0 (PB)
ISBN 978-94-024-1112-6 (HB)
ISBN 978-94-024-1113-3 (e-book)
DOI 10.1007/978-94-024-1113-3

Published by Springer,
P.O. Box 17, 3300 AA Dordrecht, The Netherlands.

www.springer.com

Printed on acid-free paper

Preface

Mass spectrometry and allied topics represent the main lectures given to the NATO-ASI (Advanced Study Institute) entitled "Molecular Technologies for the Detection of Chemical and Biological Agents", which was held at the Hotel Village La Principessa, Campora San Giovanni, Calabria, Italy, from April 9 to April 16, 2016. This ASI was a great success from both scientific and organizational points of view.

First of all, this NATO-ASI would have not been possible without the excellent help and coordination of the NATO Science Committee of the Science for Peace and Security (SPS) Programme. Special thanks are due to Professor El Mokhtar Essassi (Mohamed V University) codirector of NATO-ASI for his contribution. We were also extremely lucky to have an excellent organizing committee that simplified the tasks associated with arranging the NATO-ASI. In addition, the excellent efforts of the members of this committee, namely, Professors Richard M. Caprioli (Vanderbilt University), Giovanni Sindona (University of Calabria) and Gianluca Giorgi (University of Siena), are sincerely acknowledged.

We are particularly indebted to Professor Caprioli who was instrumental in inviting the best lecturers and securing funding from the manufacturers and to Ms. Maureen Casey, for her superb organizational skills.

We were privileged and honoured to have as lecturers for the NATO-ASI a succession of world-renowned scientists, namely, Professor Richard M. Caprioli; Stanford Moore, professor of biochemistry, director of the Mass Spectrometry Research Center and editor-in-chief of the *Journal of Mass Spectrometry* from Vanderbilt University; Professor Eric P. Skaar Department of Pathology, Immunology, and Microbiology, Vanderbilt University Medical Center; Professor Guenter Allmaier, Institute of Chemical Technologies and Analysis, Vienna University of Technology; Professor Mark Duncan of the School of Medicine, University of Colorado, USA; Professor Gianluca Giorgi, University of Siena; Professor Giovanni Sindona, University of Calabria; Professor Marc JF Suter from EAWAG, Swiss Federal Institute of Aquatic Science and Technology, Dübendorf, Switzerland; Professor Jeffrey C. Smith University of Carlton, Ottawa; and Professor Joseph H. Banoub, DFO-MUN, ST John's, NL. Canada.

The mission of NATO-ASI was to bring together leading experts in the fields of modern molecular analyses and in the biological effects of chemical and biological warfare agents. The topics covered included the detection of chemical and biological agents with state-of-the-art sensors and various mass spectrometric technologies; the effect of exposure of various agents such as toxins, bacteria, viruses, and chemical agents; and the description and protocols of new innovative molecular detection technologies and assessment of the effects of exposure on short-term and long-term health. This meeting also presented the fundamental aspects and applications of molecular technologies, especially mass spectrometry, for the detection of chemical and biological agents and for assessment of their effects on human cells.

The mission of NATO-ASI # G4915 was successfully accomplished, as it brought together leading experts in the fields of modern molecular analyses with those experts in the biological effects of chemical and biological (CB) warfare agents. The attendees were also given an opportunity to present their research through forty 10-min lectures open to all attendees. This was followed by small group discussions. The NATO-ASI meeting was advertised on the Internet and all the lectures were immediately posted online following the meeting. The relaxed atmosphere of this venue stimulated discussions between faculty and trainees concerning additional details of the lecture subjects and advanced protocols and technologies.

A major scientific message conveyed to the NATO-ASI participants was that while there is no single detection method that provides rapid and accurate detection of all CB agents, we (NATO countries) must keep abreast of the powerful new mass spectrometry state-of-the-art detection technologies that can help protect the public from chemical and biological threats.

St John's, NL, Canada Joseph H. Banoub
Nashville, TN, USA Richard M. Caprioli
Agdal, Rabat, Morocco El Mokhtar Essassi

Contents

Chapter 1
Immunorecognition of Biological Agents: An Introduction to Immunology

Alessia Ciafarone, Ilaria Bologna, Argilia D'Amico, and Mauro Bologna

Abstract The main mechanims and concepts regulating the functioning of the human immune system are illustrated, with the general aims of improving disease prevention and ecosystem dynamics knowledge. These concepts must be known for the best management of infectious diseases and in the paramount view of ecology and parasitism, which are at the basis of stage-sharing of the environment by different species of life, each defending its own uniqueness and biological existence. Man is only one of the many living species sharing the stage of the planet. These considerations are fundamental in the fields of infectious diseases and of bioterrorism which have to be efficiently contrasted.

Keywords Immunology • Parasitism • Biological defense mechanisms • Bioterrorism

1.1 Introduction

The far reaching aims of this NATO-Advanced Study Institute course allowed us to delineate a very wide landscape of scientific perspectives. Therefore, for the teaching purposes set here, we are sharing some of our studies in general immunology and some scientific and philosophical considerations concerning biological agents living in the environment and their relations with humans, in the very wide endeavours of ecological relationships, parasitism, immunological defenses and infectious

A. Ciafarone • A. D'Amico
Biologist, Post-Doctoral Fellow-Co-worker, Chair of General Pathology, L'Aquila, Italy

I. Bologna
Specialist in Occupational Medicine, MD, L'Aquila, Italy

M. Bologna (✉)
Professor of General Pathology, Chair of General Pathology, Department of Life, Health and Environmental Sciences, University of L'Aquila, 67100 L'Aquila, Italy
e-mail: mauro.bologna@univaq.it

J.H. Banoub, R.M. Caprioli (eds.), *Molecular Technologies for Detection of Chemical and Biological Agents*, NATO Science for Peace and Security Series A: Chemistry and Biology, DOI 10.1007/978-94-024-1113-3_1

disease mechanisms. All these concepts must be studied not only for the biological survival and defense of each individual in the environment but also for social prevention purposes, in the event of criminal use of biological agents (bioterrorism) aimed at harming human populations in different times and different geographical sites.

1.2 Microbes and Humans Interacting: Defense Strategies of Multicellular Live Beings

Life began on Earth *circa* 3.5 billion years ago with unicellular beings: primitive bacteria and algae (*Archean Age*) See: Archea, Wikipedia (accessed 1 Feb 2017) and De Duve [1]. Biological evolution of species progressively produced today's forms of life, as we observe them. Millions of species co-exist today and share the stage (biosphere). Evolution soon registered colonies of unicellular beings, with the advantage of wider and specialized cellular functions but with the problem of preventing undesirable aggregations (casual or purposely parasitic and potentially dangerous situations anyway). Consequently, the need of surveillance by an efficient "immune system" is always present. The basic need of this surveillance system was (and always is) to preserve the genetic integrity of the individual being and the survival of the colony (multicellular entity): any novelty cannot be "tolerated" and regular cells ("self") must be quickly recognized as such, while stranger cells ("not self") must be eliminated (rejected) efficiently [2, 3].

In the evolutionary enactment of these needs, anatomical barriers developed, like surface structures (skin, barks, cuticles) and recognition systems (surface molecules, histocompatibility antigens, intercellular communication and collaboration/ rejection functions) became very sophisticated, demanding and by necessity error-free. For a general description of the immune system, we recommend reading appropriate textbooks or specialty reference manuals [2, 4]. We can mention in this respect a recent introductory manual by Actor [3]; this book had also an Italian Edition by Pearson Publisher, 2015, "Introduzione all'immunologia", cared by our scientific team.

In presence of the enormous variety of live beings coexisting nowadays on Earth, humans (we) pretend to have absolute priority, but we have to consider that we share the stage in such a crowded environment on this planet. Therefore, we should adhere to the following principles: learn, respect, understand and prevent. LEARN the biological requirements of all the life forms, RESPECT the right of any form of life for existence and prosperity, UNDERSTAND the logic of evolution and of biological mechanisms and PREVENT the possibilities of harm and disease for humans and other animals and plants of interest, due to environmental changes (mostly man-determined).

On this planet there are millions of different species of living beings, in variable proportions in the biosphere and in different ecosystems: they are all co-existing, interacting and competing for food and survival. This encompasses the very universal phenomena of competition and of parasitism [2].

One of the most recent and precise evaluations of the number of species existing on planet earth, but still very approximate and provisional (well-illustrated and summarized in National Geographic, april 2013), finds evidence of more than five thousand species of mammals, ten thousand species of birds, twelve thousand species of reptiles, fifteen thousand species of amphibians, forty-five thousand species of fish, one hundred and fifty thousand species of crustaceans, two hundred thousand species of mollusks, six hundred thousand species of aracnids, five million species of insects and many, many millions (inestimable indeed) species of bacteria, viruses and other microrganisms. Are we numerous on Earth? Is there enough work for the immune system of each living multicellular organism to distinguish "self" from possibly harmful "not self"?

Species of microrganisms ascertained as pathogenic for humans are indeed a very small fraction of the existing ones: we come to know them better because we study the diseases connected with each of the major pathogens, but we ignore a lot about the greater number of other, presumably innocuous species of microrganisms.

In general, different species interacting may set a parasitic relationship, in which the larger biological entity (*the host*) may receive harm (food loss or disease) and the smaller one (*the parasite*) may get advantages (more food, protection). Both must preserve their identity and prevent contamination by foreign genetic material (immunologic surveillance, bilaterally). Some parasites can even live within the hosts (endoparasitism), like some bacteria and all viruses.

Three types of interactions may occur between a microrganism and a human host: (a) symbiotic relationship, in which the microrganism and the host both benefit; (b) commensal relationship, in which the microrganism gains but the host suffers no harm; and (c) a true parasytic relationship, in which the microrganism gains and the host is harmed (with or without disease) [2].

Symbiosis offers frequently mutual advantages and remains very stable in time. Pathogens (as already mentioned) are a minimal proportion of the existing microrganisms. We humans host some advantageous bacterial populations (in the intestine, on the skin, with commensal germs in the mouth and in genital mucosae): we indeed are also *made of the germs living in/on our body. Indeed,* only one cell out of ten in our body is a human cell: the rest are bacteria (the so-called microbiome). Among these, we count billions of bacteria in the intestines, definitely useful for many functions (vitamin production, competition with true pathogens, contribution to metabolism, etc) [4].

The balance between host and parasites depends on two basic forces, an aggressive force by the parasite depending on survival/proliferation/invasion capacity of the parasite itself and a defensive force by the host, depending on the immune mechanisms (phagocytosis, cellular and humoral immune reactions). In this balancing of opposite forces the parasitic relationship is played by the contendents. *If we have marked prevalence of the parasite*, we may have disease (and eventually death) of the host, but *if we have definite prevalence of the host defences,* we may have control (and eventually elimination) of parasites. For a wider general illustration of these concepts see Abbas et al. [2] and also Bologna and Lepidi [5].

1.3 Immunology: Origins and Development of a Still Young Discipline

Immunology is a rather young science (a little more than 60 years old, with this explicit name) although the first informations on the existence and validity of immune defences go back indeed to Edward Jenner with his "vaccination" practices in the 1790s and probably to some older but analogous Chinese medicine practices [6].

The first "immunological" experiment by Jenner is something that would not be ethically feasible today by any medical deontological rules! He started from the observation that English milkmaids who caught cowpox (a benign form of skin eruptions from contacts with diseased cows) did not develop the human form of the smallpox disease; therefore, he voluntarily and deliberately exposed his gardener's son (James Phipps) to biological materials from cowpox pustules (causing a fever illness in the recipient); after recovery, Jenner challenged the boy with human smallpox, verifying his attained "immunity" also to the human disease [3].

Jenner is therefore considered the father of experimental "immunology". No mechanistic explanations of that experiment were however possible at that time: only about 100–150 years later we started discovering antibodies and immune system functions allowing us to understand what was biologically happening at that time in the milkmaids (and at any other time in "immune" individuals) and in the "vaccinated" individuals like Jenner's gardener's son.

All the developing "immunological" science, enriched later through microbiology, biochemistry, physiology and pathology studies in the following decades, recorded more and more details in the functioning of the defense mechanisms of mammals and humans, having the most complex and efficient forms of immunological defense against foreign agents entering the body.

The immune system is in fact like an "eye within" the body, controlling that nothing extraneous is biologically active in each individual organism, and therefore, recognizing effectively "self" from "not-self", so that replication of self-cells are not contaminated by "foreign" biological agents and that any extraneous biological entity will be eliminated by soluble "light" weapons (antibodies) and by "heavy" killers (immune cells and macrophages, interacting and collaborating in the task).

The key steps of an immune reaction are substantially three: (a) internalization of foreign particles by macrophages (MPH) degrading and "presenting" them to T-Helper lymphocytes (TH-cells) which in turn start an elaborated attack, mostly through either (b) synthesis of soluble "light" weapons (antibodies, humoral response by B-cells) or (c) development of cellular "heavy" weapons (killer T-lymphocytes T-K cells, with cellular or cytotoxic response); in many cases we observe the activation of both responses (b + c), depending on the needs. The key steps of an immune reaction are illustrated in the following figure.

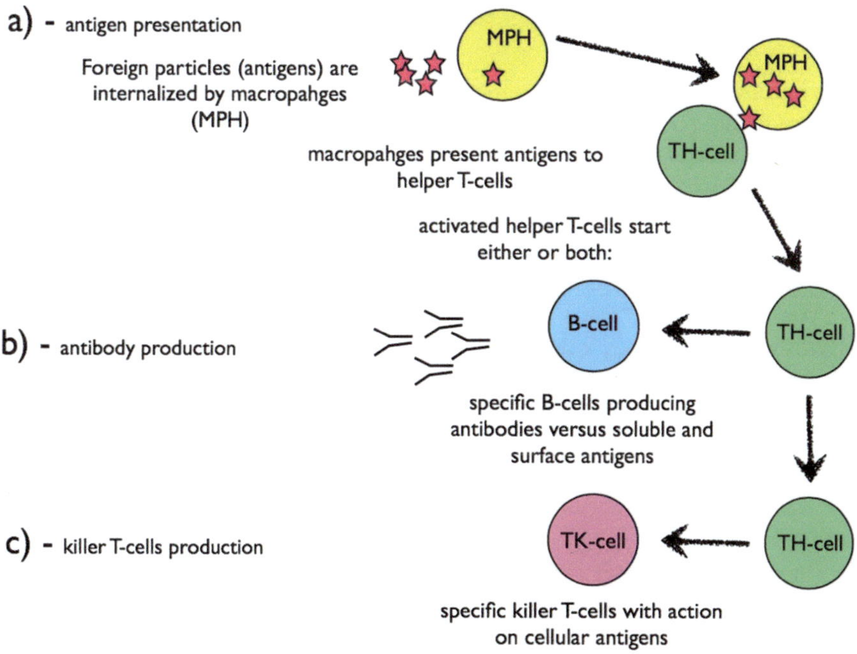

Recently, it has been shown that activated T-cells tend to aggregate, like a swarm of bees, exchanging informations, useful to coordinate the immune response (i.e. to elaborate coordinated defence plans): this is the visual demonstration of the complex molecular dialogue taking place within the various families of immune cells collaborating to the full enactment of defence mechanisms.

When invaders are present in body fluids or in the extracellular domains, like most bacteria, they can be attacked by antibodies, through specific surface recognition. However, when the invaders are instead of the intracellular type, like viruses and some bacteria (like TBC), they must instead be attacked by special killer lymphocytes (cytotoxic TK-cells) which destroy all the self-cells harboring the intruders, together with their content. Nevertheless, viruses anyway can also transit the body fluids; consequently also antibodies are produced against them.

In most responses, both humoral (antibodies) and cellular attacks are deployed. The immune response, moreover, is specific and potentiated by memory: therefore a second (or further) encounter with the same foreign agent (antigen) produces a stronger and quicker defense reaction.

Details of the immune mechanisms are continuously discovered. So this discipline is still far from being completely described and understood. New details are continuously added and better focused by immunology researchers in time [2].

1.4 Microbes and Their Attack Strategies

If germs are successful to penetrate the body, they enter a hostile territory, where they try to survive through strategic actions: rapid proliferation, identity disguise, toxic products (toxins), resistance to attacks, niche hiding, etc. If microbes prevail, they can multiply and pass to other individuals. The capacity to hide from immune attacks is exemplified by *Listeria monocytogenes*, which is able to escape vacuoles when internalized. The bacterium so hides and passes from cell to cell by endocytosis going often undetected by immune reactions and unreachable by antibodies [7].

Also tuberculosis (TBC) represents a great challenge, because the organism (*Mycobacterium tuberculosis*) can survive even after it has been internalized (phagocytosed) by macrophages. The mycobacteria can indeed hide inside the bone marrow stem cells and evade the immune reactions for many years.

But fortunately, highly pathogenic germs are very few, considering that there are millions of species of microorganisms in nature. Nevertheless, most of these non-pathogenic organism species to human are easily destroyed by the *Homo sapiens* immune system [3].

The very limited number of effective pathogens is characterized by special properties making them resistant to the immune attacks. Very virulent germs tend however to disappear in the ecosystems, because they kill their sensitive hosts (natural selection): the killer usually disappears together with the killed. This indeed is part of the natural selection of life forms able to coexist and collaborate (through non-lethal parasitic relationships). Therefore, life on Earth continues to thrive and evolve [5].

Peace and recyprocal toleration and coexistance are therefore favoured by natural selection: we may say, the biological strength of peace and cooperation (even before and without human knowledge)!!!

The attack strategies by germs are mostly implemented by their rapid proliferation, which is a key factor for microrganisms: immune response indeed requires some time (7–10 days) to be strong and effective. Another feature of attack by germs is their capacity to elaborate toxins (blocking key functions of the host): movement (tetanus), water balance (cholera), etc.

More strategies are immune evasion (antigenic change - flu viruses) (Bologna, 2010), complement inhibition, resistance to phagocytosis, or even TLy elimination (HIV). The competition grounds are therefore many and variegated [2, 3].

1.5 Prevention to Favour Defence and Immunity

To control infections and favor immune responses, therefore, we may generally try:

(a) To reduce microrganism proliferation (through the use of bacteriostatic drugs) or kill at least partially the germs (disinfection, use of bactericidal drugs);
(b) To control the vectors diffusing the infection (insects, arthropods, birds, rats, etc.); this action implies vast territory and international cooperation plans;

(c) To immunize preventively the potential hosts (vaccinations): this action requires public health planning, technology, costs, cooperation plans and time;
(d) To administer preformed antibodies (serotherapy): this also requires technology, health system costs and availability of sera (biotechnology production facilities).

Actions against germs imply the use of chemotherapeutic agents (antibiotics: bacteriostatic or bactericidal) having a multitude of action mechanisms and being suitable under precise conditions; germs tend to develop resistance to them; fewer and fewer effective antibiotics are available today, while the discovery of new molecules in this category has been scarce, lately.

Hygienic measures are very effective, well known, applicable easily almost everywhere and can be enacted at low cost, whenever possible.

Campaigns against vectors (insecticides, biological competitors - fungi, bats, genetic manipulation of vectors) can also be very effective (if they are done with a good biological and environmental knowledge of the ecological systems in play).

The consensuses on the best possible actions are mostly related to effective vaccination campaigns (if specific and effective vaccines are available and if time allows) and immune sera administration (when available and if time is critical for the disease control).

A vaccine is generally preferable, because it induces an active response in the recipient and an advantageous state of immune memory in the treated population, with minimal side effects (active immunization). A serum specific for a given antigen can be life-saving (serotherapy, immediately active), but has some side effects (heterologous proteins –horse immunoglobulins in humans are antigenic-) and does not have a lasting protection (passive immunization).

Vaccinations have resolved many infectious diseases in recent times, like diphteria, polyomielitis, pertussis, measles, (that are no longer deadly threats in many areas of the world). Vaccination against smallpox (variola) has actually extinguished the disease from the planet Earth. Unfortunately, this fact anyway presents a very dangerous problem for future potential bioterrorism actions (as we will discuss in a separate chapter in this book).

Moreover, modern life has created some extra occasions for germs and some “new” diseases: air conditioning apparatuses for instance (if not properly cared and cleaned) are a new ideal environment for bacteria; because of dirt and humidity they can foster growth of airborne bacteria never seen before as human pathogens: *Legionella pneumophyla*. The story of Legionaries disease (an often fatal lung infection by *Legionella*) is very instructive for microbiologists and epidemiologists: every scholar in this field should study the case [8, 9].

Human behaviours (sexual intercourses, exchange of syringes among intravenous drug users and frequency of air travel) have extended the contagion of formerly rare infections like HIV: also in this case the facts are dramatically instructive for medicine and epidemiology. Transfusions of unscreened blood have also diffused hepatitis C (HepC) and HIV viruses. Centralized processing of foods has sometimes diffused a contagion of food-borne infections (*E.coli*).

Airplanes have replaced ships for human travel and are at the center of attention for human communicable diseases spreading (Influenza, SARS, etc.). Many more people are traveling today to remote and tropical areas than before (forests, wilderness): this can expose more populations (even at home, on the return) to rare insects and microbes (see cases of malaria, Ebola virus, Marburg virus, etc.). In addition, economic development expands contacts: mining, forestry, agriculture in new tropical areas with recent deforestation.

And more, the increasing number of subjects with immunodeficiency diseases or immunosuppressant therapies (for transplants) increases the probability of new communicable agents to infect people, survive, and propagate in modern societies (opportunistic infections, with possible mutations in progress).

This entire panorama increases the variety of new emerging infections: here is a variety of bacterial and viral pathogens of recent discovery: for bacteria we may cite *Criptosporidium parvum*, *Legionella pneumophyla*, *Campylobacter jejuni*, *Helicobacter pylori*, and for the viruses we may cite Rotavirus, Ebola virus, Zika virus, Hantaan virus, HTLV virus, HIV virus, Herpesvirus-6 and -8, virus Guanarito, virus Sabia, nCoV-MERS virus, to name just some of the most relevant and recent ones [10, 11].

Therefore, we have a lot to learn about new infections and a lot to do to control communicable diseases, also because of the changing human ecosystem.

References

1. de Duve C (1995) The beginnings of life on earth. American Scientist. September–October
2. Abbas A, Lichtman AH, Pillai S (2014) Cellular and molecular immunology, 8th edn, Saunders, ISBN: 9780323316149
3. Actor J (2014) Introductory immunology. 1st Edition - basic concepts for interdisciplinary applications. Academic Press, ISBN: 9780124200302
4. Lynch SW, Pedersen O (2016) The human intestinal microbiome in health and disease. Phimister EG, editor. N Engl J Med 375(24):2369–2379
5. Bologna M, Lepidi A (2010) Pandemie – virologia, patologia e prevenzione dell'influenza. Bollati Boringheri, ISBN 9788833920375, www.bollatiboringhieri.i
6. Bottaccioli F (2002) Il sistema immunitario: la bilancia della vita – Com'Ã¨ fatto, come funziona in salute e in malattia. Tecniche Nuove, ISBN 8848109462, www.tecnichenuove.com
7. Soares da Costa TP, Nanson JD, Forwood JK (2017) Structural characterisation of the fatty acid biosynthesis enzyme FabF from the pathogen *Listeria monocytogenes*. Sci Report 7:39277
8. Fraser DW et al (1977) Legionnaire's disease – description of an epidemic of pneumonia. N Engl J Med 1:1189–1197
9. Jain S et al (2015) Community-acquired pneumonia requiring hospitalization among U.S. adults. N Engl J Med 373(5):415–427
10. Fauci AS, Moerens DM (2012) The perpetual challenge of infectious diseases. N Engl J Med 366(5):454–461
11. Sundman E, Olofson PS (2014) Neural control of the immune system. Adv Physiol Educ 38(2):135–139

Chapter 2
Vaccines, Sera and "New" Viruses: Ebola, Zika and Other Infectious Challenges for Human Health

Alessia Ciafarone, Argilia D'Amico, and Mauro Bologna

Abstract Following the basic description of the immunological science, we may now go in more detail into the practical applications involving immunological prevention against infectious diseases (vaccine use, requiring availability of suitable antigenic preparations, safe and effective immunization schedules and time to allow for immunity to develop) and immunotherapy of dangerous infective conditions (serotherapy, involving availability of immune sera and their use for injection in subjects needing protection against infectious agents).

All this is discussed with special regard of the many new micro-organisms described in human pathology and also in situations in which the deliberate use of them for biological warfare or related menace (bioterrorism) is realized.

Keywords Immunology • Virus • New infectious diseases • Bioterrorism

2.1 Vaccines and Sera

What is a vaccine? In a dangerous disease caused by micro-organisms and "foreign" not-self substances (an "antigen"), it may be possible to raise a vaccine, that is an innocuous preparation of that "antigen" which can induce the production of an immune response and create immune memory in the subjects we want to protect. A vaccine can be developed in many ways: killed (denatured) or non-replicating pathogens (viruses), recombinant protein antigens, live, attenuated (less harmful) strains of pathogens (for instance cowpox virus to protect against smallpox).

A. Ciafarone • A. D'Amico
Biologist, Post-Doctoral Fellow-Co-worker, Chair of General Pathology, L'Aquila, Italy

M. Bologna (✉)
Professor of General Pathology, Chair of General Pathology, Department of Life, Health and Environmental Sciences, University of L'Aquila, 67100 L'Aquila, Italy
e-mail: mauro.bologna@univaq.it

J.H. Banoub, R.M. Caprioli (eds.), *Molecular Technologies for Detection of Chemical and Biological Agents*, NATO Science for Peace and Security Series A: Chemistry and Biology, DOI 10.1007/978-94-024-1113-3_2

Vaccines can be very effective: cases of smallpox (extinction of the disease: no more cases on earth, for the success of global vaccine campaigns) and of polio (near extinction of disease, except in some nomadic populations and in critical areas of today's world, like Syria and also, surprisingly, remote Chinese provinces) are very instructive on the subject.

What is a serum? If there is no time to confer protection by means of a vaccine (which, when available, requires at least 7–10 days before sufficient amounts of specific antibodies are synthesized by the vaccinated individual) we may protect the subject exposed to the dangerous contagion by means of the serum of individuals (humans or animals) already immunized against the micro-organism or lethal toxicant in action (examples: anti-tetanus serum, anti-snake venom serum, anti-diphtheria serum, etc.).

The theme of parasitism has been treated briefly in the previous chapter by the same authors: so now we will briefly review the old and new micro-organisms posing challenges for human health in relation to Bioterrorism.

2.2 Main Diseases of Interest in the Field

In light of recent concern and interest about the potential for biological terrorism (biowarfare) there are several diseases and bacterial toxins that must be considered in particular, like anthrax [1, 2], smallpox [3, 4], plague [5], botulinum toxin [6], and tularemia [7]. A very detailed discussion of such diseases and other infectious diseases with similar risks in terms of bioterrorism goes beyond the scopes of this concise chapter, but some features of these and other infectious diseases representing important threats in the biowarfare field will be mentioned.

In this respect, we may distinguish in time diseases which are:

- Old diseases which are disappearing and sometimes returning, like smallpox and polio virus infections (which are either extinct or close to be eradicated, thanks to planetary vaccination programs);
- Diseases still active at present times, like carbuncle (anthrax), plague, tularemia, tetanus, botulinum, TBC, etc.;
- New diseases, which are appearing/spreading, like SARS (Severe Acute Respiratory Syndrome) and its more recent variety of MERS (Middle-East Respiratory Syndrome), infections by Ebola/Marburg viruses, Hantavirus, filovirus, novel Flu virus strains, Zika virus, etc.

Now we will summarize the essential facts about some of these diseases.

For a more complete medical reference to all of them, see for instance the Merck Manual of diagnosis and Therapy [8] and recent reviews which appeared on the most recent epidemics and that are listed in the references (9–25).

2.3 Definitions and History of Bioterrorism

Practice of bioterrorism goes back to very remote times of human existence and indeed to pre-historical conflicts between humans, when groups of people devised to deliberately use biological agents for conflict sustainment (damaging weapons) and for propagation of fear in the enemy populations.

Since the early times of combat between humans, some biological means have been used: for instance through the contamination of water wells in conflict areas, tainted with rotting animal remains, or through hunting-fighting arrowheads dipped in toxic plant extracts or venomous substances, before throwing them at the living target.

Usable biological agents for harm are in fact all the pathogens of biological nature, like micro-organisms (bacteria, viruses, fungi, prions), toxins, animal and plant venoms, together with the related carriers (fomites, instruments) or vectors (insects, etc.).

Terrorism is the use of violence to condition societies or governments in their political choices.

Bioterrorism is the use (or menace of use) of biological agents to enact terrorist events and induce generalized *fear* concerning negative consequences in target populations.

Here follow some examples of bioterrorism enacted in various ages:

(a) Use of poison darts/arrows (primitive populations): mostly for hunting, but also for battles against enemies.

From here derived many of the advances of toxicology, the science of toxic substances (the word "toxicology" derives indeed from the Greek words "toxon" = arch, while "farmakon" = poison; toxicon farmakon = poison for arch hunting): from such practice derives the knowledge we have of stricnin, curare, ouabain, aconite, other plant/animal poisons primarily devised for hunting and fighting.

(b) Impingement of darts in decomposing cadavers or putrefaction soil (or manure > > tetanus) before throwing at enemies (New Guinea, tribal combats; Sciites 400 B.C.)
(c) Last but not least, the use of fear that human have toward beasts. Here comes the example of Hannibal (from Carthago), leading the ships of Prussia I, king of Bithynia (West Turkey) in a battle of year 184 B.C. against Eumene II (Attalides, Pergamon); he won that naval battle because he managed to throw canisters full of reptiles at the enemy ships, causing terror and uncoordinated reactions leading to his victory. He therefore used fear as a weapon: snakes were not even harmful (not poisonous), but big was the surprise and reactions were out of control!
(d) In recent history, we can see that First World War (also named the war of Chemistry) contributed to the development and use of Nervine gases, chemical weapons banned everywhere but still existing in some countries. The Second World War (also named the war of Physics) led to the development and use of the atomic bomb.

Peace Treaties: *Geneva Protocol* ruled against chemical weapons (1925) and was followed later by additions concerning bacteriologic war. Not all the states however subscribed it. Though, most world states had banned chemical and bacteriological weapons by 1975.

Today, how scared should we be of biological and chemical terrorism? Well, since these are lethal and cheap weapons, they are of considerable concern. They may be seen as the cheap atomic bomb and represent remarkable threats to peace in local conflicts and in terroristic attacks worldwide. We should all know more on the subject and do extensive prevention.

Albert Einstein once said "I do not know by what weapons the Third World War will be fought, but for sure the Fourth will be fought with stones". Well, we do not want any more World Wars, for sure.

2.4 Main Infectious Agents in Bioterrorism and New Infections

In the following paragraphs, the main infectious agents usable for biological warfare are concisely described, together with some of the most recent biological agents appearing as world epidemics (Ebola and Zika viruses).

2.4.1 Smallpox (Variola)

Smallpox is a highly contagious disease (incubation 10–12 days) caused by the smallpox virus, an orthopoxvirus. It causes death in up to 30% of infected subjects. Indigenous infection has been eradicated (last case, Ethiopia, 1990 – WHO) [3, 4]. The main concern for outbreaks of smallpox today is from bioterrorism.

Smallpox is characterized by severe constitutional symptoms (fever, headache, extreme malaise) and a characteristic pustular rash. Treatment is supportive; prevention involves vaccination which, because of its risks (eczema, encephalitis, etc.), is done selectively [3, 4].

Pathogenesis of smallpox demonstrates that the virus is transmitted from person to person by direct contact or inhalation of droplet nuclei. Clothing and bed linens can also transmit infection. Most contagions are in the first 7–10 days after the skin rash appears. Once the crusts appear, the infectivity declines. The virus invades the oropharyngeal and respiratory mucosa, multiplies in regional lymph nodes, causing viremia and localization in small blood vessels of the skin (rash) and rarely in CNS (encephalitis) [3, 4].

Officially, smallpox is eradicated on Earth. There have no longer cases detected in the world population since 1990, but the question that often arose was: can we destroy the samples of smallpox virus existing in some virology laboratories around the world? [3]. The logical answer is certainly not, because we would lose the capability to prepare vaccine doses without the live virus samples to start from. Needless to say that without a vaccine, a small amount of wild virus could ignite a wide epidemic killing a large portion of the human population, since the vaccination is no longer mandatory in any country and a large percentage of young populations have no longer been vaccinated after the early 1990s.

2.4.2 Poliomyelitis (Infantile Paralysis)

Poliomyelitis is an acute infection caused by a poliovirus. Manifestations include a non-specific minor illness (abortive poliomyelitis), sometimes aseptic meningitis without paralysis (nonparalytic poliomyelitis) and, less often, flaccid weakness of various muscle groups (paralytic poliomyelitis) [8–10]. Diagnosis is clinical, although laboratory diagnosis is possible. Treatment is supportive. Vaccination is available, still mandatory in many countries, although soon legislations may change this. Childhood vaccination produces immunity in 95% of recipients. Declared cases worldwide have diminished remarkably, but recently some areas with particularly poor sanitary services or with conflicts preventing health services to operate are recording increased numbers of cases (Syria, 2013; China 2013).

Polioviruses have three serotypes. The virus enters the mouth via the fecal-oral route, and then enters the lymphoid tissues of the GI tract. If not contained, infection may enter the CNS with significant damage in the spinal cord and brain, specifically to nerves controlling motor and autonomic function (breathing). Spreading is through the enteric route [9–11]. The vaccine is live, attenuated virus, able to immunize many contacts in respect to the vaccinated subjects (community vaccination strategies; problems in nomad populations) [9–11].

2.4.3 Anthrax (Carbuncle)

Anthrax is caused by *Bacillus anthracis,* toxin producing, encapsulated, aerobic or facultative anaerobic organisms. Anthrax, an often fatal disease of animals, is transmitted to humans by contact with infected animals or their products (woolsorter's disease) [12].

In humans, infection typically occurs through the skin. Inhalation infection is less common; oropharyngeal, meningeal and GI infections are rare. For inhalation and GI infections, nonspecific local symptoms are typically followed in several days by severe systemic illness, shock, and often death. Empyric treatment is with cyprofloxacin or doxycycline. A vaccine is also available (antitoxin) [13].

Pathogenesis of anthrax takes place since *Bacillus anthracis* readily forms spores when germs encounter dry environment -a condition unfavorable for growth-. Spores resist destruction and can remain viable in soil, wool, and animal hair for decades. Spores germinate and multiply in favorable conditions (wet skin, tissue, blood) and can give human disease by contact (papules, black eschars, contagious also via fomites) ingestion (raw meat > fever, nausea, vomiting, diarrhea), and inhalation (flu-like illness, respiratory distress, cyanosis, shock, coma) [12, 13].

Of note is the anthrax bioterrorist attack through mail (using spores in powder form) that took place in the USA in 2001 (US Postal Service, Washington DC). An event that highly sensitized the public to the global theme of bioterroristic attacks.

2.4.4 Plague (Pestis, Black Death)

The Plague is caused by *Yersinia pestis* (formerly named *Pasteurella pestis*). Short bacillus with hairpin shape, infects wild rodents and can infect humans via tick bites [5]. Symptoms are either severe pneumonia or massive lymphadenopathy with high fever, often progressing to septicaemia. Diagnosis is epidemiological and clinical, confirmed by culture and serologic testing. Treatment is with streptomycin or doxycycline. Unfortunately, a vaccine is not available for the Plague [5].

2.4.5 Tularemia

Tularemia is a febrile disease caused by *Francisella tularensis,* it may resemble typhoid fever as symptoms are a primary local ulcerative lesion, regional lymphadenopathy, profound systemic symptoms, and, occasionally, atypical pneumonia [14]. Diagnosis is primarily epidemiological and clinical and supported by serologic tests. Treatment is with streptomycin, gentamycin and other antibiotics.

2.4.6 Tetanus

Tetanus is an acute poisoning from a neurotoxin produced by *Clostridium tetani* [15]. Symptoms are intermittent tonic spasms of voluntary muscles [15]. Spasm of the masseters accounts for the name “lockjaw” (trismus). Incubation requires 2–10 days. Diagnosis is clinical. Treatment with immune globulin and intensive support. Only unbound toxin can be neutralized. A vaccine is available, with a good extent of preventive protection.

2.4.7 Botulism

Botulism is a neuromuscular poisoning due to *Clostridium botulinum* toxin. Botulism may occur without infection if toxin is ingested [6]. Symptoms are symmetric cranial nerve palsies accompanied by a symmetric descending weakness and flaccid paralysis without sensory deficits. Diagnosis is clinical and by laboratory identification of toxin. Treatment is with antitoxin and support therapies [6].

2.4.8 Tuberculosis (TBC)

TBC is a chronic, progressive infection by *Mycobacterium tuberculosis*, often with a long period of latency following initial infection. It occurs most commonly in the lungs, with productive cough, chest pain, and dyspnea [16]. Diagnosis is most often by sputum culture and smear. TBC can involve any tissue (organ disease). Treatment is with multiple antimicrobial drugs. Forms of multiresistant TB bacteria are becoming more and more frequent [25].

2.4.9 Severe Acute Respiratory Syndrome (SARS)

Coronavirus infections in humans most frequently cause common cold symptoms; however in 2002, a relatively new coronavirus caused an outbreak of Severe Acute Respiratory Syndrome (SARS), which was much more severe than other coronavirus infections [17]. SARS is an influenza-like disease leading to progressive respiratory insufficiency with significant mortality rate. First detected in China, the SARS epidemic spread to more than 30 countries. In mid-July 2003, there were >8000 cases with >800 deaths (10% mortality) [17].

Then the outbreak subsided and no new cases have been identified from 2004 to 2012. In 2012 a new similar epidemic (sustained by the virus nCoV, novel coronavirus) started in Middle East (Arabia), with an estimated mortality above 40% (WHO, 2013). Later the nCoV epidemic has been named MERS (Middle East Respiratory Syndrome) and is being studied as a new zoonosis transmitted to humans from Dromedary camels. Studies are currently in progress, with great attention by the international sanitary authorities [17].

WHO in 2013 alarmed many countries against the new SARS-like coronavirus responsible of MERS, that has infected at the moment of this writing (April 2017; www.who.int/en) more than 1900 persons (Arabia, Great Britain, France, Germany, Tunisia, Italy, Abu Dhabi, United Arab Emirates, etc.) with reduced infective capacity as compared to SARS, but still highly lethal and communicable via close contact (family members). The latest available numbers call for 1938 ascertained diagnoses in humans, with 691 deaths (mortality, 35.7%).

Updates can be found at the following web sites: www.who.int/en; www.cdc.gov and (recommendations for clinicians) emergency.cdc.gov, emphasizing the need to consider the novel (nCoV) coronavirus when treating patients with a severe respiratory illness who have recently traveled to the Arabian Peninsula (or been in close contact with such travelers).

2.4.10 Ebola/Marburg Diseases

Marburg and Ebola are filoviruses that cause hemorrhage, multiple organ failure and high mortality rates. Diagnosis is with enzyme-linked immunosorbent assay, PCR or electron microscopy. Treatment is supportive. Strict isolation and quarantine measures are necessary to contain outbreaks. Incubation 5–10 days [18]. Marburg virus has been identified in bats and in primates. Human to human transmission occurs via skin and mucous membrane contact (humans/primates).

Filoviruses can affect intestines (nausea, vomiting, diarrhea), respiratory tract (cough, pharingitis), liver (jaudice), CNS (delirium, stupor, coma), and cause hemorrhagic phenomena (petechiae, frank bleeding) with high mortality rates (up to 90% with Ebola virus). Survivors recover very slowly and may develop long lasting complications (hepatitis, uveitis, orchitis) with only supportive care available: no specific antivirals nor vaccines are available for filovirus infections. A recent outbreak in West Africa was extremely severe in 2013–2014 [19]

2.4.11 Hantavirus, Lassa Fever, etc

Bunyaviridae contain the genus Hantavirus (4 serogroups, 9 viruses) causing hemorrhagic fevers that start with flu-like symptoms and evolve with severe renal and pulmonary consequences. Hantaviruses are single-stranded, enveloped, negative sense RNA viruses which can kill humans. Lethal in 10–15% of cases [20].

Lassa fever is an often fatal arenavirus infection occurring mostly in Africa. It may involve multiple organs, except CNS. Treated with ribavirin. No vaccinations are available so far for hantavirus infections [21].

Outbreaks of such infections have been recorded in Nigeria, Liberia, central Africa, with some rare imported cases in the USA and the United Kingdom. The animal reservoir of such viruses is in wild African rats (*Mastomys natalensis*), frequently found in African homes [19, 20]. Direct human to human transmission is documented via urine, feces, saliva or blood. Mortality (up to 45%) can be reduced by prompt ribavirin treatment. Universal hygiene precautions, airborne isolation, and surveillance of contacts are essential.

2.4.12 *Zika Virus Disease*

The original isolation of Zika virus goes back to 1947 during studies of yellow fever in macaques of the Zika forest in Uganda [22], where this new virus was first identified (a member of the flavivirus family). Then, after about 70 years, the infection spread in Africa and also diffused to South America (mainly in Brazil) where it spread substantially and showed to be highly neuropathogenic in embryos (newborns with microcephaly) and in infants (linked with Guillain-Barré syndrome) [23].

2.4.13 *Influenza Virus, with New Strains Continuously Appearing*

Last but not least, we must mention now influenza! Flu viruses are in nature among the most rapidly changing (mutating) organisms through their ability to infect a variety of hosts: birds (migrating waterfowl -ducks-, stantial poultry -chickens-), mammals (pigs, felines) and humans. In South East Asia (mostly in China, but also in Vietnam, Laos, Thailand, etc.) it is very common to have mixed farms of pigs, poultry, and ducks attended by humans [24, 25].

Every year, new strains appear in SE-Asia, favored by the reciprocal passage between migrating birds (mostly fowl), pigs and chickens, with exposure to many humans in farms, markets, rooster fighting sports, and food preparation places. A common saying in China tells that "Anything with four legs (except chairs) and anything that flies (except airplanes), can be eaten". With this philosophy, there is a lot to be desired in food safety and in general hygienic prevention in such geographical areas.

After the avian flu H5N1 of 2005–2006, highly lethal but unable to give human to human contagion, new combinations of flu strains are expected and feared, with high lethality and high human to human transmissibility.

On this widely interesting theme, of the world diffusion of new virus strains with pandemic potential, readers are encouraged to peruse the book entitled "Pandemics - virology, pathology and prevention of influenza" (Bollati Boringhieri publisher, Turin, Italy, 2010) [25].

2.5 Conclusions

In summary, we can see that a continuous surveillance is being devoted world-wide to the appearance of new strains of influenza viruses, in order to isolate as soon as possible potentially pandemic new strains and to prepare biological stocks suitable for massive vaccine preparations in due time to prevent the global spreading of potentially lethal new variants of the influenza viruses. Examples in time recall the cases of the highly lethal pandemics known as "Spanish flu" in 1917–18 (in excess of 40 million deaths worldwide), "Asian flu" in 1956 (in excess of one hundred thousand deaths worldwide) and "Hong Kong flu" in 1978 (in excess of seven hundred thousand deaths worldwide). The basic question is: when will the new pandemic strike? Sometimes soon, as international experts say. The so called "Avian flu" came close to that, but sometime in the future new mutations may emerge with the potential of being much worse.

To conclude this wide, although rapid, overview of the most frequent or alarming causes of micro-organism related human diseases with potential interest for bioterrorism, we hope to have provided sufficient matter for discussion and for further future diffusion of medical and microbiological culture that may be useful for prevention and the betterment of human social relationships and the promotion of peace.

References

1. Inglesby TV, Henderson DA, Bartlett JG, Ascher MS, Eitzen E, Friedlander AM, Hauer J, McDade J, Osterholm MT, O'Toole T, Parker G, Perl TM, Russell PK, Tonat K (1999) Anthrax as a biological weapon: medical and public health management. Working group on civilian biodefense. JAMA 281(18):1735–1745
2. Inglesby TV, O'Toole T, Henderson DA, Bartlett JG, Ascher MS, Eitzen E, Friedlander AM, Gerberding Hauer JJ, Hughes J, McDade J, Osterholm MT, Parker G, Perl TM, Russell PK, Tonat K (2002) Anthrax as a biological weapon. Updated recommendations for management. JAMA 287(7):2236–2252
3. Henderson DA, Inglesby TV, Bartlett JG, Ascher MS, Eitzen E, Jahrling PB, Hauer J, Layton M, McDade J, Osterholm MT, O'Toole T, Parker G, Perl TM, Russell PK, Tonat K (1999) Smallpox as a biological weapon: medical and public health management. Working group on civilian biodefense. JAMA 281(2):2127–2137
4. Lovinger S (2002) Addressing the unthinkable: preparing to face smallpox. JAMA 288(20):2530
5. Inglesby TV, Dennis DT, Henderson DA, Bartlett JG, Ascher MS, Eitzen E, Fine AD, Friedlander AM, Hauer J, Koerner JF, Layton M, McDade J, Osterholm MT, O'Toole T, Parker G, Perl TM, Russell PK, Schoch-Span M, Tonat K (2000) Plague as a biological weapon: medical and public health management. Working group on civilian biodefense. JAMA 283(17):2281–2290
6. Arnon SS, Schechter R, Inglesby TV, Henderson DA, Bartlett JG, Ascher MS, Eitzen E, Fine AD, Hauer J, Layton M, Lillibridge S, Osterholm MT, O'Toole T, Parker G, Perl TM, Russell PK, Swerdlow DL, Tonat K (2001) Botulinum toxin as a biological weapon: medical and public health management. JAMA 285(8):1059–1070
7. Dennis DT, Inglesby TV, Henderson DA, Bartlett JG, Ascher MS, Eitzen E, Fine AD, Friedlander AM, Hauer J, Layton M, Lillibridge SR, McDade JE, Osterholm MT, O'Toole T, Parker G, Perl MT, Russell PK, Tonat K (2001) Tularemia as a biological weapon: medical and public health management. JAMA 285(21):2763–2273
8. Porter RK (ed) (2011) The merck manual of diagnosis and therapy, 19th edn. Merck Sharp & Dohme Corp, Whitehouse Sation
9. Poliomyelitis (2012) Center for disease control and prevention
10. Wallace GS, Alexander JP, Wassilak SGF (2013) Traveler's health: poliomyelitis. Center for Disease Control and Prevention

11. Herremans T, Reimerink JHJ, Koopmans MPG (2000) Antibody responses to antigenic sites 1 and 3 of serotype 3 poliovirus after vaccination with oral live attenuated or inactivated poliovirus vaccine and after natural exposure. Clin Vaccine Immunol 7(1):40–44
12. Hendricks KA, Wright ME, Shadomy SV, Bradley JS, Morrow MG, Pavia AT, Rubinstein E, Holty JE, Messonnier NE, Smith TL, Pesik N, Treadwell TA, Bower WA (2014) Workgroup on anthrax clinical, guidelines. Centers for disease control and prevention expert panel meetings on prevention and treatment of anthrax in adults. Emerg Infect Dis 20(2). doi:10.3201/eid2002.130687. PMC 3901462.PMID 24447897
13. Turnbull P (2008) Anthrax in humans and animals, 4th edn. World Health Organization, Geneva, pp 20–36
14. Ryan KJ, Ray CG (2004) Sherris medical microbiology, 4th edn. McGraw Hill, New York, pp 488–490
15. William A (2012 May) Tetanus epidemiology and prevention of vaccine-preventable diseases, 12 edn, Public Health Foundation, pp 291–300
16. Cudahy P, Sheela VS (2016) Diagnostics for pulmonary tuberculosis. Postgrad Med J 92(1086):187–193
17. Smith RD (2006) Responding to global infectious disease outbreaks, Lessons from SARS on the role of risk perception, communication and management. Soc Sci Med 63(12):3113–3123
18. World Health Organization (2014) Ebola and Marburg virus disease epidemics: preparedness, alert, control, and evaluation, pp 123, WHO/HSE/PED/CED/2014.05
19. WHO Ebola Response Team (2014) Ebola virus disease in West Africa — the first 9 months of the epidemic and forward projections. N Engl J Med 371:1481–1495
20. Johnson KM (2001) Hantaviruses: history and overview. Curr Top Microbiol Immunol 256(256):1–14
21. Yun NE, Walker DH (2012) Pathogenesis of lassa fever. Virus 4(12):2031–2048
22. Sikka V, Chattu VK, Popli RK et al (2016) The emergence of zika virus as a global health security threat: a review and a consensus statement of the INDUSEM Joint working Group (JWG). J Global Infect Dis 8(1):3–15
23. Lyle R, Petersen LR, Jamieson D, Powers A, Honein M (2016) Zika virus. N Engl J Med 374:1552–1563
24. Mitka M (2013) Deadly mers coronavirus not yet a global concern. JAMA 310(6):569
25. Bologna M, Lepidi A (2010) Pandemie, Virologia, patologia e prevenzione dell'influenza. Bollati Boringhieri Publisher, Turin, Italy, ISBN 978–88–339-2037-5

Chapter 3
Fundamental Principles for Sensing Measuring Devices Used for the Detection of Chemical Warfare Agents

Farid Jahouh and Joseph H. Banoub

Abstract This chapter surveys the current detection technologies used in commercially available sensor detection equipment currently employed for identifying warfare chemical agents (CAs). Brief technical descriptions of these technologies are presented with emphasis placed on the principles of detection. Much of the content presented was obtained from the open-source literature and is an introduction to biosensor fundamentals.

3.1 Introduction

Since thousands of years, the use of chemical warfare as tools on the battlefields were known to man—e.g. poisoned arrows, boiling tar, dumping of dead bodies infected with plagues, arsenic smoke noxious fumes, and catapulting snakes [1].

During World War I, chlorine and phosgene gases were released from canisters on the battlefield and dispersed by the wind. The use of several different types of chemical weapons, including mustard gas (yperite), resulted in 90,000 deaths and over one million casualties during the war. Few decades ago, Iraq used chemical weapons against Iran during the war in the 1980s, and Iraq also used mustard gas and nerve agents against Kurdish residents of Halabja, in Northern Iraq, in 1988 [1].

Due to the devastating impact chemical weapons have had in the past, and the potential for the use of even more deadly chemical agents, this provided the

F. Jahouh
Department of Chemistry, Memorial University of Newfoundland,
St. John's, NF, 232 Elizabeth Avenue, A1B 3X7, Canada

J.H. Banoub (✉)
Fisheries and Oceans Canada, Science Branch, Special Projects, Chemistry Department,
Memorial University, St John's, NL, Canada
e-mail: banoubjo@dfo-mpo.gc.ca

J.H. Banoub, R.M. Caprioli (eds.), *Molecular Technologies for Detection of Chemical and Biological Agents*, NATO Science for Peace and Security Series A: Chemistry and Biology, DOI 10.1007/978-94-024-1113-3_3

imperious for the international effort to uphold the ban on such weapons and to work towards their complete global elimination [1].

However, in the last decades, chemical weapons have had an appeal to well-organized and well-funded terrorist groups. These chemicals especially the unrefined low-yield chemical weapons can be easily obtained, hidden and transported. Without a doubt, well-funded terrorist groups with a long-term vision of conflict, aiming to create destruction, always seek to inflict as much damage as possible upon unprotected populations (London, UK, Bombings, 2005 [2] and Tokyo, Japan, Sarin attack, 1995) [3]. The effects of such attacks are devastating and cannot be dismissed as tales that are too remote to contemplate. As a result, it is appropriate to think of chemicals warfare agent(s) (CWAs or CAs) can be the terrorist group private golden system, which provides all the ammunition required for their terror acts [4]. Accordingly, the uneasiness concerning the CWAs security threats has increased worldwide and has created new systematic challenges. The increased threat level of CWAs terrorism demands improved networking and coordination between relevant NATO-stakeholders [5]. It also requires common testing, analytical and certification equipment. The ability to rapidly detect, identify and monitor CWA agents is imperative for the efficient use of both military and civilian defence resources. This knowledge allows the severity and extent of a hazard to be assessed so that areas that are clean or contaminated can be identified [5].

Finally, the information acquired by the CWA detection systems provide advices to military commanders and first responders, regarding the donning of individual protective equipment (IPE), sampling, handling and analysis procedures as well as medical countermeasures, should the need arise.

Therefore, the practical front line of biodefense is to determine the presence of an attack as quickly as possible, enabling effective decontamination and treatment. While no single detection method can meet all these challenges, there are powerful new technologies in place today that can help protect the public from emerging CWA agent threats.

CWA are defined by the Organisation for the Prohibition of Chemical Weapons (OPCW) as "... *anything specifically designed or intended for use in direct connection with the release of a chemical agent (CA) to cause death or harm*". Chemical agents may be delivered either as a vapor or aerosol or in combination, depending on their physical properties, and various agents will have different persistence following an attack [6].

Immediately following the release the CAs distribution is widely determined by the characteristics of the delivery system. Accordingly, when chemical warfare agents are released in the environment, and they become distributed among four major components, namely: water, the atmosphere (air), soil and living organisms. Soon after dispersion, the environmental fate of the CWAs are principally become governed by their physical and chemical properties [4].

3.2 Various Types of Chemical Agents

Warfare agents can be classified in different groups, according to their types of action and the effects they cause on living organisms.

3.2.1 Cyanides

Sodium or potassium cyanides are white-to-pale yellow salts that can be easily used to poison food or drinks. Cyanide salts can be disseminated as a contact poison when mixed with chemicals that enhance skin penetration. Hydrogen cyanide (HCN) and cyanogen chloride (Cl-CN) are colorless-to-pale yellow liquids that will turn into a gas near room temperature. HCN has a characteristic odor of bitter almonds, and Cl-CN has an acrid choking odor and causes burning pain in the victim's eyes. These signs may provide enough warning to enable evacuation or ventilation of the targeted site before the agent reaches a lethal concentration. Exposure to cyanide may produce nausea, vomiting, palpitations, confusion, hyperventilation, anxiety, and vertigo that may progress to agitation, stupor, coma, and death. At high doses, cyanides cause immediate collapse. Medical treatments are available, but they need to be used immediately for severely exposed victims [7].

3.2.2 Nerve Agents

Sarin, tabun, and VX are highly toxic military agents that disrupt a victim's nervous system by blocking the transmission of nerve signals [8]. These agents are not commercially available, and their synthesis requires significant chemical expertise. Exposure to nerve agents causes pinpoint pupils, salivation, and convulsions that can lead to death. Medical treatments are available, but they need to be used immediately for severely exposed victims [9, 10] (Fig. 3.1).

3.2.3 Toxic Industrial Chemicals

There are a wide range of toxic industrial chemicals which are not as toxic as cyanide, mustard, or nerve agents but can be used in much larger quantities to compensate for their lower toxicity. Chlorine and phosgene are industrial chemicals that are transported in multi-ton shipments by road and rail. Rupturing the container can easily disseminate these gases [11].

The effects of chlorine and phosgene are similar to those of mustard agent. Organophosphate pesticides such as parathion are in the same chemical class as

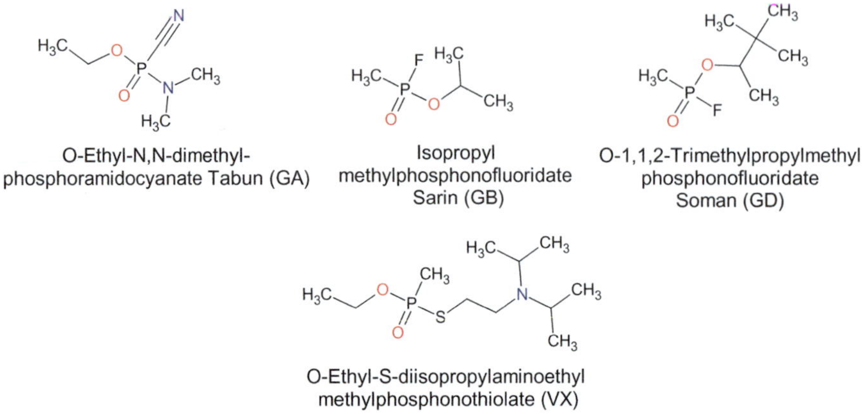

Fig. 3.1 Molecular structures of the nerve chemical agents Tabun (GA), Sarin (GB), Soman (GD) and VX

nerve agents. Although these pesticides are much less toxic, their effects and medical treatments are the same as for military-grade nerve agents.

3.2.4 Blood Agents

The name “blood agent”, like those of other groups of warfare agents, derives from its effect on victims. Blood agents are distributed via the blood and generally enter the body via inhalation. They inhibit the ability of blood cells to utilise and transfer oxygen. Thus, blood agents are poisons that effectively cause the body to suffocate. Examples of blood agents include: hydrogen cyanide (AC), cyanogen chloride (CK) and arsine (SA) [12].

3.2.5 Blister Agents

Blister agents, or vesicants, are one of the most common CW agents. These oily substances act via inhalation and contact with skin [13]. They are primarily dispersed in liquid or vapour (aerosol) form and may persist for days. Like phosgene, mustard agents have a delayed effect. Deaths typically only represent a small percentage of the casualties they cause. They affect the eyes, respiratory tract, and skin, first as an irritant and then as a cell poison. As the name suggests, blister agents cause large and often life-threatening skin blisters which resemble to severe burns [14]. Examples of the CAs include: the sulfur mustard (H, HD), the nitrogen mustard (HN), lewisite (L) and phosgene oxime (CX) (Fig. 3.2). Exposure to blister agents often results in blindness and permanent damage to the respiratory system [14, 15].

3.2.6 *Psychotomimetic Chemical Weapons*

This group of chemical agents usually includes substances which, when administered in low doses (<10 mg) cause conditions similar to psychotic disorders or other symptoms emanating from the central nervous system (loss of feeling, paralysis, rigidity, etc.) [16]. During World War II, the US military used wide range of possible non-lethal, psycho-behavioral chemical incapacitating agents, containing an indole moiety (Fig. 3.3), such as lysergic acid diethylamide (LSD) and marijuana derivatives, as well as several glycolate anticholinergics. One of the anticholinergic compounds, 3-quinuclidinyl benzilate, was developed and weaponized in the 1960s as a new chemical agent for battlefield use as a psychochemical and assigned the NATO code BZ (Fig. 3.3) [17]. The effects of this agent are transitory and cause

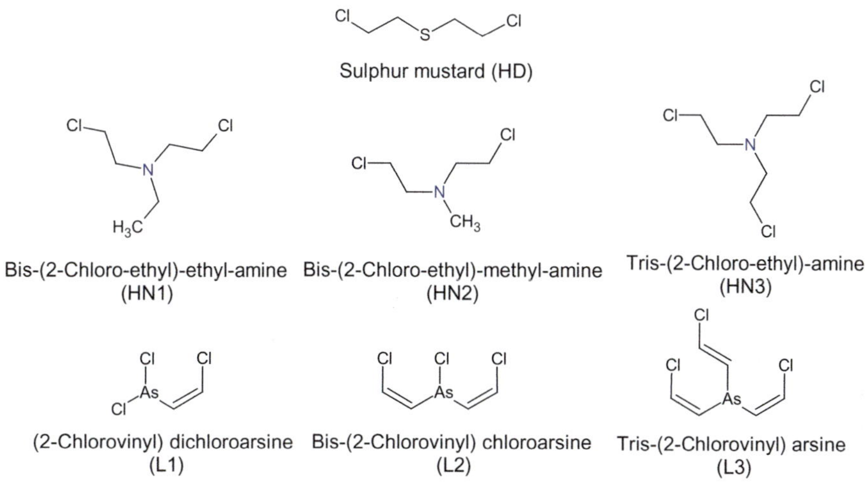

Fig. 3.2 Molecular structures of the blister agents sulfur mustard (HD), nitrogen mustard (HN1, HN2 and HN3) and lewisite (L1, L2 and L3)

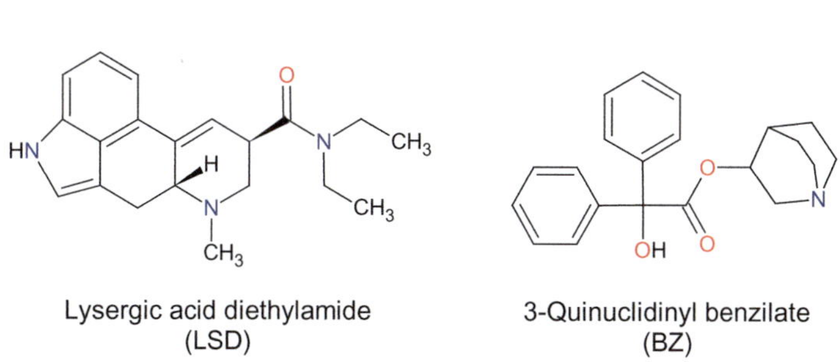

Fig. 3.3 Molecular structure of the psychotomimetic chemical agents LSD and BZ

incapacitation and inability to take decisions. Several substances may be used to achieve these objectives and only a few examples are given in this chapter.

Phencyclidine is a substance with analgesic and anaesthetic properties. Symptoms such as disturbed body-awareness, disorientation and vivid dreams are caused by this substance that is also easy to produce.

3.2.7 *Riot Control Agents*

Riot control agents cause temporary incapacitation by irritation of the eyes and irritation of the upper respiratory tract. They are often called irritants, lachrymators and harassing agents. Many tear-stimulating substances were tested as CW agents. On exposure, they cause pain in the eyes, flow of tears and skin irritation. Three types of riot control agents are recognized: lachrymators; sternutators which mainly cause sneezing and irritation of the upper respiratory tract; and vomiting agents, which additionally cause vomiting [18]. Among a long series of chemical substances only three chemical agents, viz. CN, CS and CR are usually used (Fig. 3.4).

3.3 Analytical Chemical Sensors

By definition a sensor or a detector is a device that responds to a stimulus or a form of energy. It then generates a signal that can be measured or interpreted. Analytical chemical sensors currently used for the detection and identification of chemical warfare agents are classified by the number of dimensions of information they provide [19]. An ideal chemical warfare detector, also called sensor, can be described as one that can detect both chemical warfare agents and other industrial toxicants selectively within an acceptable time, and sensitive enough to detect the toxicant agent concentrations at below levels which pose a health risk. These detectors should have a rapid reaction and recovery time whilst being portable, easy to operate and produce data that is easily interpreted.

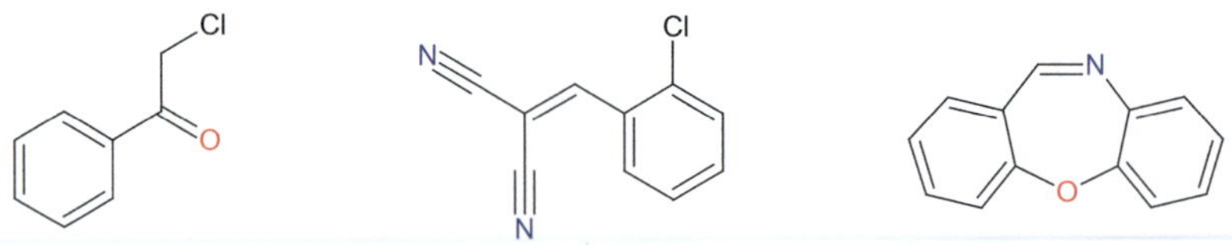

2-Chloroacetophenone (CN) 2-Chlorobenzilidenemalononitrile (CS) Dibenz-(b,f)-1,4-oxazepine (CR)

Fig. 3.4 Molecular structure of the riot control agents CN, CS and CR

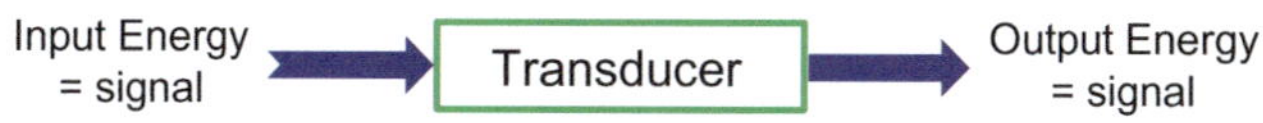

Fig. 3.5 Display of the sensing process in terms of energy conversion

A sensor detects a change in a physical stimulus and converts it into a signal which can be measured or recorded (Fig. 3.5). In addition, the terms analytical chemical detector systems can also be used to describe either a sensor and/or a transducer. A corresponding definition for the transducer dictates that it is a device that transfers power from one system to another. A sensible distinction for the use of sensor is that it senses the analyte itself, whereas the transducer is the associated circuitry for the sensing element. All transducers would thus contain a sensor and most (though not all) sensors would also be transducers (Fig. 3.5) [20, 21].

3.3.1 Sensor Classification

In order to classify a sensor, it is essential to consider all of its properties, such as: the type stimulus and specifications of its measurement; the physical phenomenon, and conversion-mechanism; the type of material used in the studied application. In this chapter will deal exclusively with optical detectors such as flame photometry, infra-red spectroscopy (IR), Raman spectroscopy, surface acoustic wave (SAW), ion mobility spectroscopy (IMS), colorimetric, photo ionization and flame ionization [22].

3.3.2 Sensor Selection

The selection of a given sensor is based on its physical property, which will be changed by the external stimulus caused by the investigated CA analyte. This will alter and change the initial steady state by either producing an electric signal and/or modifying the pre-existing external electric signal. Usually, the same stimulus caused by the analyte might be measured by using different physical phenomena, and, subsequently, by different sensors. Therefore, the selection criteria also depend on many factors such as sensor availability, cost, power consumption and environmental conditions [22].

3.3.3 Factors to Consider When Selecting Sensors

It has been shown that each sensor has certain capabilities and limitations. For this reasons, to choose the correct sensor, one must first properly define the application [20–22].

The following factors should be considered: (A) tt is essential to define the objective and to understand the instrument specification to perform the assigned task. The specifications involved should consider the ranges of the sensors and the types of gases to be measured; B) it is important to determine the background gases in the monitoring area; (C) The temperature ranges in which the sensor is to be installed should be within the sensor specifications and should be suitable for the gases to be monitored; (D) The occurrence of condensation is a function of the temperature change. In general it was observed that if the sensor transmitters are always a few degrees warmer than the environment, it avoids the formation of condensation; and (E) for applications requiring the sensors to be constantly exposed to a gas, the sensor specifications and the suppliers of the sensors may need to be consulted [20–22].

We can summarize the selection of the detector by the following: an ideal detector (or sensor) should be chosen on its ability to selectively detect both CAs and toxic industrial chemicals (TICs) within an acceptable resolution. It should be sensitive enough to detect agent concentrations below levels which pose a health risk, and it should not be affected by other environmental factors. The detector also requires a rapid reaction and recovery times, must be portable, easy to operate and produce data that is easy to interpret. In general, it is common knowledge that a universal sampling system that handles all possible CAs detection is extremely difficult to obtain. In fact, there is no general consensus that establishes which sensor is the best for a given application, and the choice of the appropriate sensor has to be made based on the analyses requirements [20–25].

3.4 Technological Requirements for the Separation and Detection of Warfare CAs

The separation and detection technologies used, usually take advantage of the physical properties of the specific warfare chemical agent, which allows its detection from other chemical compounds. Characterizing the physical attributes of that dangerous chemical includes different analyses, such as colorimetric reactions, the measurement of the gas-phase mass-to-charge ratio of that compound, absorption and scattering of electromagnetic energy, particularly in the infrared to microwave region, reactions that cause unique emissions (for exemaple: chemiluminescence), physical separations, electrochemical interactions and enzymatic reactions. In this context, a sensor is defined as a device that produces a measurable response to a change in a physical condition, such as temperature or thermal conductivity, or to a change in chemical concentration [23]. Recently, a new generation of chemical warfare agent detection system has been designed in order to detect the production of homemade explosives, like the "Mother of Satan", composed of triacetone triperoxide and also known as TATP. This explosive is a highly unstable and presents a bomb of choice for terrorist groups [24, 26]. This involves acting during the initial phase of preparation of explosives (for example, during the mixing stage), when the threat to citizens is low.

For these reasons, recently, a novel small footprint gas sensor utilizing metal oxide catalyst that was capable of detecting the heat of reaction during the catalytic decomposition of TATP was fabricated and tested, which was. As TATP possess relatively high vapor pressure compared to other explosives, it decompose readily to hydrogen peroxide and acetone, which can be successfully detected using this thermodynamic based sensor platform [25].

As a result, all known developments of integrated detector surveillance systems are beneficial for the continuous chemical background monitoring of chemical hotspots, such as bomb or biological agent workshops. Nevertheless, it is also essential to adapt the existing devices to become more user-friendly. Consequently, there is an urgent need to improve the performance of the existing technologies in terms of sensitivity, specificity, false positives detection, detection limits, distance from the source, time for detection, as well as the making the it portative to operate directly on the field [4].

3.5 Available Detection Technologies

At present there is a wide variety of commercially available detectors. However, due to their low versatility, not all of them are suitable for use in every potential threat situations. This variety of wide detection equipment can have different sensitivity and specificity responses to the targeted chemicals in a sample. In that respect, sensitivity is determined by the lowest concentration of a CA that can be detected with high confidence. It can also be referred as the limit of detection (LOD) or detection limit of the device. They also can have differences in the response time for the detector to collect and analyse a sample, determine if an agent is present, and provide interpretable data. In addition, the verification of the detectors' capacity to deliver quality data is usually required before use. This calibration process is usually conducted using a known non-toxic chemical as a simulant of the targeted compound [4, 10, 22, 23, 26].

Current chemical detection technologies have been incorporated into mobile or stationary detection platforms and can be used as a point-source detector or as a remote (stand-off) detector. The primary differences between mobile detectors and stationary detectors are the size, weight, portability, and logistical support requirements.

The following sections briefly describe the know-how of functioning and the technological basis of these detectors. Please note that no extensive description was given to the commercially available sensors that employ these technologies in their instruments. To obtain specific informations on these commercial sensors, please scrutinize the references given herein this chapter [4, 10, 22, 23, 26].

3.5.1 *Flame Photometric Detector Technology*

When burned in a flame, most elements emit a unique and characteristic wavelength light which can be detected by Flame Photometric Detection Technology (FPDT) sensors (Fig. 3.6) [27]. The FDPT sensor produces a characteristic response signal. It is composed of an optical filter that can allow light of a specific wavelength to pass through, making the photosensitive detector to produce a diagnostic signal. Since most elements emit unique and characteristic wavelength of light when burned in the flame, this device allows for the detection of specific elements (Fig. 3.6) [27]. Flame photometric detectors are commonly used in gas chromatography. For example, in the case of nerve CA that are composed of phosphorus and sulphur atoms, the FPDT detector that is equipped with optical filters can specifically detect these two elements. When phosphorus containing CA are burnt in a hydrogen flame, the excited phosphorus in the form of HPO* species is formed. Whereas when sulphur containing CA are burned, they produce the S2* excited specie. When, both the exited species return to their ground state, light is emitted near 526 nm for HPO*, and near 39,4 nm for the S2* species, allowing their detection [23–25, 27].

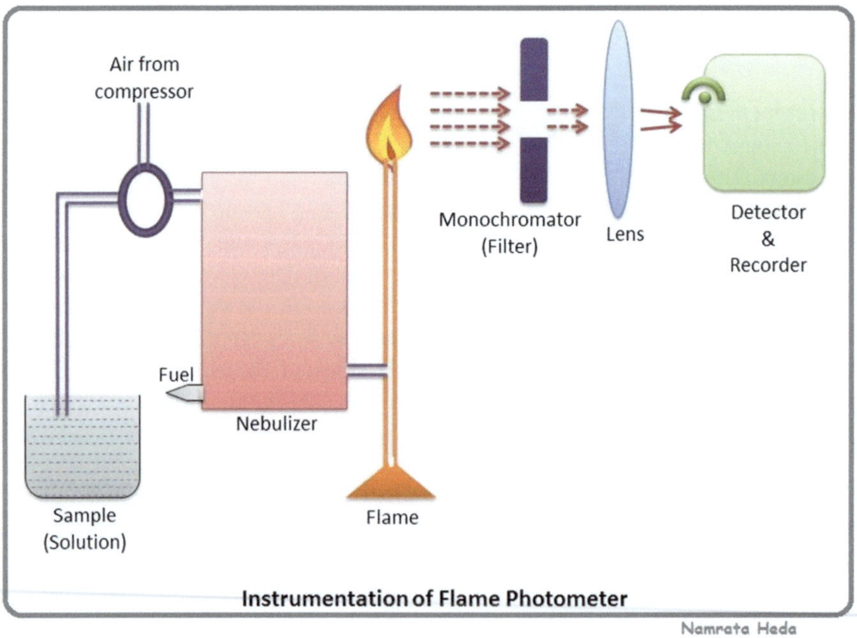

Fig. 3.6 Principle of the Flame Photometric Detector Technology (FPDT) (Adapted from Ref. [27])

It is important to realize that even when the FPDT sensors are incorporated in handheld field instruments for the detection of volatile organic compounds (VOCs) or gases (Fig. 3.6), they are only used when precise identification is not a requirement [23–25, 27]. However, FPDT detectors are very specific and sensitive to sulfur and phosphorus containing compounds at ppm (part per million) to ppb (part per billion) detection limits without the need of a separation technique to be that sensitive [23–25, 27].

3.5.2 *Infra-red Sensors*

The infrared (IR) portion of the electromagnetic spectrum is usually divided into three regions: the near-, mid- and far- infrared, named for their relation to the visible spectrum. The higher-energy near-IR, at approximately 14,000–4000 cm^{-1} (0.8–2.5 μm wavelength) can excite overtone or harmonic vibrations. The mid-infrared, at approximately 4000–400 cm^{-1} (2.5–25 μm) is used to study the fundamental vibrations and associated rotational-vibrational structure. The far-infrared (long range IR), at approximately 400–10 cm^{-1} (25–1000 μm), lying adjacent to the microwave region, has low energy and is used for rotational spectroscopy.

At the present time, there are several different detection techniques that utilise IR spectroscopy. Depending on the infrared portion range they operate, these IR techniques include: Photoacoustic IR Spectroscopy, Filter-based IR Spectroscopy, Passive IR Detection, Forward-Looking IR Spectroscopy (FLIR), and Fourier Transform IR spectroscopy [22, 23, 26, 28].

3.5.2.1 Photoacoustic Infra-red Spectroscopy (PIRS)

The photoacoustic infra-red spectroscopy (PIRS) is a highly selective technique used to identify CAs vapours. It is commonly utilised as a point detector. In this sensor, the IR beam is normally chopped or pulsed to achieve the desired frequency that allows it to pass through an optical filter and to enter in the sample cell via an optical window (Fig. 3.7) [23, 28–30]. The light transmitted by this optical filter will be selectively absorbed by the analytes in the monitored gas. As a result, the gas analyte pressure and temperature will increase. Because, the light entering the cell is pulsed, the pressure of the gas in this cell will also fluctuate. This will create and acoustic wave that is directly proportional to the gas concentration. It should be noted, that the acoustic wave frequency is similar to the chopper frequency, and it is translated into an electric signal by means of either a sensitive microphone or piezoelectric sensor, allowing further processing. Usually, two microphones are mounted inside the cell and monitor the acoustic wave signal that is subsequently sent to a control station (Fig. 3.7). PIRS technology can detect all CAs [23, 28, 30, 32].

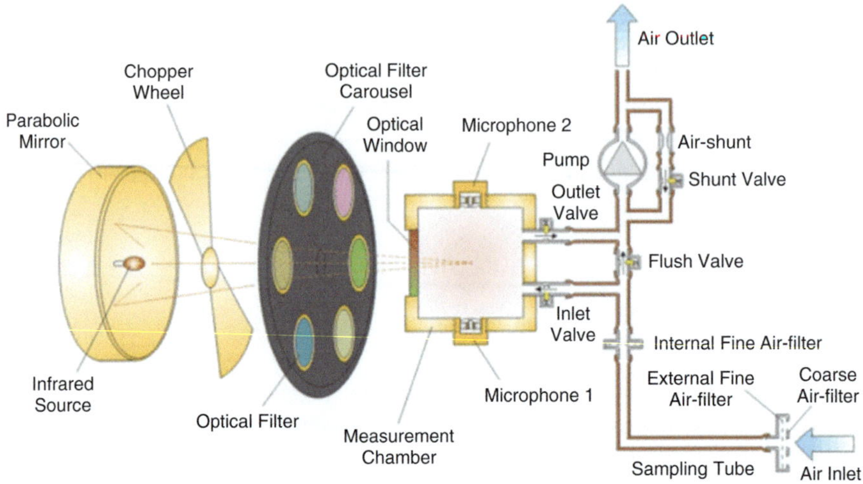

Fig. 3.7 Schematic representation of the Photoacoustic Infra-Red Spectroscopy (PIRS). (Adapted from Ref. [32])

3.5.2.2 Filter Infra-red Detectors (the Interferometer)

One of the most reliable techniques to identify gas CA species is IR absorption spectroscopy. All existing toxic compounds exhibit their strongest roto-vibrational absorption bands in the medium infrared spectral range.

Filter IR detectors act by filtering the IR wavelength so that only the desired wavelength interacts with the sample. This technique is based on a series of lenses and mirrors that direct a narrow band-pass IR beam to a pre-selected path and through the sample (Fig. 3.8). The amount of energy absorbed by the sample is measured and recorded [28, 30, 31].

The interferometer is composed of a beam splitter, a fixed mirror and a moving mirror. In this system, the beam splitter receives the incoming IR beam and divides it into two optical beams, which are reflected off the mirrors and then recombined before reaching the IR Transducer (Fig. 3.8).

It should be noted, that one of the beam path travels for a fixed length, whereas, the other beam path is constantly changing according to the mirror position. The signal that exits the interferometer is the result of these two beams 'interfering' with each other and the resulting signal is referred to as an interferogram, an example of which is shown in Fig. 3.9 [22, 30, 32, 38].

When an interferogram is measured, all IR wavelength frequencies are being measured simultaneously. The resulting interferogram signal is then converted to the frequency domain using the Fourier Transform in order to make it easier to interpret (Fig. 3.9) [22, 28, 30, 31]. Furthermore, the interferometers can collect absorption or emission spectra from a CA cloud and compare it to previously collected background spectra. For this reason, the interferometer has become a desired

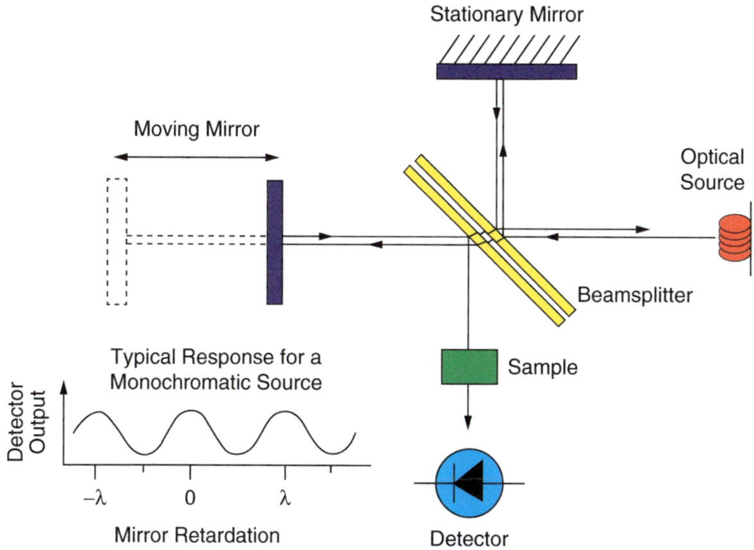

Fig. 3.8 Schematics of the interferometer (Adapted from Ref. [30])

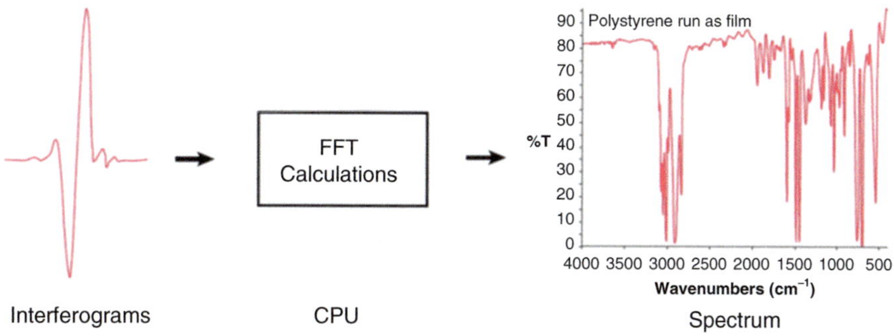

Fig. 3.9 Resulting measured interferogram signal is then converted to the frequency domain using the Fourier Transform in order to make it easier to interpret (Adapted from Ref. [30])

detector due to its ability to automatically scan the surrounding environment and subsequently give early warnings of a CA release during an attack.

3.5.2.3 Forward-Looking Infra-red Spectroscopy (FLIR)

Forward-looking infra-red spectroscopy (FLIR) usually refers to a camera that takes pictures using the mid-range of IR portion of the electromagnetic spectrum by contrast. It acquires a signal from the blackbody radiation of the scene (land, sea, sky, or combination). At ambient temperature, significant blackbody radiation in the 8 to 14 μm region is available for passive detection. Detection is therefore based on

Fig. 3.10 FLIR detection of thermal energy to create a picture in the dark (Adapted from Ref. [34]

changes in the IR energy emitted from remote objects, or from clouds formed by an agent [31(A)]. It should be noted that since FLIRs use detection of thermal energy to create the "picture" assembled for a video output, they can be used to help pilots and drivers steer their vehicles at night, and in fog, or detect warm objects against a cold background when it is completely dark as shown in Fig. 3.10 [34].

3.5.2.4 Long Wave Infrared (LWIR) Absorption Spectroscopy or Hyperspectral Imaging

Recently it was shown by Seeley and Richardson that hyperspectral imaging allows the collection and processing of information across the electromagnetic spectrum [33]. The aim of hyperspectral imaging is to obtain the spectrum for each pixel of an image in a scene, with the purpose of finding objects, identifying materials, or detecting processes [33].

Although human eye sees color of visible light in mostly three bands (red, green, and blue), spectral imaging divides the spectrum into many more bands. In the hyperspectral imaging, the recorded spectra have fine wavelength resolution and cover a wide range of wavelengths. Certain scanned objects leave unique fingerprints in the electromagnetic spectrum known as spectral signatures. These 'fingerprints' enable the identification of materials that compose a scanned object. With appropriate imaging optics, this type of sensor functions as a hyperspectral imager, providing spectral information at each pixel of the image. As a result, the hyperspectral imager generates a three-dimensional data cube, with two spatial dimensions and one spectral dimension, as described in Fig. 3.11 [33].

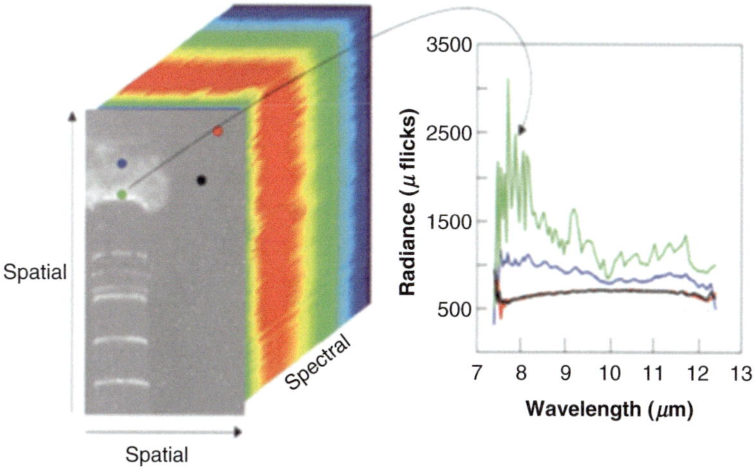

Fig. 3.11 The hyperspectral imager generates a three-dimensional data cube consisting of two spatial dimensions and one spectral dimension (Adapted from Ref. [33])

Subsequently, imaging spectrometer provides simultaneous spatial and spectral information about a scene. Therefore, when the hyperspectral data cube is obtained, it contains spectral information at every spatial point as well as an image at every spectral band. This means that we can observe one spatial slice of a cube at one particular spectral band [33]. In Fig. 3.11, the color-coded spectra of specific pixels in the image are shown on the right. The green pixel shows the spectral structure of the gas emitted from the stack. The red and black pixels correspond to cold sky and have lower radiance values [33]. In addition to the benefit of improved ACR, spatial information obtained with the use of a hyperspectral imager can add more certainty to detection statistics, potentially reducing false alarms [33].

3.5.2.5 Release of Sulfur Hexafluoride (SF6) from Five Different Locations Recorded from a Drone

Seeley and Richardson published also studied the release of sulfur hexafluoride (SF_6) near five different sites measured from a drone, as shown in Fig. 3.12 [33]. On their study, the wind was blowing from the south-southwest. The plot on the right side of the figure indicates the time of the intentional release (red) of SF_6 and the corresponding detector output measurements (blue). The first measurement with the FTIR spectrometer at the SF_6 release location shows its immediate detection (signal increase). The second SF_6 release that was directly in the sightline was immediately detected. The third SF6 release that was downwind from the line of sight was not seen, as expected from the authors. The release of SF_6 was carried out at the inlet of the building equipped with a heating, ventilation and air conditioning (HVAC) system and corresponded also at the downwind from the line of sight (Fig. 3.12) [32]. As stated

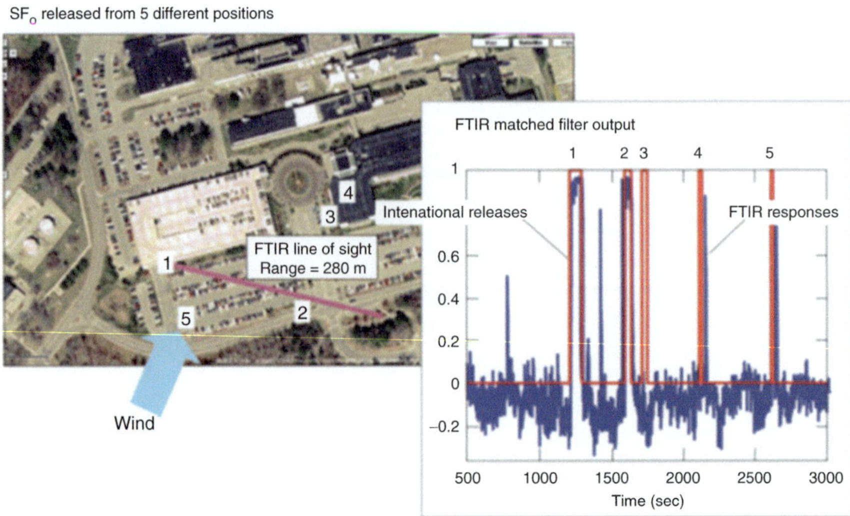

Fig. 3.12 Sulfur hexafluoride (SF_6) measurements from a drone, released near five different sites (Adapted from Ref. [33])

by the authors, this release was not expected to be seen. However, when the SF_6 gas did reach the HVAC system of the building where it was expelled out the exhaust and blown toward the line of sight, it finally got detected. The last gas release (fifth release) was done upwind from the line of sight and also detected (Fig. 3.12) [33].

The baseline of the data was not flat and the some artifact spikes were recorded prior to the releases and were explained to be probably cause by small leaks of SF_6 during experimental preparations and also by a turbulent flow inside the canyon formed by the buildings and the hill. These non-expected detections show temporal characteristics indicative of a real detection, which is the appearance and a gradual decrease of the measured gas [33].

3.6 Raman Sensors

Raman spectroscopy (RS) is a method that allows studying vibrational, rotational and other low-frequency modes in a system. In RS, a monochromatic laser light is used to illuminate the sample and the scattered light is then detected as a function of wavelength. Raman spectroscopy is based upon the observation that when radiation is passed through a transparent medium, chemical species present in that medium scatter a portion of the radiation beam in different directions [35]. RS relies on inelastic scattering, or Raman scattering of monochromatic light, usually from a laser in the visible, near infrared, or near ultraviolet range. The laser light interacts with molecular vibrations and photons resulting from excitations can be shifted up or down of their energy level. The shift in energy gives information about the vibrational states in the system [35]. When the incident light excites the system to a high-energy state and immediately recovers from this state, the scattering process occurs. If the elastically scattered light has the same energy as the incident light, this corresponds to the Rayleigh scattering. However, if the system gains energy during this process, the scattered light loses this amount of energy and the system reaches a higher energy state (higher energy level) than it had before and this is called the Stokes scattering. If the system is vibrating and excited, it loses energy during this process and the system reaches a lower energy level than it had before, this process is known as the Anti-Stokes scattering [35, 36].The difference between the scattered radiation and incident beam corresponds to wavelengths in the mid-infrared region. The degree of wavelength shift is dependent on the chemical structure of the molecules that cause the scattering. During the irradiation process, the spectrum of the scattered radiation is measured with a spectrometer.

Raman spectroscopy is a tool of choice for safely detecting CAs since it is a non-destructive technique and the sample can remain sealed in a glass container during the analysis, thereby minimising any exposure hazard to the operator. Devices using Raman spectroscopy are able to assess liquids, gases, solids and aerosols, and are not affected by water in either liquid or vapour form. Furthermore, they experience negligible variations in signatures or signal strength regardless of surface texture or reflectivity. Raman spectroscopy is a useful tool for chemical analysis as it exhibits high specificity. In addition, the Raman bands have good signal-to-noise ratio (sensitivity) and are non-overlapping (less chemical interferences) [36].

3.6.1 Differential Absorption Light Detection (LIDAR)

Differential Absorption Light Detection (LIDAR) is an IR technology largely used to track CA clouds that have already been identified. It functions by transmitting two laser pulses into the distance and then detecting the reflected IR. One of the pulses is at a frequency known to be absorbed by the CA while the other is not absorbed [37]. The difference in the intensity of the return signal is processed to measure the concentration of CA in the cloud, while the time of return is used to determine the distance from the observers. This technique is also sensitive to environmental noise but has been applied effectively to track CAs (Fig. 3.13) [37].

3.6.1.1 Radiative Transfer Sensors (RTS)

To successfully detect a chemical plume, the radiative transfer sensor (RTS) system designer must be able to anticipate its spectral signature for incorporation into the detection algorithm. Therefore, the threat cloud must also provide enough radiative signals to be detectable above the noise. Radiative transfer models predict the magnitude and spectral characteristics of the radiative signal

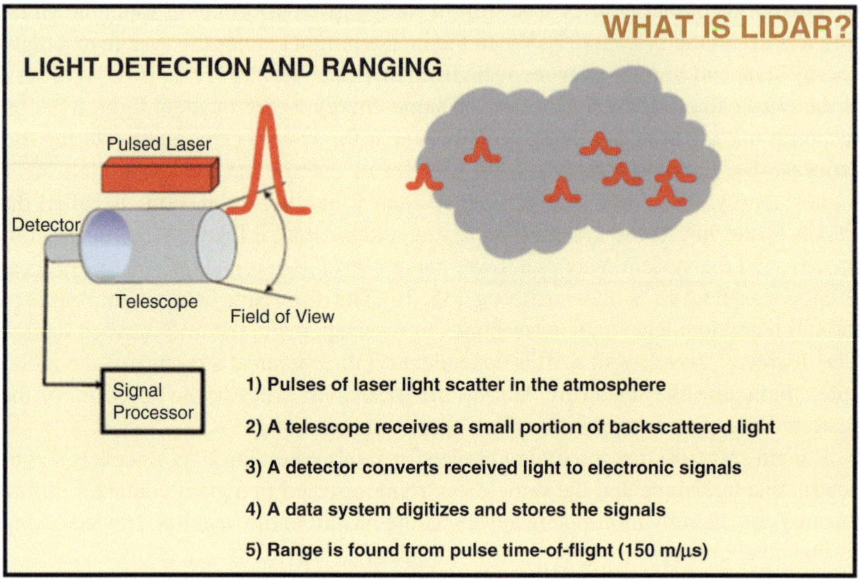

Fig. 3.13 Principle of the differential absorption Light Detection (LIDAR) (Adapted from Ref. [37])

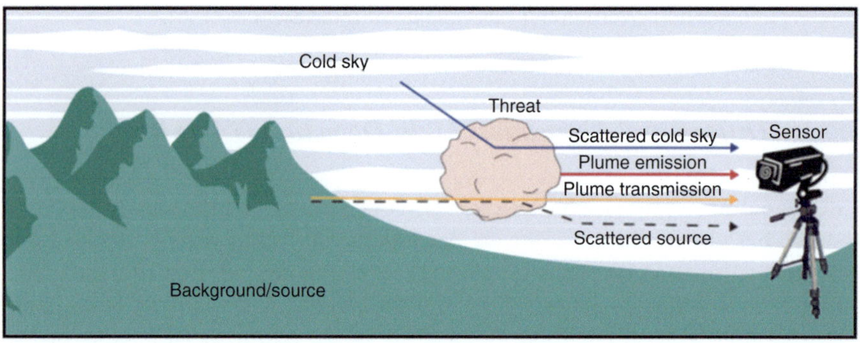

Fig. 3.14 Principle of a Radiative Transfer Sensor (RTS) applied to a cloud threat (Adapted from Ref. [33])

that a plume will produce. However, some chemical agent produce particles with sizes of the same scale or larger than the wavelength of the infrared light being used for to detect them; in that case, the dominant radiation mechanism is scattering (Fig. 3.14) [33, 38].

Long wavelengths infra-red radiation (LW-IR) emitted by the background is absorbed and emitted by all intervening materials, including a threat cloud, if present, as well as other ambient vapors or aerosols within the field of view [32].

The differential signal contributed by a threat cloud will depend upon the relative temperatures of the various absorbers/emitters, their vibrational modes, and the concentration and length of the threat cloud along the sensor line of sight.

One way to enhance area coverage rate (ACR), while maintaining high spatial resolution, is to use a detector with multiple pixels that is, a focal-plane array (FPA) [38].

3.6.2 *LaserScan Detection System*

Mid-IR laser is used to detect the chemical warfare agent methyl salicylate (MES) and explosives (RDX) on different surfaces [39]. Figure 3.15 shows the principle of a LaserScan Detection System. The lower plot shows that a detection algorithm (based on principal component analysis of the reflection spectra) can be used to clearly distinguish between three types of explosives in a luggage [39]. In addition, recently, a novel technologies such as the "Eye-safe lasers" was developed. It creates

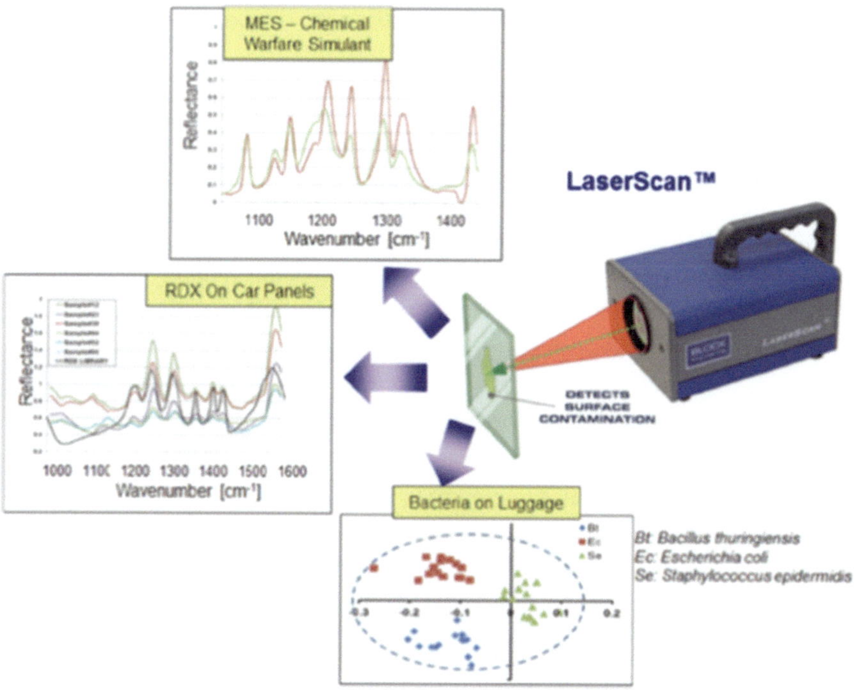

Fig. 3.15 Representation of the LaserScan system (Adapted from Ref. [39])

an invisible beam that is used to detect chemical threats in the air of airports. This system can be used for 24/7 protection against chemical attacks (Fig. 3.15) [39].

3.7 Ion Mobility-Mass Spectrometry or Ion Mobility Spectrometry (IMS)

Ion Mobility Spectrometry (IMS) functions by drawing air at atmospheric pressure into a reaction region where the constituents of the sample are ionized. The ionization is generally occurs by a collision charge exchange or an ion-molecule reaction type, resulting in the formation of low-energy stable charged molecules (or ions). The ionization source may contain the radioactive materials ^{241}Am or ^{63}Ni, in addition to the other types of atmospheric pressure ionizations such as electrospray (ESI) and atmospheric chemical ionization (APCI). IMS is a technique of choice for the analysis of all ionisable CAs and explosives [40, 41].

Usually, the IMS sensor operates by drawing the sample vapour via an inlet formed either from a membrane interface or molecular sieve packs, as these limit the entry of moisture and dust into the drift tube. A plot of the current generated over time provides the characteristic ion mobility spectrum containing a series of peaks [40]. The intensity of these signals corresponds to the relative concentration of the CA present.

IMS are used as mobile detectors for the detection of nerve, blister, and blood agents and recently for the detection of tri-cyclic acetone peroxide (TCAP) also known as the "Mother of Satan", the preferred explosive of current days terrorists [42]. In the following section, four IMS techniques will be discussed.

Recently, it was shown that there are three major analytical measurement methods for employing the IMS technique analyzers for CAs namely: ion mobility spectrometry based on drift tubes (DT IMS), aspiration ion mobility spectrometry (AIMS) and differential mobility spectrometry (DMS) [43].

3.7.1 Drift Tubes Ion Mobility Spectrometry (DT-IMS)

The graphics of a traditional drift-time IMS instrument are illustrated in Fig. 3.16 shows. After ionization, the produced ions travel through the drift tube that is connected to an electric field and contains a carrier buffer gas (drift gas) that opposes the ion motion. At the end of the tube the ions hit the detector and their intensity is monitored [43]. The ions separation is achieved through the drift gas along the axis of the applied electric field. Therefore, depending on the mass of the ion, charge, size and shape, its specific ion mobility will be measured by the migration time to reach the detector [43–45]. Obviously, larger size ions in presence of the buffer gas

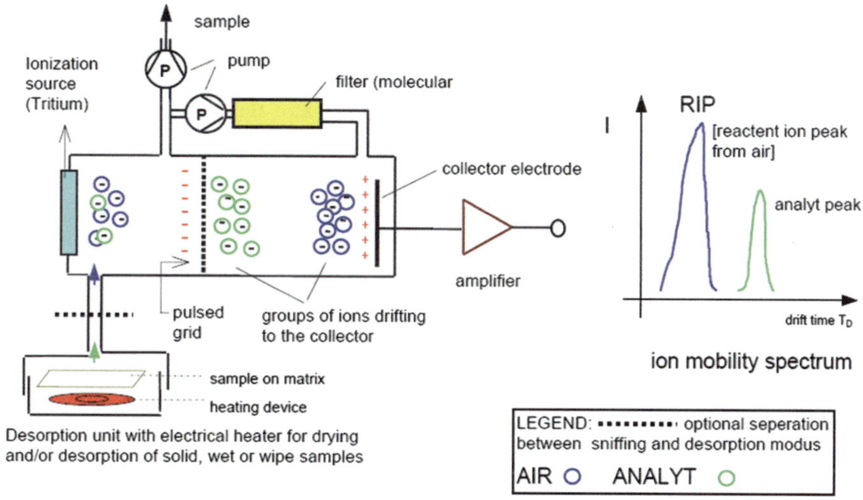

Fig. 3.16 Schematic of drift time Ion mobility spectrometer (Adapted from Ref. [43])

will collide, causing an impediment in their travel and will require a longer time to migrate through the drift tube Fig. 3.16) [43–45].

Consequently, the simplest form of an IMS system, measures how fast a given ion moves in a uniform electric field through a given atmosphere. The modus operandi of such system relies on the introduction of the sample ions in specified intervals into the drift chamber by a gating mechanism, based on a charged electrode. This electric field then drives the ions through the drift tube where they interact with the neutral drift molecules contained within the system. The time taken to traverse the device (the drift time) is correspondingly shorter (Fig. 3.15) [43–45].

3.7.2 *Aspiration Ion Mobility Spectrometry (AIMS)*

The aspiration ion mobility spectrometer (AIMS) possesses a simple and miniaturized design. A schematic figure of a typical modern aspiration type ion mobility spectrometer, which has a planar design, is shown in Fig. 3.17 [41, 46, 47]. During the operation of an AIMS instrument, a continuous laminar flow of sample gas is passed between the two parallel plates. Initially the drawn sample gas is ionized in a region composed of radioactive atmospheric pressure ionization (R-APCI) ion source (Fig. 3.16). This source may use the radioactive materials ^{241}Am or ^{63}Ni. This is followed ions transfer into the drift region, which contains different number of deflecting and collecting electrodes on opposite sides of the separation channel. Each pairs of deflecting and collecting electrodes are used to produce an electric field perpendicular to the gas flow. Consequently, the ions traveling in the laminar gas flow are pushed by the electric field towards the collecting electrodes [41, 47] Therefore, ions possessing low *m/z* values and high mobility will travel across the gas flow faster, than the ions of high *m/z* values possessing low mobility. Conversely, the higher mobility ions will travel a shorter distance in the drift region, than the ions with lower mobility. Nevertheless, as this logic may look contrary to conventional mass spectrometry theories, the positions where the collection electrodes were placed were chosen as matter of convenience. Finally, the ions are neutralized at the collecting electrodes, generating an electric current that is registered after amplification (Fig. 3.17).

As a result, a mobility histogram is formed consisting of ion intensity versus collecting electrode channel number. The mobility histogram contains analytical information and can be used for identification of CA analytes [28, 41, 58]. The resolving power of an AIMS instrument is low due to diffusion broadening of ions and the geometry and the number of the collecting electrodes [28, 41, 58]. Separation efficiency can be increased by scanning the electric field or increasing the number of collecting electrodes and variation of the electrodes shapes. It has to be noted that the AIMS instrumentation has a number of advantages, such as a high portability and the ability to perform long continuous measurements.

It should be mentioned that IMS detectors have a simple design and are quite rugged. Since they contain few moving parts, they are lightweight and can be miniaturised for field operations. In addition the detectors do not require any specialised power supplies, additional carrier gases or vacuum pumps [28, 41, 58].

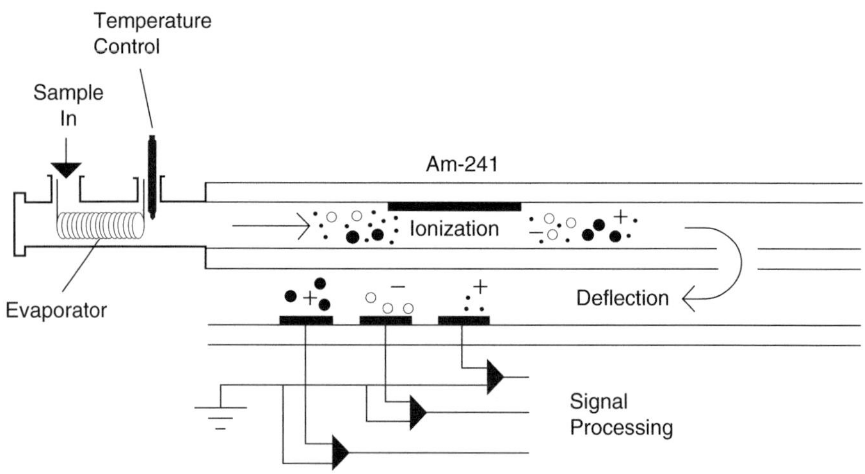

Fig. 3.17 Schematics of an aspiration ion mobility spectrometer analyzser (Adapted from Ref. [42])

3.7.3 *Travelling-Wave Ion Mobility Spectrometry (TWIMS)*

In the travelling-wave ion mobility spectrometry (TWIMS), a travelling voltage wave (T-wave) connected electrically to a series of ring electrodes (stacked ring ion guide; SRIG) pushes the ions through this detector [40, 48]. Figure 3.18. shows a schematic representation of the TWIMS principle. A radio frequency (RF) of opposite phase voltage is applied to adjacent electrodes to allow the ions confinement. When a direct transient (DC) current is superimposed on the RF that propagates to downstream at regular intervals, the ions will be pushed forward. As a result, separation will be achieved because, for a given wave speed and magnitude, the higher-mobility ions (red) will be carried forward by the wave [40, 48] (Fig. 3.18a and b).

3.7.4 *High-Field Asymmetric Waveform Ion Mobility Spectrometry (FAIMS)*

High-Field Asymmetric Waveform Ion Mobility Spectrometry (FAIMS) is another technology capable of separation of gas-phase ions at atmospheric pressure and room temperature [41, 50, 51]. FAIMS can also be operated over a wide range of pressure >1500 Torr and at different temperatures. The principle of this technology resides on the formation of ions that are then subjected to an alternating asymmetric

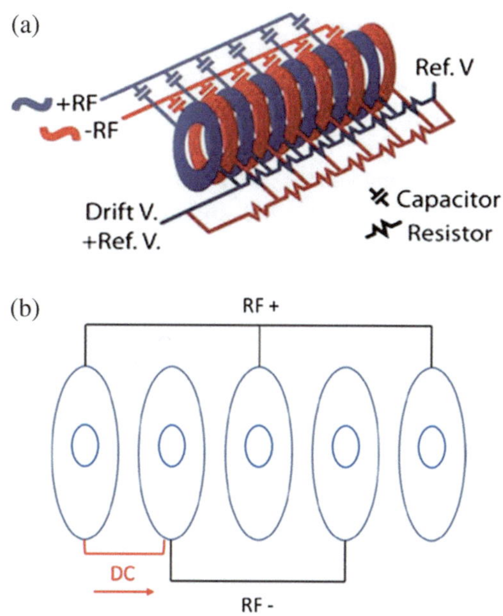

Fig. 3.18 Travelling-wave ion mobility spectrometry analyzer (TWIMS) (Adapted from Ref. [48]

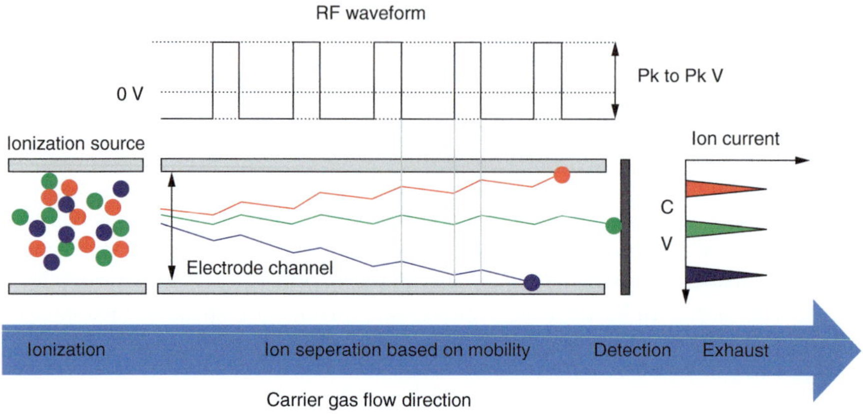

Fig. 3.19 Principle of a high-field asymmetric waveform ion mobility spectrometry (FAIMS) (Adapted from Ref. [51])

electric field (E) which will allow them to drift towards two electrodes at different rates. It should be noted, that the time at the positive voltage (t_1) is shorter than the time at negative voltage (t_2) (Fig. 3.19) [51].

Nevertheless, an equal (voltage × time) product for each part of the waveform is always maintained [40, 49, 50]. When we apply a DC voltage, termed the compensation voltage (CV) on these two electrodes, it will result in the repelling the ions and refocuses their flight through the device. Figure 3.17 shows the operation principles of FAIMS [41, 50, 51]. When the chemicals entering the system with carrier gas (nitrogen) they are ionized by the ^{63}Ni ionization source to negative or positive charged ions. After that the ions drift to the narrow gap which is formed by two electrodes, on which a high frequency asymmetric electric field is applied. This applied field, referred as dispersion field (DF), causes the oscillations of the ions perpendicularly to the gas flow [50, 51]. In order to detect all the CA analytes a compensation voltage (CV) is applied to superimpose the DF, so that the trajectories are shifted and ions with different mobility can reach the detector [51]. With the specific DF and CV the ions can be detected by the Faraday plate and the value of the ion current is proportional to the concentration of the analyte.

Nerve gases, blister agents, sulfur mustard; nitrogen mustards; lewisite; choking agents phosgene; explosives, the mother of Satan, and illicit drugs have been detected by all the different types of IMS [46–48].

3.8 Color-Change Chemistry

This technology is based on chemical reactions that occur when CAs interact with solutions and substrates. The most common indicator (for a positive response) is a color change. Detection tubes, papers, or tickets use some form of surface or substrate to which a reagent solution is applied. Many of these kits are complex and include multiple tests for specific agents or families of agents. Color change detectors can detect nerve, blister, and blood agents [22, 28, 52].

3.8.1 Dräger Tubes

Dräger tubes are commonly used for the detection of nerve and blister agents, which are phosphoric acid ester tubes and thioether tubes, respectively. The phosphoric ester tubes normally change from yellow to pink after exposure to a nerve agent, and, as agent concentration increases, the pink colour becomes more intense and tends to persist for longer time (usually greater than 60 s). The thioether tubes form an orange colour band on the yellow portion of the tube when exposed to detectable levels of HD. The intensity of the orange also increases with increased concentration of HD [22, 28, 52].

In addition, the Simultaneous Test Set, manufactured by Dräger Safety (Lübeck, Germany), consists of five tubes which are manifold together and attached to a hand pump which allows for simultaneous sampling through each of the tubes. Hence, the Test Set is capable of measuring several gases at once (Fig. 3.20) [22].

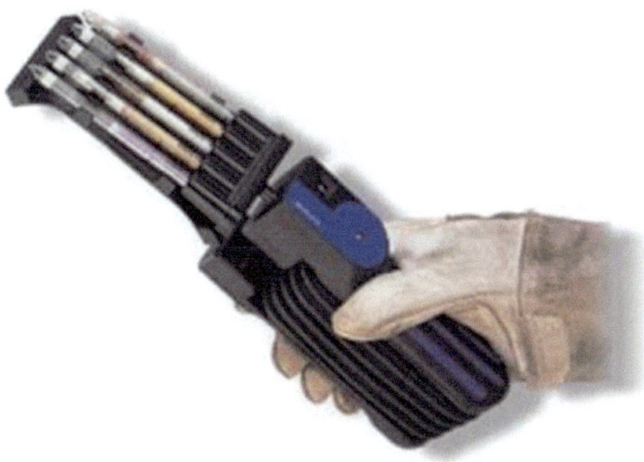

Fig. 3.20 The simultaneous test sets manufactured by Dräger safety (Lübeck, Germany) (Adapted from Ref. [53])

3.9 Photo Ionization Detectors (PIDs)

PIDs operate by passing the air sample between two charged metal electrodes in a vacuum that are irradiated with ultraviolet radiation, thus producing ions and electrons. The negatively charged electrode collects the positive ions, thus generating a current that is measured using an electrometer-type electronic circuit (Fig. 3.21) [54–56].

The measured current can then be correlated to the concentration of the molecular species. PID systems are used in portative detectors to detect nerve, blister, and mustard agents. They are highly quantitative and calibrated sensors provide excellent sensitivity levels (limits of detection reaching low ppm and ppb levels). PIDs are generally used as screening tools as they can provide fast, low-level and on-site analyses of chemicals contamination. They are able to detect and accurately measure the concentration of chemicals.

The FID detector is also not responsive to air, water. This is an extremely important advantage when volatile CAs are present in a water matrix, and therefore can be easily analyzed without any pre-treatment [23, 28, 54–56]. There is a main disadvantage in using FID-based detectors for field CA detection as they are non-selective and respond to all CAs that can generate ions via the hydrogen-air flame.

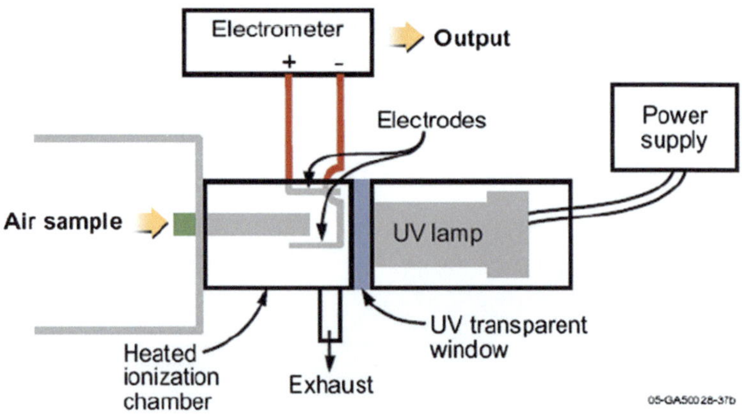

Fig. 3.21 Schematics of a photo ionization detector [56]

3.10 Conclusion

In this chapter, we have described the basic functioning principles of a series of the most important CWA detectors. Needless to say that we have omitted the presentation of commercial specific detectors used to analyse the different types of CWAs. This information coupled to the advantages and disadvantages of using these detectors can be obtained directly from the manufacturers. All of the current technologies discussed in this report have their advantages and disadvantages.

Nevertheless, the major challenge that remains is to increase the detection reliability and to reduce the frequency of false responses. Recently, research impetus has been directed towards the development of a capability for detection of both toxic chemicals and CAs. This challenge has created substantial challenges for commercial developers aimed at building improved detectors that can accurately provide advanced warning of CAs release.

References

1. Report of the OPCW on the implementation of the convention on the prohibition of the development, production, stockpiling and use of chemical weapons and on their destruction (2015) Conference of the States Parties Organisation for the Prohibition of Chemical Weapons. CS-2015-9540(E)
2. https://www.britannica.com/event/London-bombings-of-2005
3. Okumura T, Takasu N, Lshimatsu S, Minyanoki S, Mitsuhashi A, Kumada K, Tanaka K, Hinohara S (1996) Report on 640 victims of the Tokyo subway Sarin attack. Ann Energ Med 28:129–135
4. Banoub J (ed) (2014) Detection of chemical, biological, radiological and nuclear agents for the prevention of terrorism. Mass spectrometry and allied topics. IOS Press, Amsterdam and Springer, Dordrecht, in conjunction with the NATO Emerging Security Challenges Division, p 1–291
5. Hoenig SL (2007) Compendium of chemical warfare agents. Springer, New York
6. OPCW (2000) Fact sheet 4 what is a chemical weapon? OPCW, Hague
7. Borowitz JL, Kanthasamy AG, Isom GE (1992) Toxicodynamics of cyanide. In: Somani SM (ed) Chemical warfare agents. Academic Press, San Diego, pp 209–236
8. López-Muñoz F, Alamo C, Guerra JA, García-García P (2008) The development of neurotoxic agents as chemical weapons during the National Socialist period in Germany. Rev Neurol 47:99–106
9. Sidell FR, Borak J (1992) Chemical warfare agents: II. Nerve agents. Ann Emerg Med 21:865–871
10. Bajgar J (2004) Organophosphates/nerve agent poisoning: mechanism of action, diagnosis, prophylaxis, and treatment. Adv Clin Chem 38:151–216
11. Guide for the Selection of Chemical Agent and Toxic Industrial Material Detection Equipment for Emergency First Responders (2005) Volume I and II: Summary, p. 100–104
12. Walsh CJ (2008) Blood agents. In: Embar-Seddon A, Allan D, Pass AD (eds) Forensic science. Salem Press, Pasadena, p 150
13. Stuart JA, Ursano RJ, Fullerton CS, Norwood AE, Murray K (2003) Belief in exposure to terrorist agents: reported exposure to nerve or mustard gas by gulf war veterans. J Nerv Ment Dis 191:431–436

14. Pechura CM, Rall DP (1993) Chapter 3: history and analysis of mustard agent and lewisite research programs in the United States. In: Veterans at risk: the health effects of mustard gas and lewisite. National Academies Press, Washington, DC
15. Le HQ, Knudsen SJ (2006) Exposure to a First World War blistering agent. Emerg Med J 23:296–299
16. https://www.opcw.org/about-chemical-weapons/types-of-chemical-agent/psychotomimetic-agents/
17. Ketchum JS, Salem H (2008) In Incapacitating Agents, Medical Aspects of Chemical Warfare Tuorinsky SD (ed), pp 411–440
18. Olajos EJ (2004) Riot control agents issues in toxicology, safety, and health. CRC Press, Boca Raton/London/New York/Washington, DC, p 353
19. Norige A, Thorton J, Schiefelbein C, Rudzinski C (2009) High-density distributed sensing for chemical and biological defense. Lincoln Lab J 18:27–40
20. Carrano J (2007) Chemical and biological sensor standards study, Technical report, Defense Advanced Research Projects Agency, Arlington, VA, August 2007
21. Braden CG, Greg EC (2007) Synthetic methods applied to the detection of chemical warfare nerve agents. Curr Org Chem 11:255–265
22. Sferopoulos RA (2009) Review of Chemical Warfare Agent (CWA) Detector Technologies and Commercial-Off-The-Shelf Items. DSTO-GD-0570, Human Protection and Performance Division DSTO Defence Science and Technology Organisation, Victoria 3207 Australia, p 90
23. Duffy LM, Downing E, Huey BM, Mckone TM (2000) National Research Council, Commission on Life Sciences, Commission on Engineering and Technical Systems, Division of Military Science and Technology and Board on Environmental Studies and Toxicology, National Academies Press, p 272
24. Cripping JB (ed) (2005) Explosives and chemical weapons identification, Forensic science techniques series. CRC Press/Francis and Taylor, Boca Raton, p 288
25. Amani M, Chu Y, Waterman KL, Hurley CM, Platek MJ, Gregory OJ (2012) Detection of Triacetone Triperoxide (TATP) using a thermodynamic based gas sensor. Sens Actuators B Chem 162:7–13
26. http://news.nationalpost.com/news/world/mother-of-satan-the-highly-unstable-bomb-of-choice-for-terrorists-likely-used-in-brussels-attacks
27. http://namrataheda.blogspot.fr/2016/05/spectrophotometry-flame-photometry.html
28. Sun Y, Ong KY (2005) Detection technologies for chemical warfare agents and toxic vapors, 1st edn. CRC Press, Boca Raton/Florida, p 272
29. Budzier H, Gerlach G (2011) Thermal infrared sensors: theory, optimisation and practice. Wiley, Hoboken, p 324
30. Davis G. CBRNE – Chemical Detection Equipment. http://www.emedicine.com/emerg/topic924.htm
31. Luma Sense. Available online: www.lumasense.dk/INNOVA-1412.gas_monitoring4.0.html/
32. Photoacoustic Detection (PAS) In: Innova, vol 2007
33. Seeley JA, Richardson JM (2007) Early warning chemical sensors. Lincoln Lab J 17:85–99
34. http://www.darkgovernment.com/news/flir-forward-looking-infrared/flir-2-soldiers/
35. Gardiner DJ (1989) Practical Raman spectroscopy. Springer, Berlin
36. Lombardi JR, Birke RL (2008) A unified approach to surface-enhanced Raman spectroscopy. J Phys Chem C 112:5605–5617
37. Robinson R (2015) The Application of Differential Absorption Lidar (DIAL) for Pollutant Emissions Monitoring, Environmental Measurements Group – Analytical Science Division National Physical Laboratory Teddington, http://www.npl.co.uk/environment
38. Mudaliar S (2013) Remote sensing of layered random media using the radiative transfer theory. Radio Sci 48:535–546
39. Kotidis P, Deutsch E, Goyal A. Standoff detection of chemical and biological threats. Defense & security: http://spie.org/newsroom/5844-standoff-detection-of-chemical-and-biological-threats

40. Kano AB, Dwivedi P, Tam M, Matz L, Hill HH Jr (2008) Ion mobility-mass spectrometry. J Mass Spectrom 43:1–22
41. Eiceman GA, Karpas Z, Hill HH Jr (2013) Ion mobility spectrometry third edition. CRC Press/ Francis and Taylor, p 444
42. Smith J (2015) Brodie's Bombs and Bombings A Handbook to Protection, Disposal and Investigation for Industry, Police and Fire Departments Fourth Edition, Charles C Thomas Publisher LTD, p 297
43. Zolotov YA (2006) Ion mobility spectrometry. J Anal Chem 61:519
44. Srebalus CA, Li JW, Marshal WS, Clemmer DE (1999) Gas-phase separations of electrosprayed peptide libraries. Anal Chem 71:3918–3927
45. Trimpin S, Plasencia MD, Isailovic D, Clemmer DE (2007) Resolving oligomers from fully grown polymers with IMS-MS. Anal Chem 79:7974
46. Zhang W, Quernheim M, Joachim Räder M, Müllen K (2016) Collision-induced dissociation ion mobility mass spectrometry for the elucidation of unknown structures in strained polycyclic aromatic hydrocarbon macrocycles. Anal Chem 88:952–959
47. (a) Buryakov IA (2011) Detection of explosives by Ion mobility spectrometry. J Anal Chem 66:674 (b) Puton J, Namiesnik J (2016) Ion mobility spectrometry: current status and application for chemical warfare agents detection. Trends Anal Chem 85:10–20
48. Hill HH, Simpson G (1999) Capabilities and limitations of ion mobility spectrometry for field screening applications. Field Anal Chem Technol 13:119–134
49. Smith DP, Knapman TW, Campuzano I, Malham RW, Berryman JT, Radford SE, Ashcroft AE (2009) Deciphering drift time measurements from travelling wave ion mobility spectrometry-mass spectrometry studies. Eur J Mass Spectrom (Chichester) 15:113–130
50. Hsi P, Sheikham B, Negre T, Mulhem M, Stohecker J (2103) The PID handbook-theory and applications of direct-reading photoionization detectirs, 3rd ed. RAE Systems by Honeywell, RAE Systems Inc, San Jose. www.raesystems.com
51. http://www.geo.uni-tuebingen.de/studium/studentische-projekte/wiss-praesentieren-ss2012/applied-environmental-geoscience/yu-ye.html
52. Smith PA, Lepage JC, Harrer KL, Brochu PJ (2007) Handheld photoionization instruments for quantitative detection of Sarin vapor and for rapid qualitative screening of contaminated objects. J Occ Env Hyg 4:729–738
53. Dräger Tube Sets: Simultaneous Test Sets. In Dräger safety
54. Zrodnikov Y, Davis CE (2012) The highs and lows of FAIMS: predictions and future trends for high field asymmetric waveform ion mobility spectrometry. J Nanomed Nanotechol 3:5
55. Buryakov IA, Krylov EV, Nazarov EG, Rasulev UK (1993) A new method of separation of multi-atomic ions by mobility at atmospheric-pressure using a high-frequency amplitude-asymmetric strong electric field. Int J Mass Spectrom Ion Process 128:143–148
56. Daum KA, Watrous MG, Neptune MD, Michael DI, Hull KJ, Evans JD (2006) Data for frist responder use of photoionization detectors for vapor chemical constituents. INL Idaho National Laboratory, SAIC, pp 78

Chapter 4
Fundamental Principles for Luminescence Sensing Measuring Devices Used for the Detection of Biological Warfare Agents

Joseph H. Banoub and Farid Jahouh

Abstract This chapter surveys the current detection technologies used in commercially available luminescence biosensor detection equipments currently employed for identifying warfare biological agents (BAs). Brief technical descriptions of these technologies are presented with emphasis placed on the principles of detection. Much of the content presented was obtained from the open-source literature and is an introduction to biosensor fundamentals

4.1 Introduction

Biological agents (BAs) are widely found in the natural environment (work place, hospitals, HVAC, etc.) and as a result of voluntary release of biological warfare or terrorism agents [1, 2]. Biological agents include series of different virulent bacteria, viruses, fungi (yeasts and moulds) and parasites. All of these agents possess irreversible threat to potentially cause ill health and death to soldiers/humans. The incidents of anthrax-laced letters, the emergence of Severe Acute Respiratory Syndrome (SARS), and repeated occurrences of illnesses caused by food-borne pathogens highlight the need for rapid and sensitive identification of the responsible biological agents [3, 4].

Biological agents are usually invisible and possess the ability to infect in very small doses [1, 2]. They also have the cunning ability to replicate rapidly and require minimal resources to survive. For these reasons, it is impossible to initially feel their

J.H. Banoub (✉)
Fisheries and Oceans Canada, Science Branch, Special Projects, Chemistry Department, Memorial University, St John's, NL, Canada
e-mail: banoubjo@dfo-mpo.gc.ca

F. Jahouh
Department of Chemistry, Memorial University of Newfoundland, St. John's, NF, 232 Elizabeth Avenue, A1B 3X7, Canada

J.H. Banoub, R.M. Caprioli (eds.), *Molecular Technologies for Detection of Chemical and Biological Agents*, NATO Science for Peace and Security Series A: Chemistry and Biology, DOI 10.1007/978-94-024-1113-3_4

presence or to predict the risks they present. Recently, it has been demonstrated that bioterrorism raises the specter of exposure to toxins by devising new deliveries by which BAs can be weaponized [5–7]. It has become apparent that BAs weapons pose a real and potentially immediate threat as they are relatively cheap to manufacture and employ, and they have tremendous potential impact as terror weapons [2]. These features make biological weapons attractive to rogue states and terrorist organizations. In this article we briefly describe the threat of biological weapons [5–7].

4.1.1 Classification of Biological Agents

Biological Agents are categorized according to the Code of Practice to the Safety, Health and Welfare at Work (Biological Agents) Regulations, 2013 [8]. This classification system is based on whether: the biological agent is pathogenic to humans, represents a hazard to soldiers, it is transmissible to nearby community and if there is a possible effective treatment available [8–11]. Consequently, BAs are classified into four risk groups (RG), namely, RG1, RG2, RG3 and RG4, as follows:

- RG1 that includes BAs not linked with diseases in healthy adult humans. Examples of RG1 agents include *Bacillus subtilis* or *Bacillus licheniformis*, *Escherichia coli*-K12, and adeno-associated virus (AAV) types 1 through 4 [9, 10].
- RG2 that includes BAs connected with human diseases which is rarely serious and for which preventive or therapeutic interventions are often available. Examples of RG2 agents include *Salmonella* sp., *Chlamydia psittaci*, measles virus, and hepatitis A, B, C, D, and E viruses [9, 10].
- RG3 that contains BAs associated with serious or lethal human diseases for which preventive or therapeutic interventions may be available (high individual risk but low community risk). Examples of RG3 agents include *Brucella*, *Mycobacterium tuberculosis*, *Coccidioides immitis*, yellow fever virus and human immunodeficiency virus (HIV) types 1 and 2 [9, 10].
- RG4 includes BAs that are likely to cause serious or lethal human disease for which preventive or therapeutic interventions are *not usually* available (high individual risk and high community risk). RG4 agents only include viruses. Examples of RG4 agents consist of Crimean-Congo hemorrhagic fever virus, Ebola virus and herpesvirus simiae (B-virus). Under the classification system, These Risk Group 1 agents are the least hazardous whilst Group 4 are the most hazardous [9, 10].

4.2 Analytical Measurement of Biological Agents

Field BAs and urban public health surveillance systems usually provide rapid determination of the presence of BAs in the atmosphere and may in time provide an indication of when and where the biological agent was released [11–15]. Once, the point source is revealed and the BA identified, rapid clean-up effort could be initiated, after the release of the compound. Moreover, it is essential to monitor the presence of BAs in the environment, to provide people at risk with the means of rapidly identifying contaminated air, water, food and equipment [11–15]. It is well known that antibodies specifically targeting proteins or pathogens, can be generated to a wide variety of target cell bacterial analytes and are the most popular choices for the recognition element in many biosensors. Typical immunoassay formats include competitive, displacement, sandwich, and enzyme-linked immunosorbent assays (ELISA) [14]. ELISA, for example, uses colorimetric or chemiluminescent enzyme substrates for signal transduction and is more suited to automated instruments because of the multiple incubations and washes required.

The classical approach to detect the bacteria species type or microbe involves the use of differential metabolic assays, monitored colorimetrically and on immunochemical routine tests. In addition, the use of cell culture and electron microscopy are vital for the diagnosis of viruses, bacteria and intracellular parasites. Consequently, all samples retrieved from the affected environment must ne cultured in order to obtain sufficient numbers of various cell types for reliable identification. Unfortunately, one of the major drawbacks is that the time required for the microbial outgrowth is long and some bacteria obtained are not culturable, as a result of genetic mutation. This chapter has been drafted as a "Comptes Rendus" and general *modi operandi* of novel biodetection approaches using luminescence biosensing approaches, published recently, for the detection of warfare potential BA weapons; it will exclude all the conventional approaches. The readers are encouraged to peruse the references included within the text, to search for the applied commercial used biosensors, that for simplification reasons, we did not discuss herewith.

4.2.1 What Are Biosensors?

According to a recently proposed IUPAC definition a biosensor is "a self-contained integrated device which is capable of providing specific quantitative or semi-quantitative analytical information using a biological recognition element (biochemical receptor) which is in direct spatial contact with a transducer element [15]. A biosensor should be clearly distinguished from a bioanalytical system, which requires additional processing steps, such as reagent addition. The term "biosensor" is short for "biological sensor." The device is made up of a transducer and a biological element that can be an enzyme, an antibody or a nucleic acid. The bioelement interacts with the analyte being tested and the biological response is converted into an electrical signal by the transducer (Fig. 4.1) [16, 17].

Fig. 4.1 Schematic representation of the detection of a bioelement using a biosensor

4.2.2 Overview of Predominant Sensing Techniques

Sensing can be explained as the use of recognition elements (biological in origin) for binding to the biothreat molecule of interest. The binding event must be transduced in a manner that signals the presence of the targeted analyte. Biosensor probes are becoming increasingly sophisticated, mainly owing to combination of advances in two technological fields: microelectronics and biotechnology. Biosensors are highly valuables devices in measuring a wide spectrum of BA analytes [18, 19].

Ideally each sensing detection technology should contain the following characteristics:

- Specific and able to discriminate between closely related pathogenic and non-pathogenic organisms or toxins.
- Sensitive and able to detect small amounts of target within a high background matrix.
- Possess high affinity and being able to maintain binding even through repeated washing steps.
- Stable enough to allow long-term use.

It should be understood that luminescence biosensing technology is totally distinct from other physiochemical methods, such as mass spectrometry (MS) or Fourier transform infrared spectroscopy (FTIR) and Raman based analysis. These methods of course, have their merits and are very sensitive and specific. Therefore, in the following sections, we will provide an overview of the state-of-the-art prime molecular sensing technologies for the detection of BAs.

4.3 Electrical Detection of Pathogenic Bacteria via Immobilized Antimicrobial Peptides

It should be noticed that the current methods for detecting pathogenic bacteria, which include ELISA and PCR [20, 21] are assays that exploit antibodies as molecular recognition elements due to their highly specific targeting of antigenic sites. Nonetheless, these antibodies lack the stability needed to detect pathogenic species under harsh environments, and requires a one-to-one pairing of antibody-based sensors for each target to be detected. Whereas, nucleic acid probe-based techniques such as PCR can reach single-cell detection limits, they still require the extraction of nucleic acids and are limited in portability [21, 22].

McAlpine and coworkers developed a robust and portable biosensor for the detection of pathogenic bacteria that could impact water quality monitoring for bacterial contamination [23–25]. The particular interest of the developed biosensor was that it combined the natural specificity of biological recognition to label-free sensors providing sensitive electronic readout. Thus, McAlpine et al. reported the selective and sensitive detection of infectious agents via electronic detection based on antimicrobial peptide-functionalized microcapacitive electrode arrays [25].

The semi-selective antimicrobial peptide magainin I, which occurs naturally on the skin of African clawed frogs, was immobilized on gold microelectrodes via a C-terminal cysteine residue. Significantly, exposing the sensor to various concentrations of pathogenic *Escherichia coli* revealed detection limits of approximately 1 bacterium/μL, a clinically useful detection range. The peptide-microcapacitive hybrid device was further able to demonstrate both Gram-selective detection as well as interbacterial strain differentiation, while maintaining recognition capabilities toward pathogenic strains of *E. coli* and *Salmonella* [25].

It is well known that the synthesis of antimicrobial peptides (AMPs) and their resulting intrinsic stabilities render them particularly interesting candidates for the use as molecular recognition elements in electronic biosensing platforms [23–25]. AMPs do exist in nature and are located either in the skin of higher organisms and/or in the extracellular milieu of bacteria [25]. The replacements of current antibody-based affinity probes, with more stable and durable AMPs in biological sensors, have a major advantage as recognition elements. This advantage stems from the AMPs semi-selective binding nature to target cells of a variety of pathogens.

The bioactivity of AMPs toward microbial cells can be classified into groups according to their secondary structures [24, 25]. Many AMPs adopt amphipathic conformations that spatially shield the hydrophobic group of the cationic amino acids, thereby targeting the negatively charged head groups of lipids in the bacterial membrane. In contrast, the membranes of plants and animals separate negative charges to the inner leaflet and contain cholesterols that reduce AMP activity [25]. The AMPs, linear cationic peptides such as magainins, are particularly attractive for microbial sensing applications because of their small molecular size and intrinsic stability [26, 27]. In particular, the positively charged AMP magainin I (GIGKFLHSAGKFGKAFVGEIMKS) binds most selectively to the bacterial cell *E. coli* O157:H7 as a precursor to bactericidal activity [25–27].

4.3.1 Development of an AMP-Based, Label-Free Electronic Biosensor

Magainin I displays a broad-spectrum activity toward other Gram-negative bacteria, which comprise the majority of pathogenic infection in humans [25–28]. The first step toward the development of an AMP-based, label-free electronic biosensor consisted of targeting the microbial cells by magainin I using impedance spectroscopy.

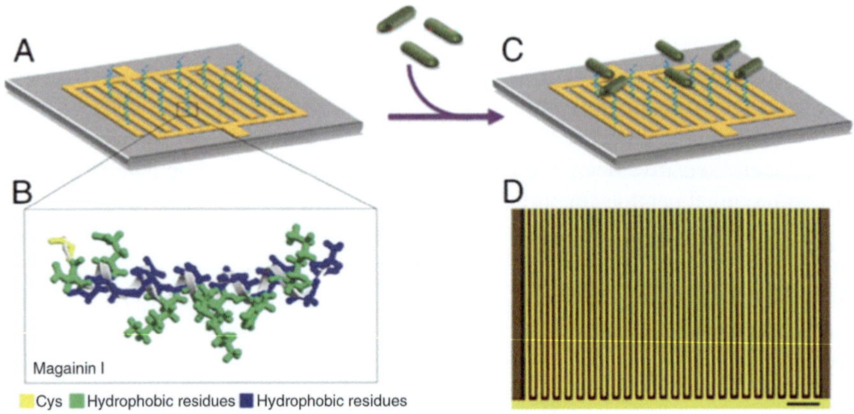

Fig. 4.2 AMP-based electrical detection of bacteria (Adapted from Ref. 25).

It should be repeated that electrical impedance measures the total opposition to a circuit or a part of the circuit presented to an electric current. Usually the impedance is the results for both resistance and reactance. It is important to remember that the resistance component arises from collisions of the current-carrying charged particles with the internal structure of the conductor. Furthermore, the reactance component is an additional opposition to the movement of the electric charge that result from the changing magnetic and electric fields in circuits carrying alternating current.

Figure 4.2 outlines the sensing platform. First, the AMPs are immobilized on the micro-fabricated interdigitated gold electrodes (Fig. 4.2a). It should be specified that magainin I contains a cysteine residue on the C terminus (Fig. 4.2b), which allows the facile site-specific covalent attachment to the gold electrodes. Next, the heat-killed bacterial cells are injected and incubated on the AMP-modified electrodes [25]. When, the bacteria are recognized by the AMPs, binding will ensue (Fig. 4.2c), causing the dielectric property to changes that can be monitored by a spectrum analyzer. Usually, the impedance is measured over a frequency range of 10 Hz to 100 kHz. Figure 4.2d shows an optical micrograph of the device, which is made using standard microfabrication techniques.

The results of measurements performed after incubation of the immobilized AMPs with pathogenic *E. coli* O157:H7 cells concentrations ranging from 10^3 to 10^7 cfu/mL are shown in Fig. 4.3. When, a blank device with no immobilized AMPs was also tested for comparison, it was found that there was no change in the impedance of the blank device without immobilized AMPs, upon exposure to various bacterial concentrations. Figure 4.3a shows that at low frequencies, the different concentrations of bacterial cells have the effect of increasing the impedance in proportion to the number of cells present in the sample. As the frequency increases, the contribution to the impedance from the bacterial cells decreases, leaving only the dielectric relaxation of small dipoles including water molecules in the buffer solution to affect the measured impedance. Figure 4.3b depicts the impedance change at a fixed frequency of 10 Hz. The variation in the impedance is directly proportional

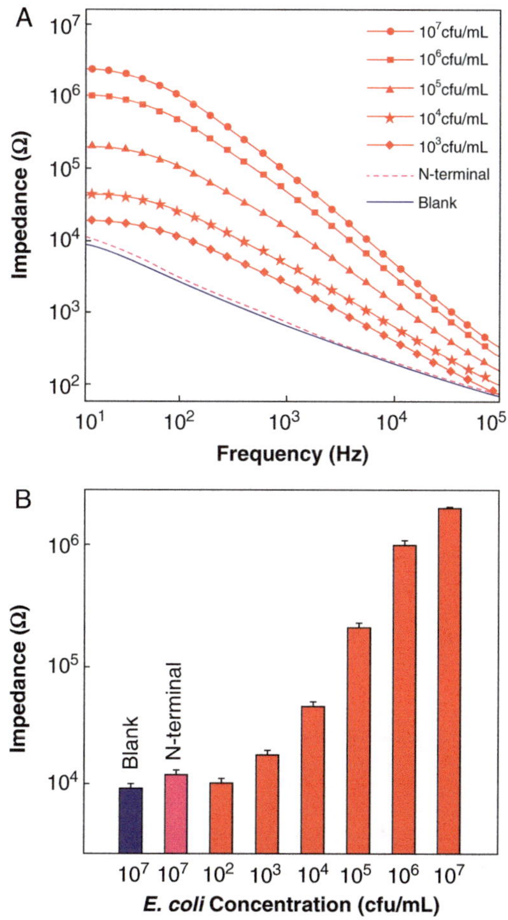

Fig. 4.3 Impedance spectra and sensitivity of the AMP electronic biosensor (Adapted from Ref. 25)

to the number of bacterial cells bound to the immobilized AMPs and manifested in a logarithmic increase with respect to serially diluted bacterial concentrations. Significantly, the detection limit of response of the hybrid AMP-microelectrode device to *E. coli* was found to be 10^3 cfu/mL (1 bacterium/μL). This lowest limit of detection appears to be limited by the presence of impedance due to the electrical double layer resulting from the electrode polarization effect at low frequencies. Notably, these sensitivity limits is clinically relevant [30] and compares favorably to AMP-based fluorescent assays [26], antibody-based impedance sensors [27], and to the LAL test [25].

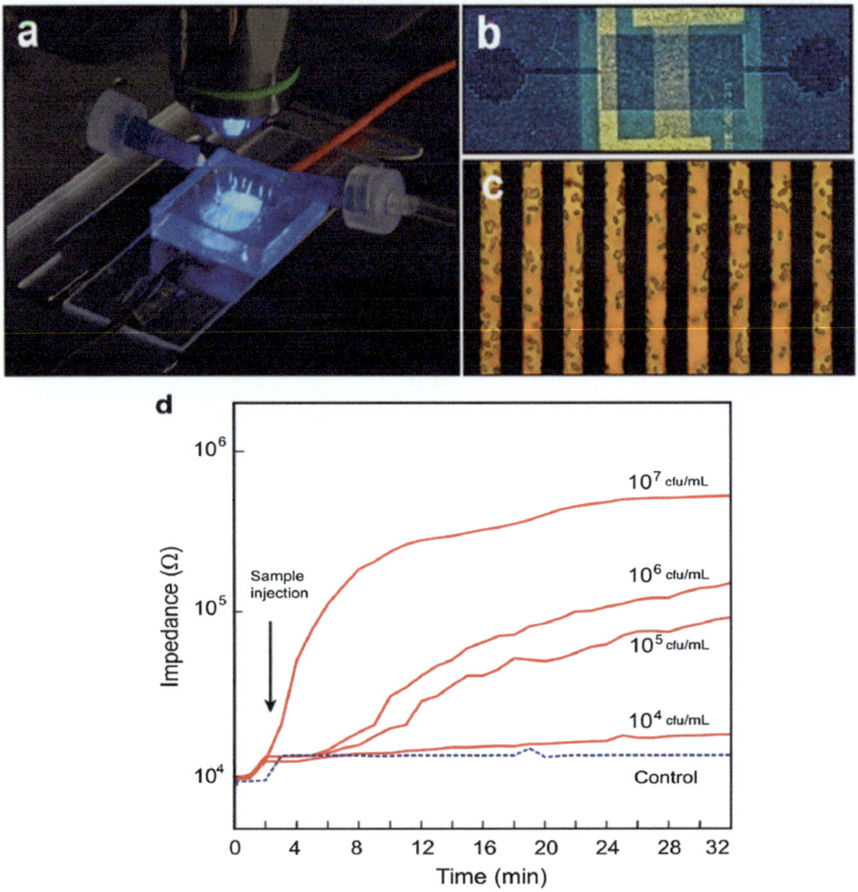

Fig. 4.4 Real-time binding of bacteria to AMP biosensors (Adapted from Ref. 25)

4.3.2 Pathogenic Bacteria Real-Time Detection

To simulate the use of the AMP microelectrodes in everyday applications, such as direct water sampling, the biosensor response was investigated in real time, as shown in Fig. 4.4. First, a microfluidic cell was bonded to the interdigitated biosensor chip (Fig. 4.4a), such that the electrodes were perpendicular to the direction of the sample flow (Fig. 4.4b) [25].

Next, the fluid was injected using a syringe pump connected to the inlet port and allowed to flow through to the outlet port at a flow rate of 100 μL/ min. The flow cell was first flushed with buffer to establish a baseline. This was followed by injection of various dilutions (10^4–10^7 cfu/mL) of the pathogenic *E. coli* cells in PBS to the channel at a reduced flow rate of 5 μL/ min for 30 min. For example, Fig. 4.4c shows the microelectrode array after exposure to 10^7 cfu/mL bacterial cells.

Simultaneously, the impedance response was continuously monitored during the sample flow-through process (Fig. 4.4d).

All samples produced a measurable response relative to the control sample within 5 min, with the highest concentration sample yielding a response within 30 s; the responses saturated after after 20 min. These results augured well for the implementation of this sensor in continuous monitoring of flowing water supplies.

4.3.3 Selectivity Measurements

McAlpine also examined the selectivity o the AMP-functionalized biosensors toward the following various bacterial species (*i*) Gram-negative pathogenic *E. coli* O157 : H7, (*ii*) the nonpathogenic *E. coli* strain American Type Cell Culture (ATCC) 35,218, (*iii*) Gram-negative pathogenic *Salmonella typhimurium*, and (*iv*) the Gram positive *Listeria monocytogenes*. The selectivity was first investigated using fluorescent microscopy methods, by staining bacterial cells and optically mapping their binding density to gold films hybridized with AMPs. The discriminative binding patterns of immobilized magainin I to and the surface density of the various bacterial cells (all 10^7 cfu/mL) stained with propidium iodide (PI) nucleic acid stain are showed in Fig. 4.5. In summary, coupling of AMPs with microcapacitive biosensors has resulted in the implementation of a portable, label-free sensing platform for the detection of infectious agents. The achievable sensitivity approached 1 bacterium/μL-a clinically relevant limit-and the AMPs allowed for sufficient selectivity to distinguish pathogenic and Gram-negative bacteria, while retaining broadband detection capabilities [25–28]. In addition, the simulated water sampling chip which consisted of a microfluidic flow cell integrated onto the hybrid sensor, demonstrated the potential of real-time on-chip monitoring of the interaction of *E. coli* cells with the antimicrobial peptides [25].

4.4 NRL Array Biosensor for Toxin Detection

This following part discusses the progress made with the NRL Array Biosensor, which is a portable instrument for rapid and simultaneous detection of multiple targets which was developed, automated and miniaturized for operation at the point-of-use by the US Navy Research Laboratories. This Array Biosensor has been used for the quantitative immunoassays against an expanded number of toxins and toxin indicators in food and clinical fluids, and for its usefulness as semi-selective molecules which can be used as alternative recognition moieties [29–31]. In this sensor, the antibodies or other capture molecules are immobilized in a two-dimensional array on an optical waveguide (as either stripes or spots) and standard fluoroimmunoassays are performed within the channels of a multi-channel flow cell, which is placed on the waveguide surface (Fig. 4.6, left). The spots are interrogated using

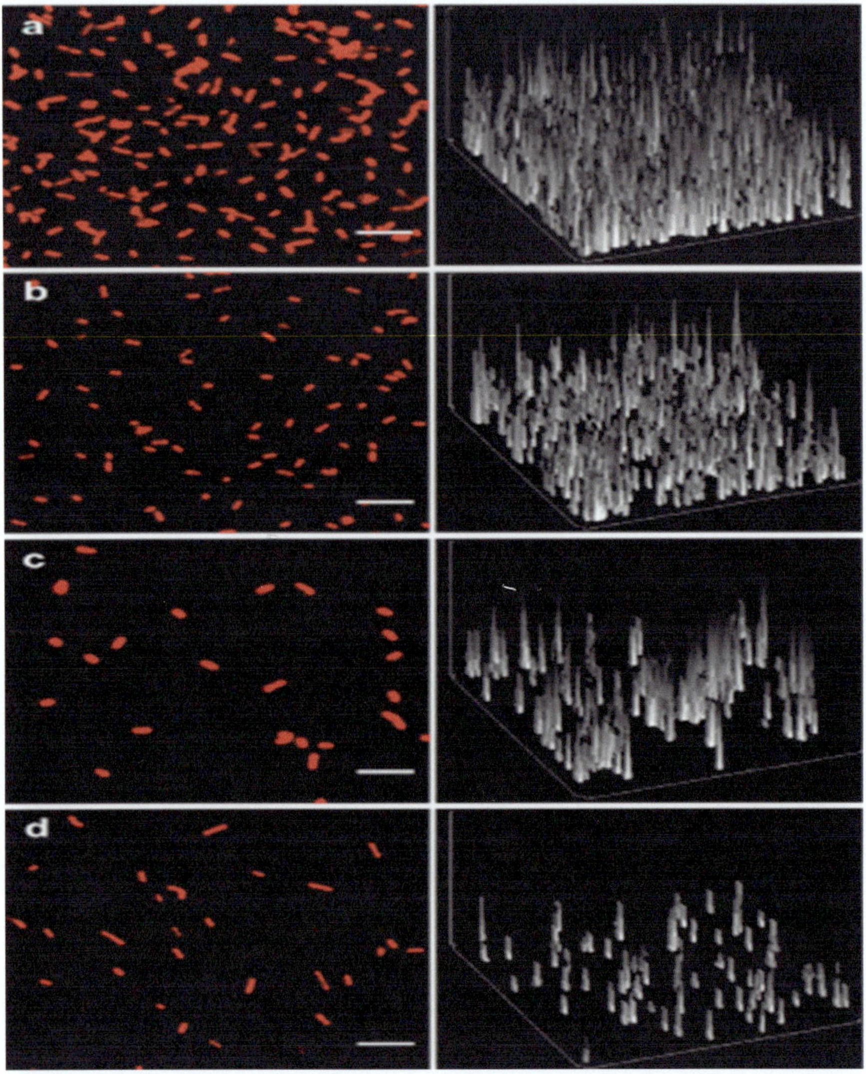

Fig. 4.5 Optical microscopy of the selectivity of AMPs (Adapted from Ref. 25)

evanescent wave technology: light from a 635 nm diode laser is focused into the edge of the patterned slide/waveguide and after propagation and mixing within the waveguide; the confined beam produces an evanescent field within the sensing portion of the waveguide [29]. In the NRL array biosensor the spots are interrogated using evanescent wave technology that is specifically: a light issued from a 635 nm diode laser that is focused into the edge of the patterned slide/waveguide and after propagation and mixing within the waveguide, the confined beam produces an evanescent field within the sensing portion of the waveguide [29, 32]. The definition of

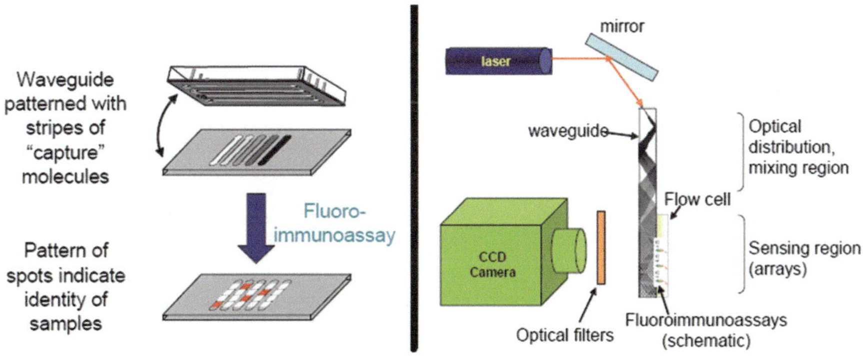

Fig. 4.6 NRL array biosensor (Adapted from Ref. 29)

an evanescent field or wave is: an oscillating electric and/or magnetic field, which does not propagate as an electromagnetic wave but whose energy is spatially concentrated in the vicinity of the source (oscillating charges and currents) [32].

The Surface bound molecules labeled with fluorophore are excited by this evanescent field, producing a fluorescence signal; this fluorescence is then detected using a CCD camera fitted with appropriate bandpass and longpass filters (Fig. 4.6, right). Since the penetration depth of the evanescent field is limited, only surface-bound fluorophores are excited, enabling analysis of non-homogeneous or turbid samples. The locations and intensities of the fluorescent spots indicate the identity and concentration of the target sample in each lane [29, 32].

The NRL Array Biosensor was used successfully for the detection of toxins and for multiple toxins simultaneously in multiple samples, it could also detect toxin levels as low as 500 pg/mL and quantify the toxin concentration, and finally it could perform toxin assays in clinical, food, and environmental samples [29]. Furthermore, both sandwich immunoassays for protein toxins (e.g., staphylococcal enterotoxin B [SEB] and ricin) and competitive immunoassays for low molecular weight toxins (e.g., trinitrotoluene and fumonisin B1) were reported (Fig. 4.6) [29].

4.4.1 Toxins Environmental Testing in Food and in Air

Determination of bacteria and large toxins in foods and air by the NRA Array Biosensor Assays normally employ a sandwich immunoassay format. However, mycotoxins are smaller in size and are therefore better assayed using an indirect competitive immunoassay [32–35].

The validity of the NRL Array Biosensor was demonstrated for the detection of mycotoxins (ochratoxin A, deoxynivalenol and aflatoxin B1) in various food matrices and in air [32–35]. This competitive assay protocol involved attaching the biotinylated mycotoxin derivatives onto the waveguide, this was followed by incubating

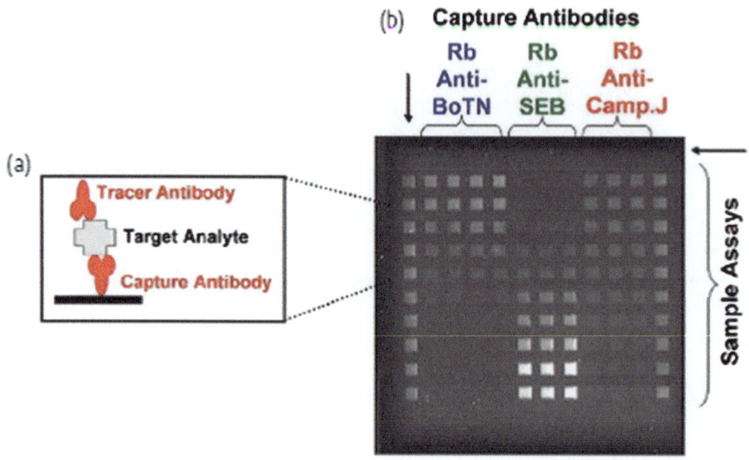

Fig. 4.7 Immunoassay with the NRL array biosensor using the sandwich immunoassay format (Adapted from Ref. 29)

the test sample with cyanine 5 (Cy5)-labeled anti-toxin antibodies and then passing the pre-incubated mix over the immobilized mycotoxin derivatives (Fig. 4.7). Since, the immobilized mycotoxin derivatives competed with the toxin in the test sample for binding to the fluorescent antibodies; it was found that the resulting fluorescent signal of the immunocomplex on the waveguide surface was inversely proportional to the concentration of toxin in the sample (decrease in signal with increasing concentration [32–35].

This type of Array Biosensor has been automated and miniaturized for operation as point-of-use quantitative immunoassay arrays. This methodology was utilized to measure an expanded number of toxins and toxin indicators in food and clinical fluids. In addition, semi-selective recognition molecules were also used to expand the repertoire of toxins that can be detected on a single array. In the automated system, up to 6 samples can be analyzed simultaneously while the non-automated system can test up to 12 samples [30B].

4.4.2 Use of Antimicrobial Peptides for Toxin Detection

Taitt and coworkers investigated the possibility to use antimicrobial peptides (AMPs) for the detection of inactivated botulinum toxins A, B, and E as well as other toxins in assays analogous to the AMP-based bacterial assays [36]. They observed clear differences in the patterns of binding between botulinum neurotoxoids A, B, and E. It was found that the detection limits were improved when immobilized AMPs were used for target capture [36].

4.5 Surface Acoustic Wave (SAW) Sensors

The Love wave (LW) physical effect was originally discovered by the mathematician Augustus Edward Hough Love [37, 38]. Typically, LW sensors consist of a transducing area and a sensing area. The transducing area consists of the interdigital transducers (IDTs), which are metal electrodes, sandwiched between the piezoelectric substrate and the guiding layer [39–41]. The input IDT is excited electrically (applying an rf signal) and launches a mechanical acoustic wave into the piezoelectric material which is guided through the guiding layer up to the output IDT, where it gets transformed back to a measurable electrical signal (Fig. 4.8).

It should be understood that the sensing area, is the area of the sensor surface, located between the input and output IDT, which is exposed to the analyte. Consequently, to permits the use of a SAW device as a biosensor, the device has to be coated with a biospecific layer corresponding to the analyte. The immobilization chemistry strongly depends upon the underlying SAW substrate with or without a guiding layer and hence on the chemical environment available. Gold surfaces, for example, allow the use of functionalized thiols, whereas quartz or SiO2 surfaces enable the use of various silanes [42]. Analyte-specific molecules (e.g., antibodies) are immobilized on the SAW device to catch analyte molecules (e.g., antigens) from the sample stream. Analytes binding to the immobilized capture molecules will influence the velocity of the SAW and hence the output signal generated by the driving electronics [39].

SAW detectors have the ability to identify and measure many BAs simultaneously and are relatively inexpensive, making them a popular choice amongst civilian response units [39–42]. SAW detect changes in the properties of acoustic waves as they travel at ultrasonic frequencies in piezoelectric materials. The basic transduction mechanism involves interaction of these waves with surface-attached matter. Multiple sensor arrays with multiple coatings and pattern recognition algorithms provide the means to identify agent classes and reject interferant responses that could cause false alarms [39–43].

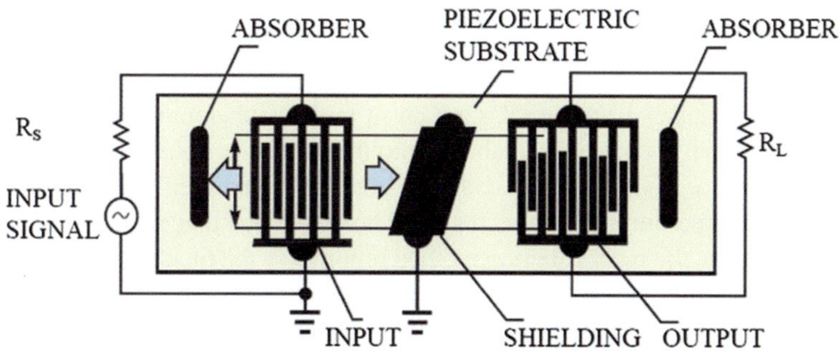

Fig. 4.8 Surface Acoustic Wave Sensor (SAW) (Adapted from Ref. [40])

Recently, SAW immunosensors were successfully applied to detect *E. coli*, *Legionella*, the anthracis simulant B8 *Bacillus thuringiensis* (B8) and M13 bacteriophage (M13) acting as model analyte for bacteria or viruses. Tamarin et al. used Love wave sensors based on quartz substrate with a SiO2 wave-guiding layer. Antibodies against M13 were immobilized on the sensor surface to detect the bacteriophage directly [43].

Stubbs et al. developed a SAW immunoassay for the detection of analytes in the gas phase, e.g., cocaine plumes [44]. For this purpose, SAW devices based on quartz were used. Antibodies were coupled to the SAW device via adsorbed protein A and coated with a hydrogel layer to overcome the problem of hydration of the biomolecule [44]. Benzoylecgonine, the major metabolite of cocaine, could be detected in vapor [45]. In general, it can be specified that SAW based biosensors offer the possibility of observing real-time binding events of proteins at relevant sensitivity levels [45, 46].

4.6 Biosensing with Luminescent Semiconductor Quantum Dots

Recently, luminescent semiconductor nanocrystals or quantum dots (QDs) have become a successful novel nanomaterial possessing unique photophysical fluorescent properties, which have helped create a new generation of robust fluorescent biosensors [47–53]. It should be stated that the fluorescent properties of QDs have overcome most of the liabilities of conventional organic and protein-based fluorophores. The biosensing QD properties of interest include high quantum yields, broad absorption spectra coupled to narrow size tunable photo-luminescent emissions and exceptional resistance to both photo-bleaching and chemical degradation. In this section, we will examine the progress in adapting QDs for several predominantly in vitro biosensing applications including use in immunoassays, as generalized probes [47–53].

4.6.1 Immunoassays Using Quantum Dots

The unique advantages of using QDs are owed to their inherent photostability, their improved sensitivity, and size tunable photoluminescence coupled to their broad absorption spectra. These unique advantages have allowed them to serve as multicolor or multiplexed immunoassays. It should be mentioned that in terms of coupling QDs to antibodies, the most common method reported in the literature utilizes biotin-avidin interactions [48]. The avidin/streptavidin coated QDs are commercially available, while biotin-labeling of antibodies are usually prepared in-house. QDs presenting available carboxylic acids from their capping agents may also be covalently attached

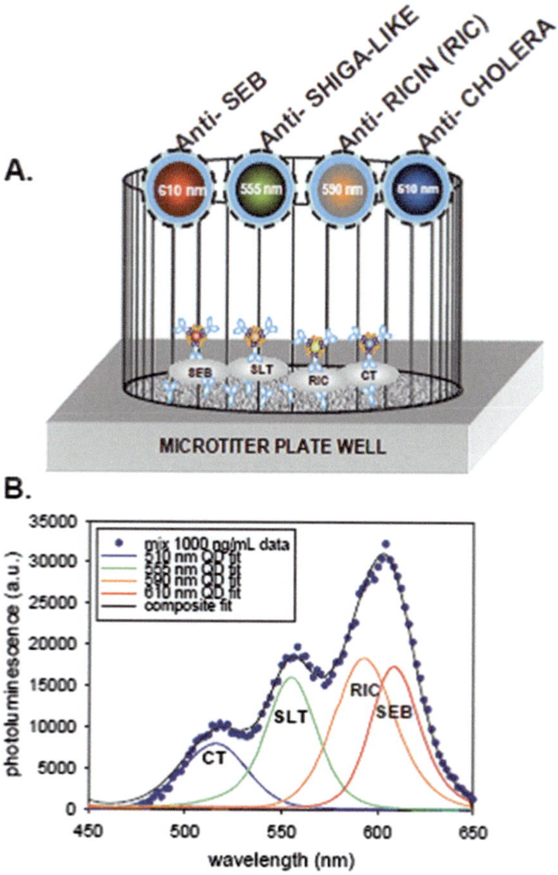

Fig. 4.9 QD multiplexed biosensing (Adapted from Ref. 54)

to the epsilon amine of an antibody's lysine residues by using EDC/NHS coupling chemistry [47–53]. Alternatively, simple electrostatic interactions can be used depending on the overall protein charge at the pH of conjugation [48].

Goldman et al. used sandwich immunoassays for the simultaneous detection of four toxins: cholera toxin, ricin, shiga-like toxin 1 and staphylococcal enterotoxin B (SEB), in a single microtiter well, see Fig. 4.9 [47, 48, 54].

In this assay capture antibodies immobilized in a microtiter well plate were first exposed to the mixed toxin sample. Antibodies specific for each of the toxins coupled to a different color QD were then added to the microtiter well plate. The resulting signal from the mixed toxin samples was then deconvoluted using a simple algorithm. Similarly, QD-antibody bioconjugates were used to identify and differentiate between diphtheria toxin and tetanus toxin proteins which were non-specifically immobilized onto poly-L-lysine coated cover slips and for the simultaneous detection of *Escherichia coli* O157:H7 and *Salmonella typhimurium* bacteria using different colored QDs as immunoassay labels [55].

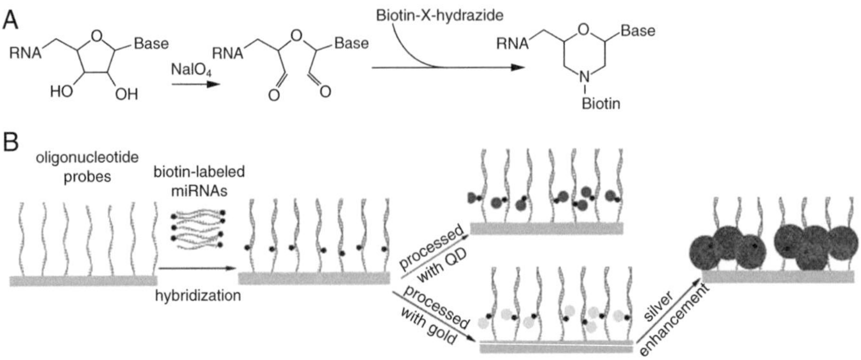

Fig. 4.10 Schematic principles of the miRNA profiling microarray (Adapted from Ref. 53)

4.6.2 Nucleic Acid Detection

Liang et al. applied QDs to miRNA microarray assays by using streptavidin QDs probes to label biotinylated miRNA targets derived from rice, see Fig. 4.10 [53]. Initially, miRNAs were oxidized with sodium periodate to oxidize the ribose 2′- and 3′- hydroxyl groups into aldehydes. The resulting dialdehyde was then reacted with biotin-X-hydrazide resulting in biotinylated miRNA. This was followed by immobilizing the 5′ amine-modified oligonucleotide probes antisense to miRNAs on the amine-reactive glass slides. The biotinylated miRNAs were captured on the microarray by oligonucleotide probes in hybridization. Quantum dots were labeled on the captured miRNAs through the strong specific interaction of streptavidin and biotin. As QDs have a high extinction coefficient and a high quantum yield, so trace amounts of miRNAs are easily detected with a laser confocal scanner. In addition as alternative, the colorimetric gold–silver detection method was used in which captured miRNAs were labeled with streptavidin-conjugated gold followed by silver enhancement. During silver enhancement, the gold nanoparticles bound to miRNAs catalyzed the reduction of silver ions to metallic silver, which further autocatalyzed the reduction of silver ions to form metallic silver precipitation on gold, resulting in a signal enhancement [53]. This process allowed straightforward detection of the microarray with an ordinary charge-coupled device (CCD) camera mounted on a microscope.

They found that QD probes provided good sensitivity down to sub-femtomolar concentrations and dynamic range over several orders of magnitude. This was far better than other dye-based methods and further obviated the use of amplification while allowing a semi-quantitative comparison of the amount of miRNA in different samples [53].

Figure 4.11 displays a set of images for various concentrations of miRNA (21 nt siRNA) detected by QD [53]. It should be noted that the signals become gradually weaker with the decrease in miRNA concentration (Fig. 4.11a). When the miRNA concentration was as low as 39 pM, the fluorescence signal could be detected, indi-

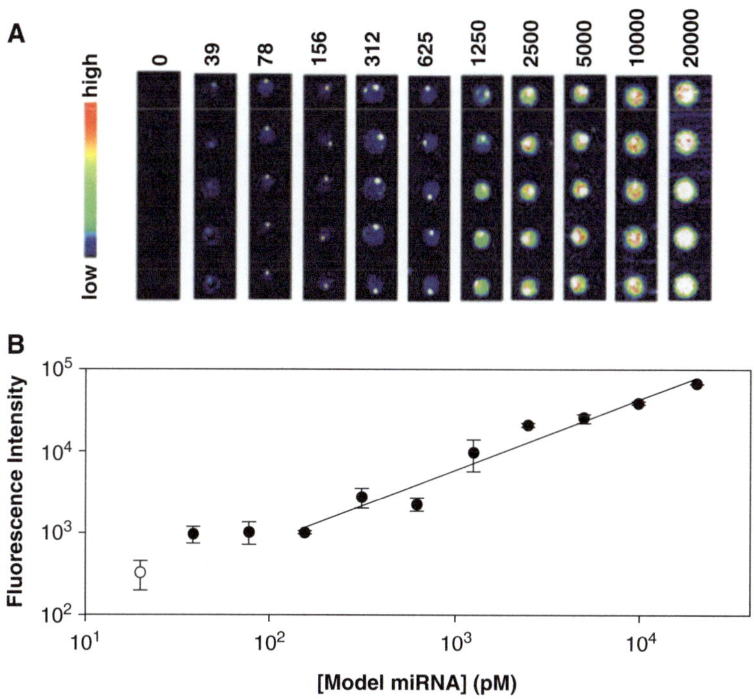

Fig. 4.11 Detection limit and dynamic range of the model miRNA detection microarray. (**a**) Image sets of microarrays hybridize with various concentrations of miRNAs from 20 nM to 39 pM and the backgound. The 50 μM concentration of oligonucleotide probes printed on slides pentaplicately. The volume of model miRNA needed to hybridize with microarray was 10 μl. (**b**) Correlation between fluorescence intensity of spots and concentrations of model miRNA. The values were calculated from image in (**a**). Open circle represents the background (Adapted from Ref. 53).

cating that the lower detection limit of miRNA microarrays is at least 0.4 fmol. As shown in Fig. 4.11b, the fluorescence intensity of the spots is linear to the model miRNA in a logarithmic fashion from 156 to 20,000 pM, and the dynamic range is about 2 orders of magnitude. This implies that this method can be used to quantify miRNAs with broad concentration range [53].

4.6.3 Sensing Based on FRET with Quantum Dot Bioconjugates

Fluorescence Resonance Energy Transfer (FRET) is a physical radiationless transmission of energy phenomenon which relies on the distance-dependent transfer of energy from a donor molecule to an acceptor molecule. FRET has been extensively used in biophysical and biochemical studies to probe ligand-receptor binding and molecular structural changes [56–58]. In the following example, the authors

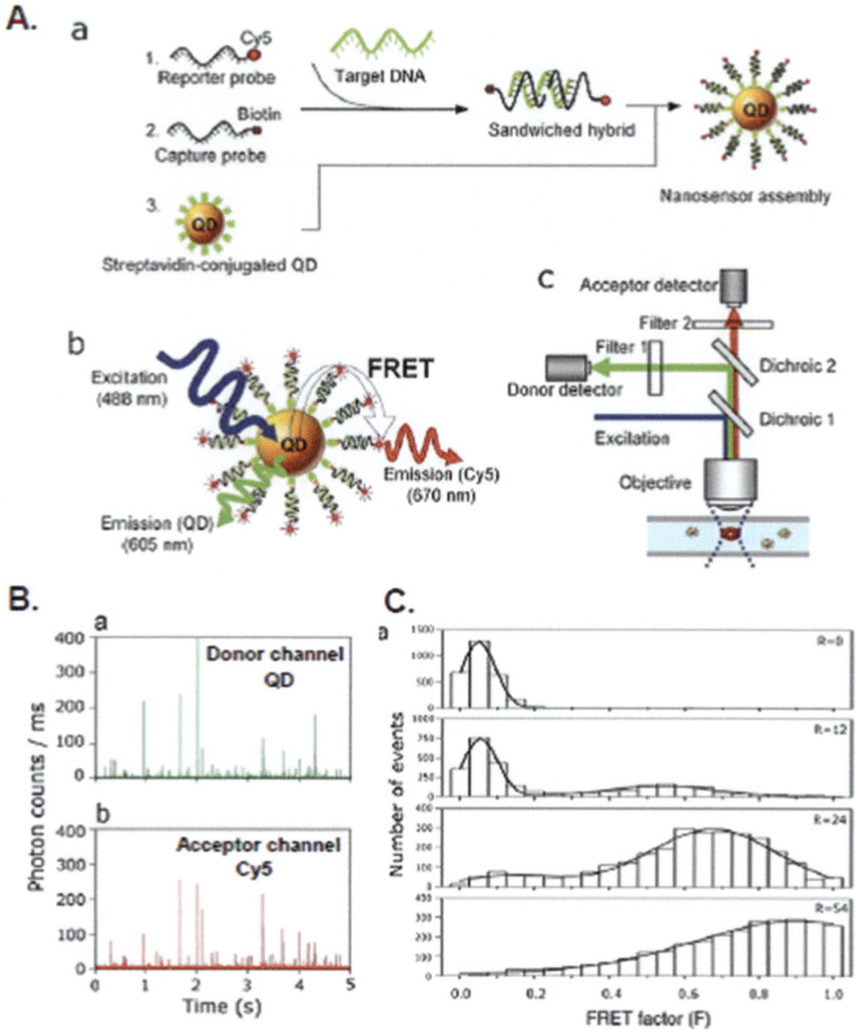

Fig. 4.12 Single QD-based DNA nanosensor (Adapted from Ref. 54)

incubated the dye-labeled DNA targets with biotinylated capture DNA probes, which were allowed to be conjugated to streptavidin QDs, only when the two DNA sequences hybridize. The resulting hybridization was then detected via FRET between the QD and the dye acceptor (Fig. 4.12) [54]. It should be noted that the additional background caused by the acceptor direct excitation is virtually eliminated through the choice of an appropriate excitation wavelength; this led to a 100-fold improvement in sensitivity compared to single organic dye molecular beacon-based detection. This type of sensing schemes can also be amenable to use in a multiplex format. The narrow and symmetric QD emissions allow easy spectral deconvolution and the most straightforward configuration relies on several QD populations interacting with the same dye acceptor, rather than the opposite [54].

4.7 Engineered Cell-Based Sensors: The CANARY System

Petrovick et al. have developed a novel inexpensive genetically engineered white-blood cells biosensor for the rapid identification of warfare BA pathogens and toxins [59]. This new sensor was named and abridged as CANARY for "Cellular Analysis and Notification of Antigen Risks and Yields". CANARY sensors are capable to detect soluble protein toxins, which are an important class of potential bioweapon, and can also be used for sequencing DNA and RNA [60].

4.7.1 CANARY Bioelectronic Sensor

The CANARY technology is based on genetically engineered white blood B cells, which has the ability to bind to and to recognize pathogens quickly and assists other parts of the immune system to fight the infection. It is well known that B cells are the fastest known pathogen identifiers (intrinsic response less than 1 sec). The B-lymphocytes recombinantly express cytosolic aequorin, a Ca-sensitive bioluminescent protein from the jellyfish *Aequoria victoria,* which emits light in response to elevations of intracellular Ca, along when membrane-bound by antibodies. Binding of the pathogen to the cell surface antibodies causes an increase in intracellular Ca levels, resulting in the emission of light from the cytosolic aequorin [60–63].

Two routine genetic modifications enable engineered B-cell lines to express cytosolic aequorin, a calcium sensitive bioluminescent protein, as well as membrane bound antibodies specific for pathogens of interest [62, 63]. This was achieved by crosslinking the membrane-bound antibodies to a polyvalent antigen that induces a signal-transduction cascade. This latter, sequentially involves tyrosine kinases, phospholipase C and inositol triphosphate (IP3). The IP3 activates calcium channels, thereby increasing cytosolic calcium from both internal stores and the extracellular medium, which activates the aequorin, causing it to emit light (Fig. 4.13) [59, 62–64].

The CANARY sensor can detect less than 50 colony-forming units (cfu) of pathogen in less than 3 min, which include the time required to concentrate the sample [65]. It should be mentioned that state-of-the-art immunoassays take at least 15 min, whereas polymerase chain reaction (PCR) takes longer than 30 min. The novel genetic-engineering system developed by Petrovick et al. have consisted of the efficient production of B-cell lines that can react specifically and rapidly to a variety of pathogens [59]. The antibody genes were cloned from hybridomas and inserted into expression vectors. These were transfected into a parental B cell line that expresses active aequorin, and the cells are screened for their response to patho-

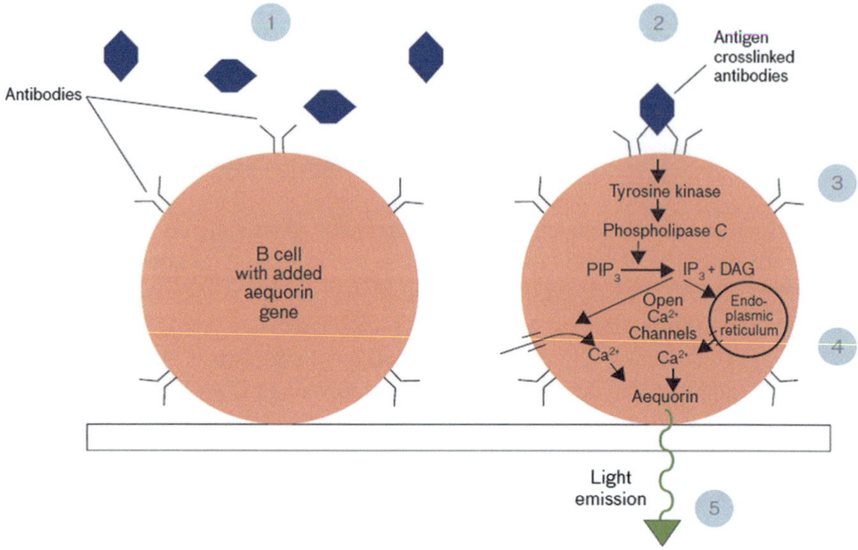

Fig. 4.13 The CANARY bioelectronic sensor (Adapted from Ref. 59)

gen. The genetically engineered CANARY cells can be used separately in a single identification assay, or as many as three can be combined to achieve a multiplexed assay. Alternatively, several antibodies can be expressed in a single cell line to provide a classification assay [59]. It is also feasible to create B cells that emit at different wavelengths of light, enabling multiplexed assays that simultaneously distinguish among several targets [59].

4.7.2 *Toxin Detection*

A novel effective method described by Petrovick et al. for the immobilization of a toxin was to capture it on beads coated with antibodies against that specific toxin [59]. The antibody-coated beads were then incubated in a solution containing the suspected toxin, and washed to remove the contaminating proteins and other materials. The toxin- obtained decorated beads are excellent candidates for the purification and immobilization of toxin analytes by the CANARY cells, as illustrated in Fig. 4.14 [59].

It should be mentioned, that the CANARY cells should express an antibody that binds to the toxin at a different site from that of the capture antibody. Because the toxin is immobilized on the bead, the antibodies on the CANARY cell that bind to the toxin are also immobilized, and therefore light emission is stimulated [59]. This approach has been used to develop a very effective CANARY assay for botulinum neurotoxin type A (BoNT/A) which is the most poisonous toxin known to man, with

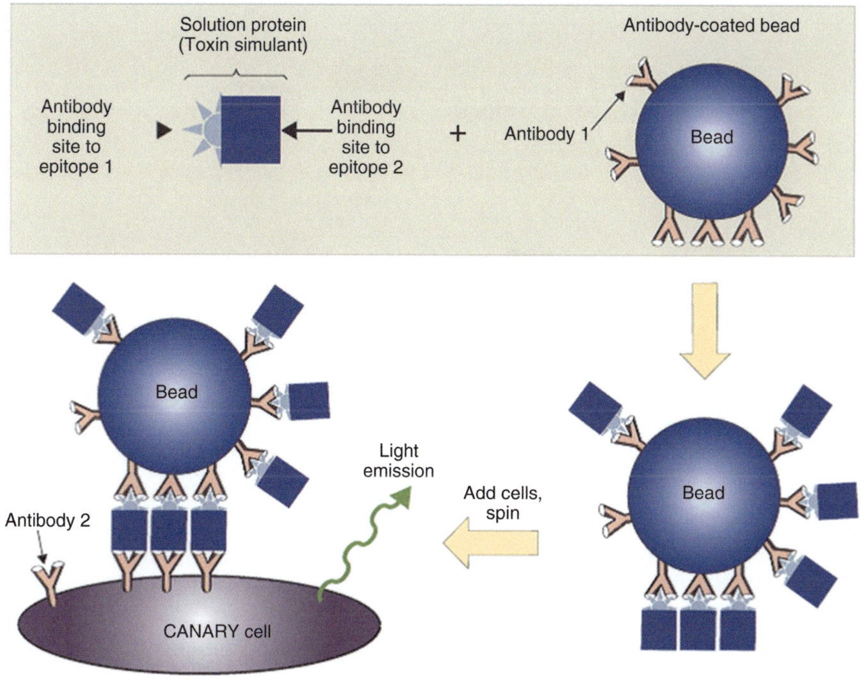

Fig. 4.14 Two antibodies assay for toxins detection. (Adapted from Ref. 59)

an LD50 for a 55 kg adult of about 550 ng by inhalation. Under ideal conditions, the assay sensitivity is currently 16 pg (1.6 ng/mL) [59].

4.7.3 DNA Sequence Detection

The detection of soluble macromolecules has a second interesting application: identification of DNA and RNA sequences. The ability to identify nucleic acid (NA: DNA or RNA) sequences is central in the deduction of the first hard NA sequence information, concerning a new or genetically modified pathogen [59]. It is important to develop assays that have the flexibility to respond quickly to new threats. For this reason, developing NA probes allow for the rapid and quick examination of the actual genetics of the target organism [59]. Once the NA sequence of a pathogen is determined, multiple short probes are synthesized that bind adjacent to each other along a specific sequence on the target NA [59]. Consequently, Petrovick et al. developed a novel assay that uses a single CANARY cell line that expresses an antibody against digoxigenin. Each of these probes is labeled with a single digoxigenin molecule. If these probes are added to solution containing the target NA sequence, the binding of multiple digoxigenin-containing probes produces a tight

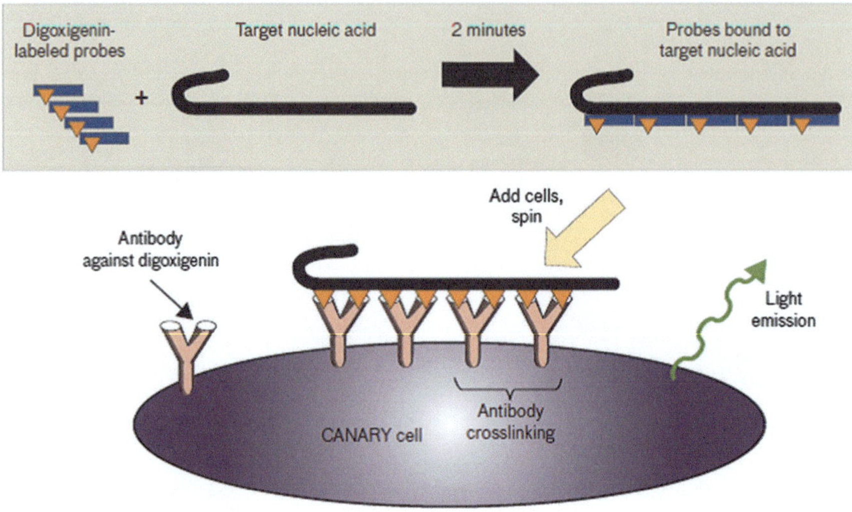

Fig. 4.15 DNA probes designed that bind to a specific region on a single-target nucleic acid (NA) (Adapted from Ref. 60)

cluster of immobilized digoxigenin molecules, which will stimulate light production from the CANARY cell (see Fig. 4.15) [59].

In the absence of target NA, each digoxigenin-labeled probe remains monomeric, and therefore cannot crosslink antibodies on the surface of CANARY cells. There are advantages to detecting RNA as compared to DNA. Primarily, while there is only one copy of genomic DNA per bacterium, there can be thousands of copies of a single RNA; so the number of target molecules per bacterium is much higher. Furthermore, since probe binding requires that the target NA be single stranded, a denaturing step must be performed to separate the two constituent strands of DNA. RNA, however, is normally single stranded [59].

4.7.4 CANARY Detection of **Francisella tularensis** *and* **Yersinia pestis**

The CANARY approach takes advantage of a receptor that binds to the constant region of antibodies, leaving the antigen-binding region of the antibody free. After binding bacteria with the captured antibodies, the receptor initiates a signal cascade, similar to the one induced by the crosslinking of membrane-bound antibodies on B cells, which activates aequorin [59].

The excellent combination of speed and sensitivity of the CANARY system was demonstrated with cell lines expressing an antibody specific for the F1 antigen of *Yersinia pestis (Yp)*, shown in Fig. 4.16 [59]. When concentrated in the centrifuge

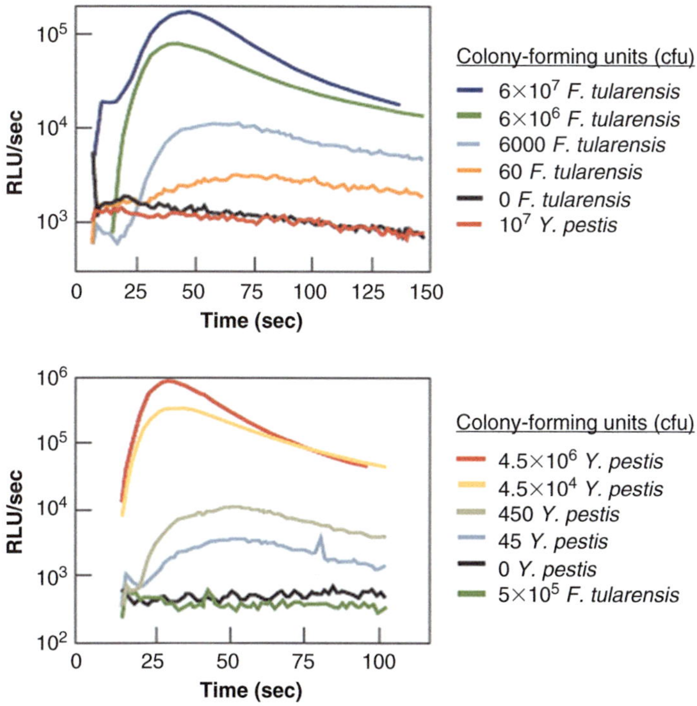

Fig. 4.16 The dose-response curves for inactivated *Francisella tularensis* (*top*) and *Yersinia pestis* (*bottom*) (Adapted from Ref. 60)

luminometer, as little as 45 cfu of formalin-inactivated Yp are detected. However, there was no response to relatively large numbers of *F. tularensis* [59].

For biological defense applications, the CANARY technology was incorporated into a flexible biological-aerosol sensor platform called PANTHER that can form the core of a family of mission-specific bio-aerosol identification sensors useful as standalone sensors for site/building protection, emergency response, rapid screening, and environmental monitoring [59].

To summarize, the CANARY's capabilities open possible applications in pathogen genotyping, virulence testing, antibiotic resistance screening, and viability assessment. Moreover, using these cells demonstrate the best known combination of speed and sensitivity. Other applications of CANARY technology include biological aerosol sampling, point-of-care diagnostics, pre-symptomatic diagnosis in the aftermath of a biowarfare attack, detection of agricultural pathogens at ports of entry, or screening of perishable food supplies medium [62], which activates the aequorin, causing it to emit light [59, 60].

4.8 High-Density Microsphere-Based Fiber Optic DNA Microarrays

It is well known that an optical fiber or optical fibre is a flexible, transparent fiber made by drawing glass (silica) or plastic to a diameter slightly thicker than that of a human hair [66, 67]. Optical fibers consist of an inner core which is surrounded by a clad material of lower refractive index. Because of the differences in refractive index light is totally reflected. A fiber optic bundle consists of thousands of individual fibers fused together such that each fiber retains its ability to transmit light independently of its neighbors (Fig. 4.17) [68].

Pantano and Walt showed that by selectively etching the fiber core, an array of microwells can be formed [69]. These microwells can be filled with oligonucleotide-functionalized microspheres. The array dimensions can be tailored to suit any size of oligonucleotide-functionalized microsphere. The well diameters are equal to those of the fiber cores, and the depths are dependent on the etchant concentration, the exposure time, and the fiber composition. Because each microsphere is optically wired to a fiber, the specific interactions on each microsphere surface can be independently monitored. Walt and coworkers developed a high-density fiber optic DNA microarray consisting of oligonucleotide-functionalized, 3.1-μm-diameter microspheres randomly distributed on the etched face of an imaging fiber bundle [69].

Usually these fiber bundles are composed of around 6000/50000 fused optical fibers, in which each fiber –contains an etched well [70]. The desired oligonucleotide sequences are attached to individual microspheres, than added to each etched wells on the fiber optic bundle face. The produced microwell arrays are capable of casing complementary-sized microspheres, each containing thousands of copies of a unique oligonucleotide probe sequence. Walt and coworkers showed that the array fabrication process resulted in random microsphere placement. It should be understood that the determination of the position of microspheres in the random array, essentially required an optical encoding scheme. The detection schemes, which are combined the intrinsic recognition abilities of nucleic acids, are usually measured with fluorescence-based detection methods. This "optical bar code" obtained is simply a combination of fluorescent dyes with different excitation and emission wave lengths and intensities that allows each bead to be independently identified [70].

Nonetheless, additional degrees of freedom are available in the fiber-optic format, that allows additional number of excitation and emission wavelengths that can be used. It should be mentioned that the optically bar-coded arrays, can be decoded in a matter of seconds. This occurs byh conventional image processing software that collect series of fluorescence images at different excitation and emission wavelengths and then analyze the relative intensities of each bead. Alternatively, preformed oligonucleotides may be added directly to surface-activated microspheres [70].

Needless to say those fluorescence-based assays are more desirable than traditional radiolabeled methods due to their increased safety and experimental versatil-

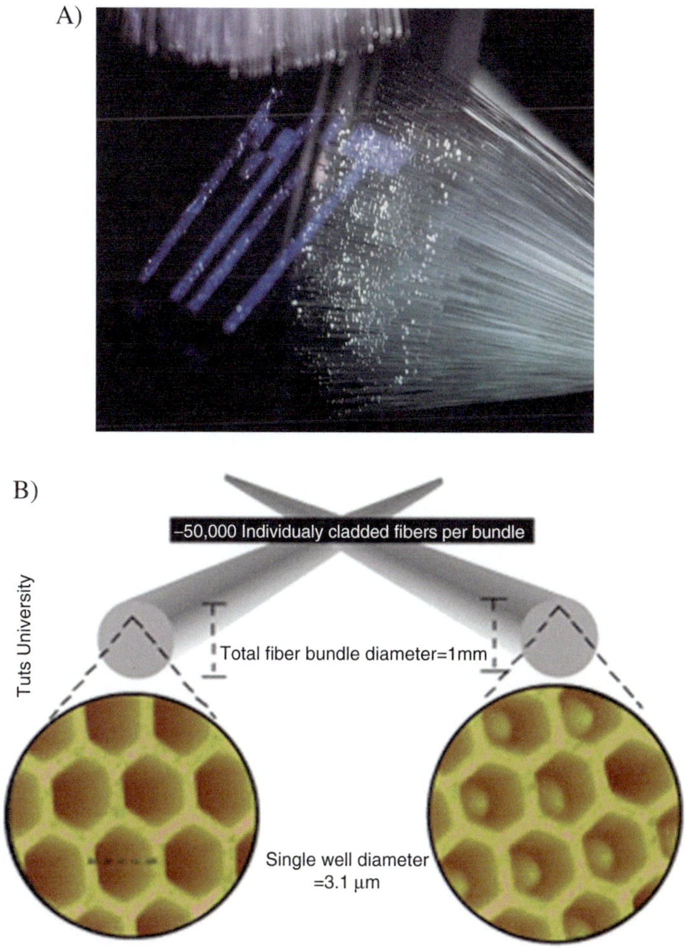

Fig. 4.17 (**a**) Individual optical fibre is a flexible, transparent fiber and a fiber optic bundle consists of thousands of individual fibers fused together bundle. (**b**) Scanning force images of the etched face of an optical fiber imaging bundle. The first (*left*) is empty, and the second (*right*) contains complementary-sized microspheres. Each well diameter is approximately 3.1 μm [68]

ity. Fluorescence can be incorporated into microarray assays by fluorescent intercalating dyes, fluorescently labeled targets, or label-less methods employing fluorescence resonance energy transfer (FRET). Fluorescence-based assays enable the measurement of multiple wavelengths independently and simultaneously. The use of multiple fluorophores enables parallel interrogation schemes [70–73].

4.8.1 Fiber Optic DNA Biosensors

A general array protocol entails immobilizing a probe sequence (primer) that can hybridize to its fluorescently-labeled complementary target. The fluorescent tag is commonly incorporated into the target molecules, via polymerase chain reaction (PCR) [74]. This primer labeling method is convenient when amplification is needed for detection, or when it is used to construct a cDNA library from a genomic RNA pool via reverse transcription. In addition, the derivatized fiber core experiments are able to detect unlabeled (non-fluorescent) target solutions 74]. This is achieved by competitive hybridization with fluorescent target samples. In this method, the fluorescent synthetic target complements are synthesized and initially hybridized to the array to saturate the array probe elements [68]. The unlabeled target solution is then hybridized to the same array, competing with the prehybridized synthetic targets. The presence of unlabeled target is determined by a fluorescence decrease caused by displacement of the fluorescent synthetic target by the unlabeled species. This procedure eliminates the need to incorporate fluorescence into the target and allows quantitative measurements to be performed. These fiber optic platforms are the basis for microsphere array designs that improved array fabrication and allowed extremely high-density sensor placement [68].

4.8.2 Analysis Setup and Protocol

The imaging system consists of a light source, an inverted microscope, and a modified Olympus epifluorescence microscope/charge-coupled device camera (Photometrics PXL). A fiber chuck held the imaging fiber in a fixed position while electronically controlled filter wheels switched between the analytical wavelength and the encoding wavelengths, enabling complete analysis and identification of the microspheres within minutes. Excitation light is sent into the proximal tip of the imaging fiber, and emission from the fluorescing molecules captured and directed onto the CCD camera detector (Fig. 4.18) [72].

The multiplex analysis images were acquired for 1 and 0.5 s at wavelengths specific to each encoding dye. A 365-nm excitation filter and a 600-nm long-pass emission filter were used for the Eu-dye. A 620-nm excitation filter and a 670-nm emission filter were used for the Cy5 dye. A 530-nm excitation filter and a 580-nm emission filter were used for TAMRA. That is detection of amplified DNA fragments which were incubated with a fluorescein-labeled sequencing primer in the presence of the two allelic ROX-ddNTP or TAMRA-ddNTP terminators. All targets were labeled with fluorescein. It was shown that the fluorescence intensity was proportional to the extent of hybridization at each probe position [72A]. In addition, the camera was equipped with an internal chip that provides megapixel resolution (1280/1024). This megapixel chip was able to resolve the arrays miniatured feature sizes (3 mm) and provided multiple pixels for each optical channel in the fiber bundle (Fig. 4.19).

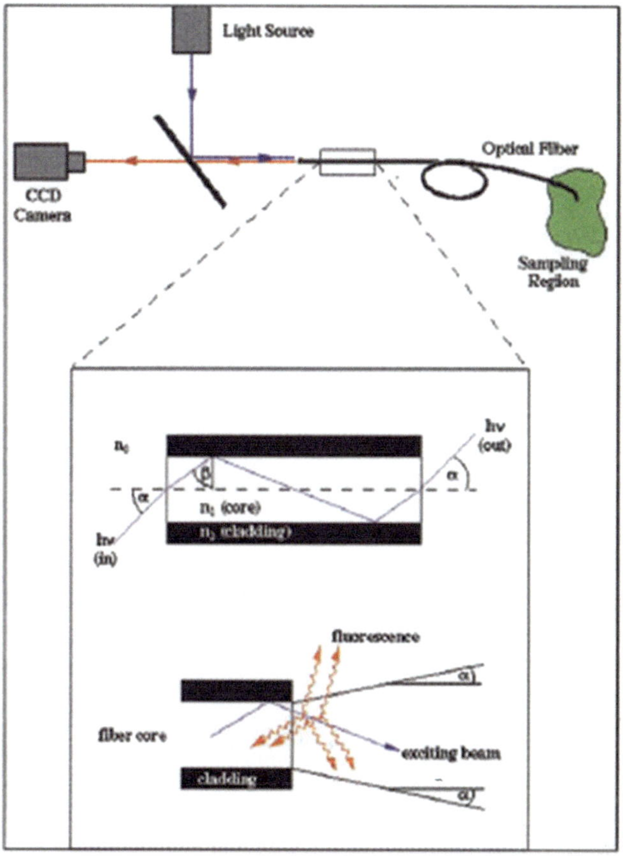

Fig. 4.18 Schematic diagram of the fiber optic imaging system (Adapted from Ref. 73)

Finally, it can be surmised that the described fiber optic microsphere-based biosensor is a versatile platform that possesses micro-scale features and an overall array size that enables rapid analysis and extremely low detection limits. Redundant detection elements in the array increase the signal-to-noise ratio and avoid the potential for false positive and false negative results. Microsphere-based arrays are reusable, and are easy to fabricate. The fiber optic platform has also been applied to other applications including artificial olfaction and cell-based array sensing [68, 73].

4.9 Surface Plasmon Resonance

Recently, numerous strategies for protein labeling were developed and they allowed the characterization of proteins regarding their structure, folding, or interaction with other proteins [75]. Surface plasmon resonance (SPR) is a label-free detection

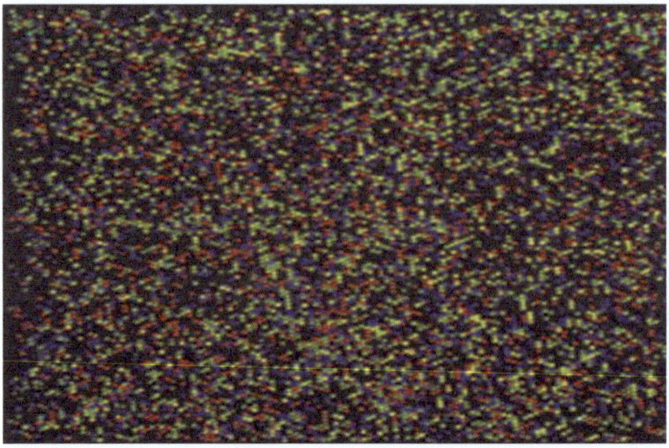

Fig. 4.19 An image of a 13,000-well square fiber bundle microarray, with a diameter of −1.2 mm. This entire array is about the size of a single spot on a traditional spotted array (Adapted from Ref. 73).

method which is a suitable and reliable platform in clinical analysis for biomolecular interactions. Undeniably, SPR has been proven to be one of the most powerful technologies to determine the specificity, affinity and kinetic parameters displayed during the binding of macromolecules. These binding of macromolecules include, but are not limited to, protein-protein, protein-DNA, enzyme-substrate or inhibitor, receptor-drug, lipid membrane-protein, protein-polysaccharide and cell or virus-protein [76–85]. SPR is an optical technique which measures the refractive index changes in the vicinity of thin metal layers (i.e., gold, silver, or aluminum films) in response to biomolecular interactions.

4.9.1 SPR Principle

When the photon of an incident light strikes a metal surface, typically a gold surface, surface plasmon resonance occurs. When a portion of the light energy excites the electrons of the metal surface layer, at a certain angle of incidence, it creates electron movements which propagate parallel to the metal surface. This electric oscillation is termed as a "plasmon". This oscillation, in turns, generates an electric field whose range is around 300 nm from the boundary between the metal surface and sample solution [76, 79]. In a commercial SPR biosensor configuration, the incident light employed is generated by a high-reflective index glass prism of the Kretschmann geometry of the attenuated total reflection (ATR) method (Fig. 4.20) [79].

The defined SPR angle, at which resonance occurs, relies on the refractive index of the material coating the metal surface, and the constant light source wavelength. When there is no change in the reflective index of the sensing medium, the plasmon

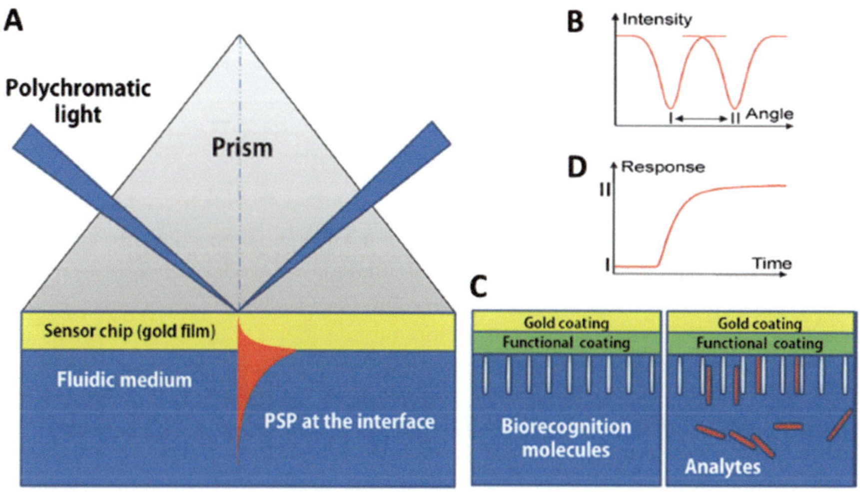

Fig. 4.20 Concept of a surface plasmon resonance (SPR) biosensor: (**a**) Kretschmann geometry of the ATR method; (**b**) spectrum of reflected light before and after refractive index change; (**c**) analyte-biorecognition elements binding on SPR sensor surface and (**d**) refractive index changes caused by the molecular interactions in the reaction medium (Adapted from Ref. [79]).

oscillation cannot be formed (Fig. 4.20a). In addition, it should be noted that when the surface of the sensing material has been coated through biomolecule attachment only, there will be perhaps an unnoticed small change in the reflective index of the sensing medium, and as a result, the plasmon oscillation cannot be formed (Fig. 4.20c, left side) [79]. However, when the metal surface has been coated with an analyte-biorecognition couple of biomolecules, detection is accomplished by measuring the changes in the reflected light obtained on a detector (Fig. 4.20c, right side). In addition, the amount of surface concentration can be quantified by monitoring the reflected light intensity or tracking the resonance angle shifts. Typically, an SPR biosensor has a detection limit of 10 pg/mL [76–79].

In all commercial SPR biosensors, probe molecules are firstly immobilized on to the sensor surface. When the solution of target molecules is flown into contact with the surface, a probe-target binding via affinity interaction occurs, which consequently induces an increase in the refractive index at the SPR sensor surface (Fig. 4.20d) [79]. In SPR experiments, resonance or response units (RU) are used to describe the signal change, where 1 RU is equivalent to a critical angle shift of 10^{-4} degree [80–84]. At the start of the experiment whereas probe target interactions have not occurred, the initial RU value corresponds to the starting critical angle. The change in refractive index Δnd arisen within a layer of thickness h can be calculated as:

$$\Delta n_d = (dn / dc)_{vol}\ \Delta\Gamma / h \quad (4.1)$$

where $(dn/dc)_{vol}$ is the increase of refractive index n with the volume concentration of analyte c, and $\Delta\Gamma$ is the concentration of the bound target on the surface [76].

4.9.2 High-Throughput Screening (HTS)

There are several different formats of SPR biosensors. These comprise the array format, multi-channel unit format, and SPR imaging format, which allow simultaneous and continuous detection to analyze the performance of hundreds to thousands of affinity binding events on a chip surface [81–83]. In SPR imaging, the incidence angle remains fixed, and the binding of biomolecules on a gold surface is measured as the change in reflectivity (or reflectance) in relation to the incident ray intensity, unlike SPR sensors that depend on the measurement of the absorption dip in the SPR angle or SPR wavelength [80–84]. Recently, it was shown that SPR imaging technology using a multi-analyte biosensor allows the measurement of a high-throughput approach, and it achieves a similar degree of sensitivity of conventional SPR biosensors (Fig. 4.21) [83].

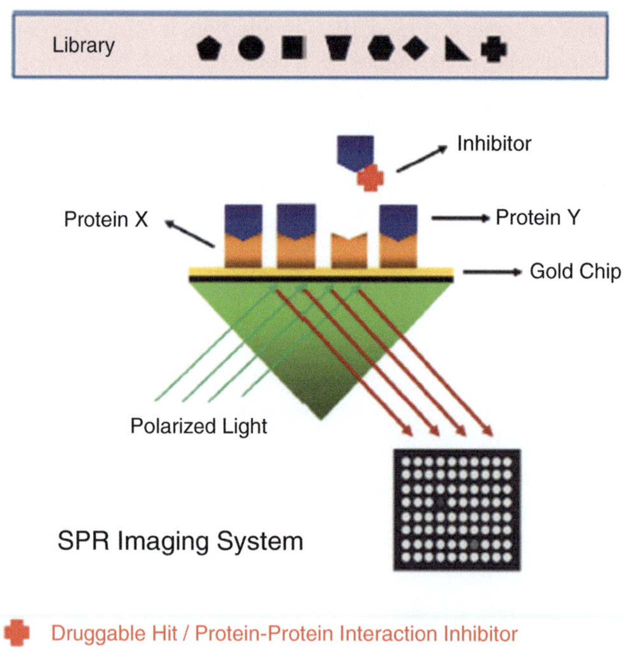

Fig. 4.21 High-throughput drug screening using an SPR imaging protein chip system. The bright image indicates protein-protein interaction on a gold surface. Upon the binding of an inhibitor to the target protein, protein-protein interactions are disrupted, resulting in changes in SPR imaging signal intensity and a darker image (Adapted from Ref. [83])

Consequently, SPR imaging systems without any labeling requirements can be used for high-throughput screening (HTS), particularly in drug discovery, than any other optics-based detection techniques [83, 84].

4.9.3 Surface Plasmon Resonance Sensing of Biological Warfare Agent Botulinum Neurotoxin A

Dhaked and coworkers developed a label free real time method for the detection and quantification of botulinum neurotoxin A (BoNT/A) using surface plasmon resonance (SPR) [85]. The authors used antibody against rBoNT/A-HCC fragment and synaptic vesicles (SV), which were immobilized on a carboxymethyldextran modified gold chip. The immobilization of BoNT/A antibody and interaction of BoNT/A with immobilized antibody were characterized in-situ by SPR and electrochemical impedance spectroscopy. A sample solution containing BoNT/A antigen in concentrations ranging from 0.225 to 4.5 fM and 0.045 to 5.62 fM was interacted with the immobilized antibody and immobilized SV, respectively [85].

4.9.3.1 Immobilization of BoNT/A Antibody and SV Protein on CM5 SPR Sensor Chip

Dhaked and coworkers described the stepwise immobilization of BoNT/A antibody and SV protein on a CM5 chip, respectively in nine steps. These steps involved the stabilization of the baseline, the activation of carboxyl groups on the CM5 chip, and converting them into activated carboxymethylated groups on the sensor chip for future bounding to the free amino groups of BoNT/A antibody. This was followed by washing with PBS and measuring the SPR angle that has shifted nearly to the baseline [85]. When the BoNT/A antibodies are injected on the CM5 chip an increase in SPR angle was observed. Finally, washing with 1000 mM ethanolamine was performed to prevent non-specific binding and to block the unreacted NHS-ester groups on CM5 chip. From Figs. 4.22 and 4.23, a net angle change of 95.44 m° and 48 m° are observed and this ascribes the attachment of 0.79 ng/ mm2 of antibody and 0.4 ng/mm2 SV protein on CM5 chip, respectively [87].

4.10 Conclusion

Most of the sensing measuring devices used for the detection of biological warfare agents are based on luminescence immunoassay signal transduction mechanisms which are optical. In this chapter, we have discussed the following subjects: impedance spectroscopy, evanescent wave technology and internal reflection fluorescence

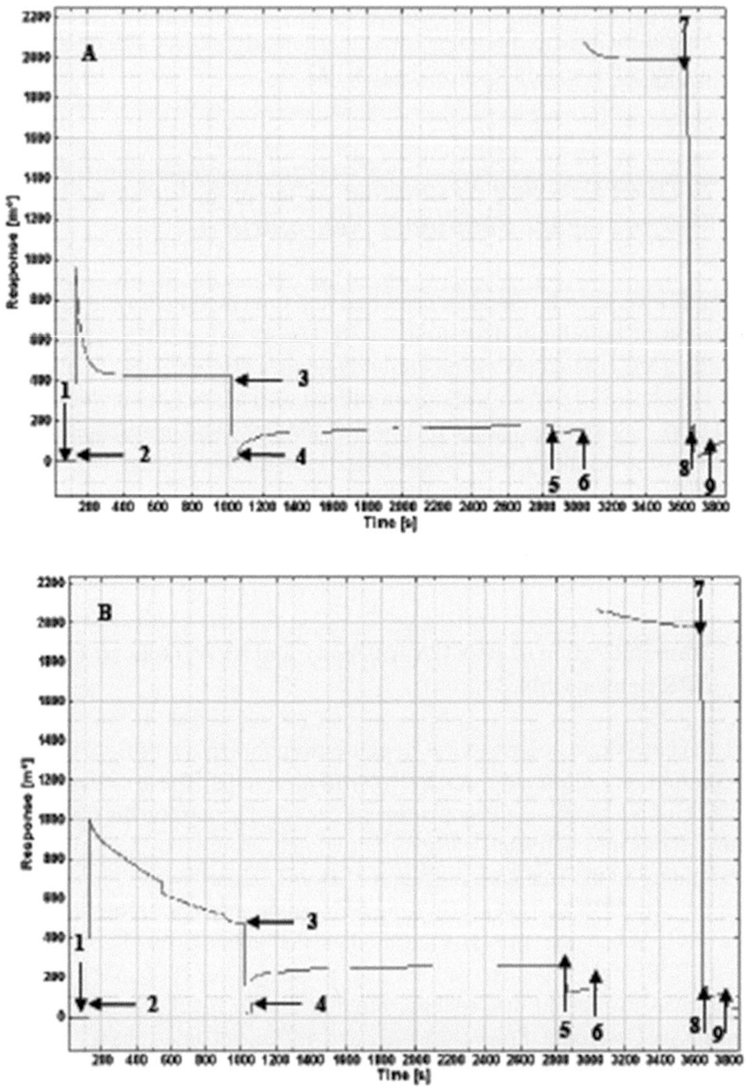

Fig. 4.22 Sensorgram showing different steps [(1) Baseline (2) EDCNHS activation (3) Washing (4) Antibody coupling (5) Washing (6) Deactivation (7) Washing (8) Regeneration and (9) Back to baseline] involved in the (**a**) immobilization of BoNT/A antibody (1: 500 dilution); (**b**) immobilization of SV on CM5 chip [87]

(TIRF) excitation, surface acoustic wave sensors, fluorescent biosensors, Fluorescence Resonance Energy Transfer (FRET), light emission (CANARY), microsphere-based fiber optic biosensors, and surface plasmon resonance. It should be noted that although fluorescence detection is the most popular, surface plasmon resonance (SPR)-based immunoassay formats with surface imaging capabilities are growing in popularity

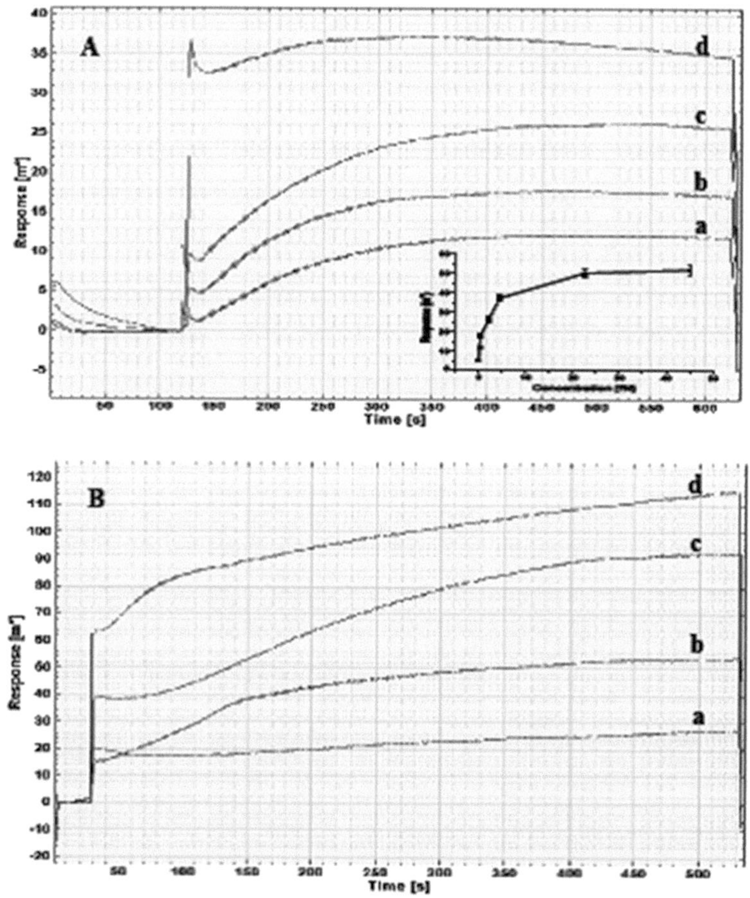

Fig. 4.23 SPR sensor response for the interaction of different concentration of BoNT/A antigen with (**a**) immobilized BoNT/A antibody (*a*) 0.225 fM, (*b*) 0.45 fM, (*c*) 2.25 fM and (*d*) 4.5 fM (**b**) immobilized SV (*a*) 0.045 fM, (*b*) 0.225 fM, (*c*) 1.12 fM and (*d*) 5.62 fM Temperature: 25o C and pH: 7.5 [87]

[80–86]. Multiplexing, or simultaneous detection of multiple analytes, is one of the most important prerequisites for biothreat agent detection. As mentioned in our introduction, the presented work is more a "Comptes Rendus" and an introduction to this series of sensing measuring devices used for the detection of biological warfare agents that were presented to the participants of this NATO-ASI meeting.

References

1. Eitzen EM Jr, Takafuji ET (1997) Historical overview of biological warfare. In: Sidell FR, Takafuji ET, Franz DR (eds) Medical aspects of chemical and biological warfare. Office of the Surgeon General, Borden Institute, Walter Reed Army Medical Center, Washington, DC, pp 415–423
2. Ridel R (2004) Biological warfare and bioterrorism: a historical review. Proc (Baylor Univ Med Cent) 17(4):400–406
3. WHO. Severe acute respiratory syndrome (SARS) (2003) Wkly Epidemiol Rec, 78: 86.92
4. Gray C (2007) Another bloody century: future warfare, pp 265–266, Phoenix
5. Heymann DL (2001) Strengthening global preparedness for defense against infectious disease threats. Committee on foreign relations, United States Senate. Hearing on the threat of bioterrorism and the spread of infectious diseases
6. Hoffman RE, Norton JE (2000) Lessons learned from a full-scale bioterrorism exercise. Emerg Infect Dis 6:652–653
7. US Commission on National Security in the 21st Century. New world coming: American security in the 21st century, supporting research and analysis. September 15, 1999
8. Szinicz L (2005) History of chemical and biological warfare agents. Toxicology 214(3):167–181
9. CDC. Biological and Chemical Terrorism: Strategic Plan for Preparedness and Response (2005) Recommendatons of the CDC Strategic Planning Workgroup. MMWR Recomm Rep 49(RR-4):1–26
10. Cenciarelli O, Rea S, Carestia M, D'Amico F, Malizia A, Bellecci C, Gaudio P, Gucciardino A, Fiorito R (2013) Bioweapons and bioterrorism: a review of history and biological agents. Defence S & T Tech Bull 6(2):111–129
11. Barrett JA, Bowen GW, Golly SM, Hawley C, Jackson WM, Laughlin L, Lynch ME (1998) Assessment of biological agent detection equipment for emergency responders, June 1, 1998. Chemical Biological Information Analysis Center (CBIAC), P.O. Box 196, Gunpowder, MD 21010–0196
12. Chemical and biological terrorism: research and development to improve ivilian medical response to chemical and biological terrorism incidents, National Academy of Sciences, 1999. National Academy Press, 2101 Constitution Avenue, N.W., Box 285, Washington, DC 20055
13. Barrett JA, Bowen GW, Golly SM, Hawley C, Jackson WM, Laughlin L, Lynch ME (1998) Final report on the assessment of biological agent detection equipment for emergency responders, U.S. Army Chemical and Biological Defense Command (CBDCOM), CBIAC, P.O. Box 196, Gunpowder, MD 21010–0196
14. Marquette CA, Blum LJ (2006) State of the art and recent advances in immunoanalytical systems. Biosens Bioelectron 15;21(8):1424–1433
15. IUPAC (1992) 64: 143
16. Sapsforda KE, Bradburneb C, Delehantyb JB, Medintzb I (2008) Sensors for detecting biological agents (2008). Mater Today 11(3):38–49
17. Thevenot R, Toth K, Durst RA, Wilson GS (1999) Electrochemical biosensors: recommended definitions and classification. Pure Appl Chem 71:2333–2348
18. Thevenot R, Toth K, Durst RA, Wilson GS (2010) Electrochemical biosensors: recommended definitions and classification. Biosens Bioelectron 16:121–131
19. Thevenot R, Toth K, Durst RA, Wilson GS (2001) Electrochemical biosensors: recommended definitions and classification. Anal Lett 34:635–659
20. Daly P, Collier T, Doyle S (2002) PCR-ELISA detection of *Escherichia coli* in milk. Lett Appl Microbiol 34:222–226
21. Tsai W-L, Miller CE, Richter ER (2000) Determination of the sensitivity of a rapid *Escherichia coli* O157:H7 assay for testing 375-gram composite samples. Appl Environ Microbiol 66(9):4149–4151
22. Nicolas P, Mor A (1995) Peptides as weapons against microorganisms in the chemical defense system of vertebrates. Annu Rev Microbiol 49:277–304

23. Zasloff M (2002) Antimicrobial peptides of multicellular organisms. Nature 415:389–339
24. Meng H, Kumar K (2007) Antimicrobial activity and protease stability of peptides containing fluorinated amino acids. J Am Chem Soc 129:15615–15622
25. Mannoor MS, Zhang C, Link J, McAlpine MS (2010) Electrical detection of pathogenic bacteria via immobilized antimicrobial peptides. Proc Natl Acad Sci U S A 107(45):19207–19212
26. Zasloff M, Martin B, Chen HC (1988) Antimicrobial activity of synthetic magainin peptides and several analogues. Proc Natl Acad Sci U S A 85:910–913
27. Zasloff M (1987) Magainins, a class of antimicrobial peptides from Xenopus skin: isolation, characterization of two active forms, and partial cDNA sequence of a precursor. Proc Natl Acad Sci U S A 84:5449–5453
28. Matsuzaki K, Sugishita KI, Harada M, Fujii N, Miyajima K (1997) Interactions of an antimicrobial peptide, magainin 2, with outer and inner membranes of Gram-negative bacteria. BBA-Biomembranes 1327:119–130
29. Taitt CR, Shriver-Lake LC, Ngundi MM, Ligler FS (2008) Array biosensor for toxin detection: continued advances. Sensors 8:8361–8377
30. Ligler FS, Taitt CR, Shriver-Lake LC, Sapsford KE, Shubin Y, JPb G (2003) Array biosensor for detection of toxins. Anal Bioanal Chem 377(3):469–477
31. Feldstein MJ, Golden JP, Ligler FS, Rowe CA (2001) Reflectively coated optical waveguide and fluidics cell integration. US Pat 6:192, 168
32. Golden JP, Taitt CR, Shriver-Lake LC, Shubin YS, Ligler FS (2005) A portable automated multianalyte biosensor. Talanta 65(5):1078–1085
33. Ngundi MM, Qadri SA, Wallace EV, Moore MH, Lassman ME, Shriver-Lake LC, Ligler FS, Taitt CR (2006) Detection of deoxynivalenol in foods and indoor air using an array biosensor. Environ Sci Technol 40(7):2352–2356
34. Sapsford KE, Taitt CR, Fertig S, Moore MH, Lassman ME, Maragos CA, Shriver Lake LC (2006) Indirect competitive immunoassay for detection of aflatoxin B-1 in corn and nut products using the array biosensor. Biosens Bioelectron 21(12):2298–2305
35. Ngundi MM, Taitt CR (2006) An array biosensor for detection of bacterial and toxic contaminants of foods. In Diagnostic Bacteriology Protocols, 53–68
36. Shriver-Lake LC, Erickson JS, Sapsford KE, Ngundi MM, Shaffer KM, Kulagina NV, Hu JE, Gray SA, Golden JP, Ligler FS, Taitt CR (2007) Blind laboratory trials for multiple pathogens in spiked food matrices. Anal Lett 40(16–18):3219–3323
37. Milne EA (1941) Augustus Edward Hough Love. 1863-1940. Obituary Notices of Fellows of the Royal Society 3(9):466
38. O'Connor JJ, Robertson EF. "Augustus Edward Hough Love", MacTutor History of Mathematics archive, University of St Andrews
39. Länge K, Rapp BE, Rapp M (2008) Surface acoustic wave biosensors: a review. Anal Bioanal Chem 391:1509–1519
40. A) Grate JW, Frye GC (1996) Acoustic wave sensors. In: Baltes H, Göpel W, Hesse J (eds) Sensors update, vol 2. Wiley-VCH. B) Https://wiki.metropolia.fi/display/sensor/Surface+acoustic+wave+(SAW)+sensors
41. Weinheim E, Wohltjen H, Dessy R (1979) Anal Chem 51:1458–1464
42. Collings AF, Caruso F (1997) Rep Prog Phys 60:1397–1445
43. Tamarin O, Comeau S, Déjous C, Moynet D, Rebière D, Bezian J, Pistré J (2003) Biosens Bioelectron 18:755–763
44. Stubbs DD, Hunt WD, Lee SH, Doyle DF (2002) Biosens Bioelectron 17:471–477
45. Stubbs DD, Lee SH, Hunt WD (2003) Anal Chem 75:6231–6235
46. Sapsford KE, Berti L, Medintz IL (2005) Fluorescence resonance energy transfer concepts, applications and advances. Minerva Biotech 16:253–279
47. Goldman ER, Medintz IL, Mattoussi H (2006) Luminescent quantum dots in immunoassays. Anal Bioanal Chem 384:560–563
48. Goldman ER, Balighian ED, Mattoussi H, Kuno MK, Mauro JM, Tran PT, Anderson GP (2002) Avidin: a natural bridge for quantum dot-antibody conjugates. J Am Chem Soc 124:6378–6638

49. Wang SP, Mamedova N, Kotov NA, Chen W, Studer J (2002) Antigen/Antibody immunocomplex from CdTe nanoparticle bioconjugates. Nano Lett 2:817–822. 70
50. Sun BQ, Xie WZ, Yi GS, Chen DP, Zhou YX, Cheng J (2001) Microminiaturized immunoassays using quantum dots as fluorescent label by laser confocal scanning fluorescence detection. J Immunol Methods 249:85–89
51. Hoshino A, Fujioka K, Manabe N, Yamaya S, Goto Y, Yasuhara M, Yamamoto K (2005) Simultaneous multicolor detection system of the single-molecular microbial antigen with total internal reflection fluorescence microscopy. Microbiol Immunol 49:461–470
52. Zhu L, Ang S, Liu WT (2004) Quantum dots as a novel immunofluorescent detection system for *Cryptosporidium parvum* and *Giardia lamblia*. Appl Environ Microbiol 70: 597-598. Liang RQ, Li W, Li Y, Tan CY, Li JX, Jin YX, Ruan KC (2005) An oligonucleotide microarray for microRNA expression analysis based on labeling RNA with quantum dot and nanogold probe. Nucleic Acids Res 33: e17
53. Sapsford KE, Thomas P, Medintz IL, Hedi M (2006) Biosensing with luminescent semiconductor quantum dots. Sensors 6:925–953
54. Zhang CY, Ma H, Nie SM, Ding Y, Jin L, Chen DY (2000) Quantum dot-labeled Trichosanthin. Analyst 125:1029–1031
55. Jares-Erijman E, T b J (2003) FRET Imaging. Nat Biotechnol 21:1387–1395
56. Sapsford KE, Berti L, Medintz IL (2005) Fluorescence resonance energy transfer concepts, applications and advances. Minerva Biotechnol 16:253–279
57. Sandros MG, Shete V, Benson DE (2006) Selective, reversible, reagentless maltose biosensing with core-shell semiconducting nanoparticles. Analyst 131:229–235
58. Petrovick MS, James D, Harper JD, Frances E, Nargi FE, Eric D, Schwoebel ED, Mark C, Hennessy MC, Todd H, Rider TH, Hollis MA (2007) Rapid sensors for biological-agent identification. Lincoln Laboratory J 17:63–84
59. Cormie MJ, Prasher DC, Longiaru M, McCann RO (1989) The enzymology and molecular biology of the Ca2+–activated photoprotein, aequorin, photochem. Photo-Dermatology 49(4):509–512
60. Shimomura O, Musicki B, Kishi Y (1989) Semi-synthetic aequorins with improved sensitivity to Ca2+ ions. Biochem J 261:913–920
61. Wilson HA, Greenblatt D, Poeni M, Finkelman FD, Tsien RY (1987) Cross-linkage of B lymphocyte surface immunoglobulin by anti-Ig or antigen induces prolonged oscillation of intracellular ionized calcium. J Exp Med 166:601–606
62. Tsuji FJ, Inouye S, Goto T, Sakaki Y (1983) Site-specific mutagenesis of the calcium-binding photoprotein aequorin. Proc Natl Acad Sci U S A 83:8107–8111
63. Shimomura O, Johnson FH (1978) Peroxidized coelenterazine, the active group in the photoprotein aequorin. Proc Natl Acad Sci U S A 75(6):2611–2615
64. Rider TH, Petrovick MS, Nargi FE A B cell–based sensor for rapid identification of pathogens. Science 301:213–215
65. Senior JM, Yousif JM (2009) Optical fiber communications: principles and practice, Pearson Education
66. Epstein JR, Leung APK, Kyong-Hoon L, Walt DR (2003) High-density, microsphere-based fiber optic DNA microarrays. Biosens Bioelectron 18:541–546
67. http://spie.org/newsroom/decoding-dna
68. Pantano P, Walt DR (1996) Ordered nanowell arrays. Chem Mater 8:2832–2835
69. Fodor SPA, Read JL, Pirrung MC, Stryer L, AT L, Solas D (1991) Light-directed, spatially addressable parallel chemical synthesis. Science 251:767–773
70. Schena M, Shalon D, Davis RW, Brown PO (1995) Quantitative monitoring of gene expression patterns with a complementary DNA microarray. Science 270:467–470
71. Epstein JR, Lee M, Walt DR (2002) High-density fiber-optic genosensor microsphere array capable of zeptomole detection limits. Anal Chem 74:1836–1840
72. A) Ferguson JA, Steemers FJ, Walt DR (2000) High density fiber optic DNA random microsphere array. Anal Chem 72: 5618–5624; B) Olivas SJ, Arianpour A, Stamenov I, Morrison R,

Stack RA, Johnson AR, Agurok P, Ford JE (2015) Image processing for cameras with fiber bundle image relay. Applied Optics 54(5): 1124–1137
73. Mullis KB (1994) Polymerase chain reaction (Nobel Prize). Angew Chem 106:1271–1276
74. Nguyen HH, Park J, Sebyung Kang S, Kim M (2013) Surface plasmon resonance: a versatile technique for biosensor applications. Sensors 15:10481–10510
75. Stephanopoulos N, Francis MB (2006) Choosing an effective protein bioconjugation strategy. Nat Chem Biol 7:876–884
76. Tugarinov V, Kanelis V, Kay LE (2006) Isotope labeling strategies for the study of high-molecular-weight proteins by solution NMR spectroscopy. Nat Protoc 1:749–754
77. Phelan ML, Nock S (2003) Generation of bioreagents for protein chips. Proteomics 3:2123–2134
78. Šípová H, Homola J (2013) Surface plasmon resonance sensing of nucleic acids: a review. Anal Chim Acta 773:9–23
79. De Feijte JA, Benjamins J, Veer FA (1978) Ellipsometry as a tool to study the adsorption behavior of synthetic and biopolymers at the air-water interface. Biopolymers 17:1759–1772
80. Smith EA, Corn RM (2003) Surface plasmon resonance imaging as a tool to monitor biomolecular interactions in an array based forma. Appl Spectrosc 57:320A–332A
81. Steiner G (2004) Surface plasmon resonance imaging. Anal Bioanal Chem 379:328–331
82. Jung SO, Ro HS, Kho BH, Shin YB, Kim MG, Chung BH (2005) Surface plasmon resonance imaging-based protein arrays for high-throughput screening of protein-protein interaction inhibitors. Proteomics 5:4427–4431
83. Kim M, Han SH, Shin Y (2007) Surface plasmon resonance biosensor chips. Biochip J 1:81–89
84. Tomar A, Gupta G, Singh MK, Boopathi M, Singh B, Dhaked RK (2016) Surface plasmon resonance sensing of biological warfare agent botulinum neurotoxin A. J Bioterror Biodef 7(2):142–154
85. Tsai WC, Li IC (2009) SPR-based immunosensor for determining staphylococcal enterotoxin A. Sensors Actuators B Chem 136:8–12
86. Stenberg E, Persson B, Roos H, Urbaniczky C (1991) Quantitative determination of surface concentration of protein with surface plasmon resonance using radiolabeled proteins. J Colloid Interface Sci 143:513–526
87. Tomar A, Gupta G, Boopathi M, Singh B, Dhaked RK (2016) Surface plasmon resonance sensing of biological warfare agents. J Bioterorr Biodef 7(1), 1–12

Chapter 5
Mass Spectrometry and High Resolution Mass Spectrometry: An Overview

Gianluca Giorgi

Abstract In this chapter, a brief overview of fundamentals in mass spectrometry and some applications are presented. The main concepts in ionization techniques, *classical* techniques and *ambient mass spectrometry* are followed by an overview on analyzers.

Particular attention has been paid to high resolution mass spectrometry and accurate mass measurements, highly debating topics in modern mass spectrometry, with some examples and applications. Finally tandem mass spectrometry, collision-induced dissociations (CID), interactions of ions with electrons (ECD, ETD), photons (IRMPD) and with surfaces (SID) are also briefly presented.

5.1 Mass Spectrometry: An Overview

Mass spectrometry (MS) is an important and very powerful methodology for identifying and structurally elucidating unknowns, and for determining their quantity. It is characterized by high specificity, selectivity, sensitivity and speed. Thus it is possible to carry out analyses on compounds at trace level on food, environmental samples, in chemistry, clinics, biosciences, proteomics, metabolomics, homeland security, and in many other branches of science [1–5].

Mass spectrometry is based on production of ions in the gas phase, occurring in the ion source, on their separation according to their mass-to-charge (*m/z*) ratios, done by the analyzer, and finally on their detection (Fig. 5.1).

Modern mass spectrometry offers different tools for obtaining a plethora of information on the sample [5]. Thus it is possible to determine the molecular weight of an analyte; its separation from isobaric species and its chemical formula through *high resolution mass spectrometry* (HRMS) and *accurate mass measurements*; its

G. Giorgi (✉)
Department of Biotechnology, Chemistry & Pharmacy, University of Siena,
Via Aldo Moro, I-53100 Siena, Italy
e-mail: gianluca.giorgi@unisi.it

J.H. Banoub, R.M. Caprioli (eds.), *Molecular Technologies for Detection of Chemical and Biological Agents*, NATO Science for Peace and Security Series A: Chemistry and Biology, DOI 10.1007/978-94-024-1113-3_5

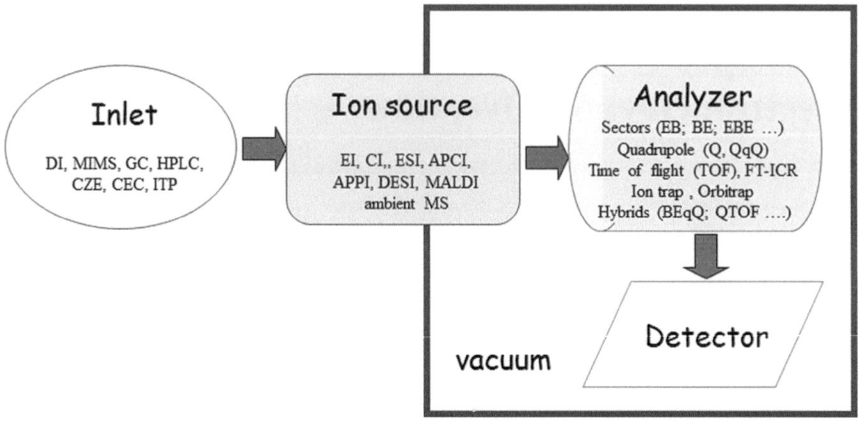

Fig. 5.1 The diagram of a mass spectrometer

molecular structure and structural information by *tandem mass spectrometry*, inducing ion decompositions through collision induced dissociations (CID), interactions with electrons (electron capture dissociation (ECD) [6], electron transfer dissociations (ETD) [7], interactions with photons (infrared multiphoton dissociation (IRMPD), ion spectroscopy) or with surface (surface induced dissociations (SID)) [8], mapping the distribution of an analyte on a surface (tissue, leaf, etc.) by *mass spectrometry imaging*; obtaining information on shape and conformation by using *ion mobility mass spectrometry*.

5.2 Ionization Techniques

Mass spectrometry can study wide classes of molecules, ranging from small volatile up to polar and very large molecules. As they have different properties, *e.g.* molecular weight, volatility, polarity, etc., all of them cannot be ionized by just one ionization technique, but a wide range of ionization techniques are nowadays available.

There are classical techniques which ionize volatile molecules in the gas phase (electron and chemical ionization (EI, CI)), desorption techniques (such as matrix assisted laser desorption (MALDI), secondary ion mass spectrometry (SIMS), …) and spray ionization techniques in which sample solutions are sprayed forming small droplets and then ions (electrospray (ESI), atmospheric pressure chemical ionization (APCI) [9], atmospheric pressure photoionization (APPI), …..) [10].

There is a quite new branch of mass spectrometry, called *ambient mass spectrometry* consisting of direct sampling and ionization of the analytes at ambient conditions with no or minimal effort for sample preparation. Many ionization techniques, and their number is continuously increasing, belong to this group. Some examples are: desorption electrospray ionization (DESI), direct analysis in real time

(DART), extractive electrospray ionization (EESI), low temperature plasma probe (LTP), paper spray ionization (PSI), and many others [11–18].

While electron ionization and chemical ionization occur under vacuum, most of all the other ionization techniques operate at atmospheric pressure. In the case of MALDI, the ionization generally occurs under vacuum, but it can be also at atmospheric pressure (atmospheric pressure MALDI (AP-MALDI)) [1–5].

Dependently on the technique used, ionization can occur through one electron removal from the molecule in the gas phase, thus forming a radical cation (the molecular ion), as it occurs in electron ionization, but in most of cases ionization consists of protonation or deprotonation reactions, thus forming even-electron ions $[M+H]^+$or $[M-H]^-$. In the case of electrospray, if the molecule is sufficiently large, multiply charged ions $[M+nH]^{n+}$ or $[M-nH]^{n-}$ are formed. Adducts formation can also occur yielding species $[M+Na]^+$, $[M+K]^+$, $[M+Cl]^-$, $[M+HCOO]^-$ [1–5].

Finally, ionization techniques can be grouped into *hard* and *soft*. In the case of hard ionizations, mostly represented by electron ionization, a large amount of energy is deposited onto the molecule, yielding a molecular ion with a high energy excess. This causes its abundant fragmentation and, as a consequence, lot of information on structural characterization of the analyte. In the case of *soft* ionizations, none or very few fragment ions are formed, so no any structural information can be obtained and tandem mass spectrometry is required for producing fragmentation in the analyzer region [1–5].

5.3 Analyzers

Once the ions are formed in the ion source, they are accelerated towards the mass analyzer where their separation according to their *m/z* ratios occurs. To fulfil their aim, analyzers use forces which influence ion movement, and in particular (i) a magnetic field; (ii) an electric field; (iii) a radiofrequency. An analyzer can use just one or a combination of two or three of them [1–5].

Analyzers can be divided into two main groups: those based on ion separation *in space*, sometimes also referred to as *beam-type* analyzers, and those separating ions *in time* (Table 5.1).

Double sector instruments, having a magnetic and an electric field, quadrupole and time of flight belong to the first group. In this case ions travel for some centimeters (roughly 30–100 cm in quadrupole) or meters (roughly 2–5 m in sectors or 2–17 m in ToF) and along the way they are submitted to one or two of the forces described above.

Analyzers separating ions *in time* are trapping devices: ion traps (linear or tridimensional ion trap), Orbitrap, and Fourier Transform Ion Cyclotron Resonance (FT-ICR). In these cases the ion path may be from cm to meters long, depending on trapping time [1–5].

Analyzers can make a scan (sectors, quadrupole, ion trap), measure a flight time (ToF) or measure a frequency (frequency of axial oscillations (Orbitrap), cyclotron

Table 5.1 Analyzers and their main features

Ion separation *in space*					
Analyzer	**Force**	**Separation based on**	***m/z* range**	**Resolving power**	**Mass accuracy**
Sectors (EB, BE) 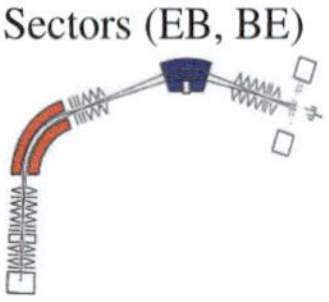	Magnetic and electrostatic fields	Ion momentum and kinetic energy	10,000	10,000	< 1 ppm
Quadrupole 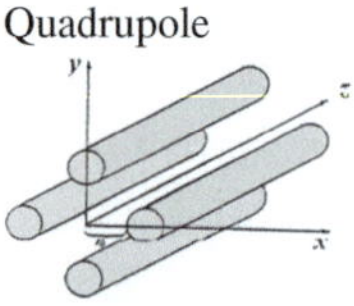	Electric field and radiofrequency	Stability/ Instability of the orbits	2000–4000	Unit (0.2 u FWHM)	No (>20 ppm)
Time of flight 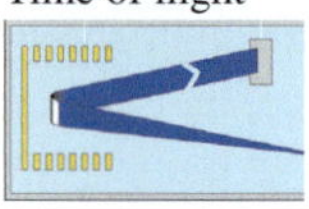		Speed	>100,000	>20,000	2–5 ppm
Ion separation *in time*					
Analyzer	**Force**	**Separation based on**	***m/z* range**	**Resolving power**	**Mass accuracy**
3D and 2D ion trap  	Electric field and radiofrequency	Frequency of the orbits	<6000	<500	no
Fourier Transform Ion Cyclotron Resonance (FT-ICR) 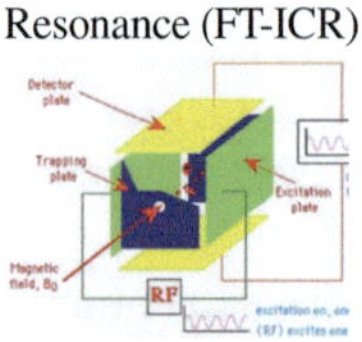	Electric field, radiofrequency, magnetic field	Frequency of the orbits	>10,000	>500,000	< 1 ppm
Orbitrap 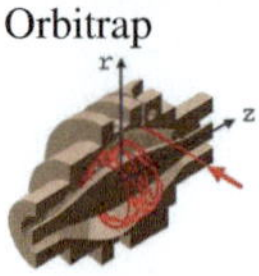	Electric field	Frequency of the harmonic oscillations	<6000	>100,000	2–5 ppm

frequency (FT-ICR)). In the last case, the Fourier transform mathematical operations are applied for generating mass spectra from time domain transients produced by the image current.

Main topics concerning analyzers are resolving power and resolution, accurate mass measurements and tandem mass spectrometry.

5.4 Resolving Power, Resolution and Accurate Mass Measurements

One of the most important performances of analyzers is their resolving power, *i.e.* the ability to distinguish between ions differing in the quotient mass/charge by a small increment [19]. A given resolving power yields a given resolution in the mass spectrum. Resolving power and resolution are the most debated terms in mass spectrometry.

Two main definitions can be used for resolution (Fig. 5.2):

(a) *Valley definition*: it considers two peaks of equal height in a mass spectrum at masses m and $(m + \Delta m)$ be separated by a valley (generally defined at 5–10%) which, at its lowest point, is at x% of the height of either peaks;
(b) *Peak width definition*: for a single peak made up of singly charged ions at mass m in a mass spectrum, the resolution may be expressed as $m/\Delta m$, where Δm is the width of the peak at a height which is a specified fraction of the maximum peak height. Generally the 50% peak height is considered, so a resolution at full width at half maximum (FWHM) is defined.

From above it results that resolution and peak width are strictly related each to the other. Depending on the analyzer used, resolution or peak width can be constant or they vary in the scan. In particular:

1. Quadrupole and ion trap: the peak width is constant over all the mass spectrum. Thus if FWHM is 0.5u: at *m/z* 40 the resolution will be 40/0.5=80, but at *m/z* 1000 the resolution will be 1000/0.5=2000. In other words, resolution increases by increasing the *m/z* value. In any case, quadrupole and ion trap do not have high resolving power.
2. TOF and sector analyzers: the resolution remains constant over all the mass spectrum, *i.e.* the peak width increases by increasing the *m/z* value. Thus if Res $_{FWHM}$ is 20,000: at *m/z* 40 the peak width at 50% of height will be 40/20,000=0.002 u, but at *m/z* 4000 FWHM will be 4,000/20,000 = 0.2u. In other words, peaks are

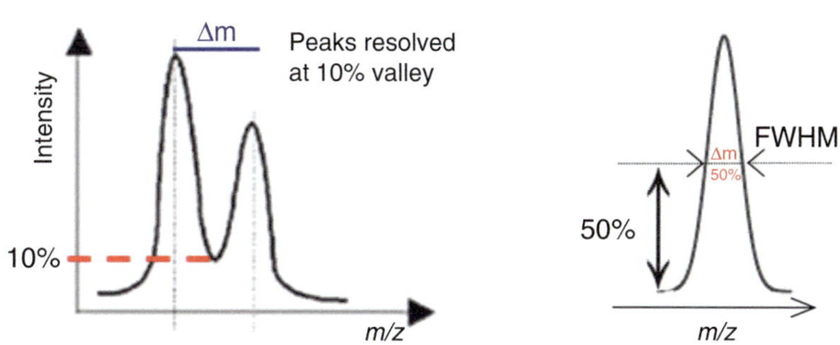

Fig. 5.2 Definitions of resolution: valley definition (*left*) and full width half maximum (FWHM, *right*)

wider by increasing the *m/z* value. Modern TOF with reflectron have resolving powers around 30,000–40,000 while for sector instruments resolving power is about 10,000–15,000.

3. FTICR: at constant detection time, the resolving power is inversely proportional to *m/z*. Thus if resolution is 1×10^6 at *m/z* 100, it will be 1×10^5 at *m/z* 1000.
4. Orbitrap: at constant detection time, the resolving power is inversely proportional to the square root of *m/z*. Thus if resolution is 100,000 at *m/z* 100, at *m/z* 1000 it will be $100{,}000\times(100/1000)^{1/2}=31{,}646$.

Resolution is strictly related to sensibility (i), specificity (ii) and scan speed (iii).

(i) Resolution *vs* sensitivity.
In general an increase of resolution causes a drastic decrease of sensitivity. So in sector instruments, increasing resolution involves a narrower ion spread which is obtained by closing the slits on the ion path. Passing from low resolution to resolution 5000 (at 10% valley) causes a decrease of sensitivity of about 50–70%. A similar situation occurs in ToF analyzers, passing from linear mode to reflectron mode, in Orbitrap and FT-ICR analyzers. A drastic decrease of sensitivity can have negative effects on accurate mass measurements (see below).

(ii) Resolution *vs* specificity.
An increase of resolution increases specificity by distinguishing and separating isobaric ions, i.e. ions having the same nominal mass, but different chemical compositions. As an example CO, N_2 and C_2H_4 have the same nominal mass, 28 Da, but different chemical compositions. So a high resolution regime allows one to eliminating species interfering with the analytes of interest (Fig. 5.3).

(iii) Resolution vs scan speed.
In general, an increase of resolution causes longer scan rates or longer detection times in FT analyzers. This is a very important topic especially when the mass spectrometer is coupled to a chromatographic system. In fact a good reproducibility of a chromatographic peak profile requires making at least 12–15 scans (Fig. 5.4). If the peak width is 5 sec, the cycle time cannot be longer than 0.4 sec, *i.e.* ~3 Hz. Thus in some cases, resolution must be reduced to a value in agreement with peak width and scan rate.

High resolution is a mandatory requirement for accurate mass measurements, *i.e.* measuring the *m/z* value with significance on the fourth decimal digit [20]. So when working in low resolution regime, calculations on masses can be done by using the *nominal mass*, *i.e.* the mass of the ion calculated using the mass of the most abundant isotope of each element rounded to the nearest integer value, *e.g.* H=1, C=12, N=14, O=16, etc. But when working in high resolution with accurate mass measurements, the *monoisotopic mass*, *i.e.* the mass of the ion calculated using the mass of the most abundant isotope of each element, has to be used. As an example, monoisotopic masses for hydrogen, carbon and nitrogen are 1.007825, 12.000000 and 14.00307, respectively.

By using the accurate mass value it is possible to obtain information on the chemical formula of the ion [21]. This is done via software starting from an input

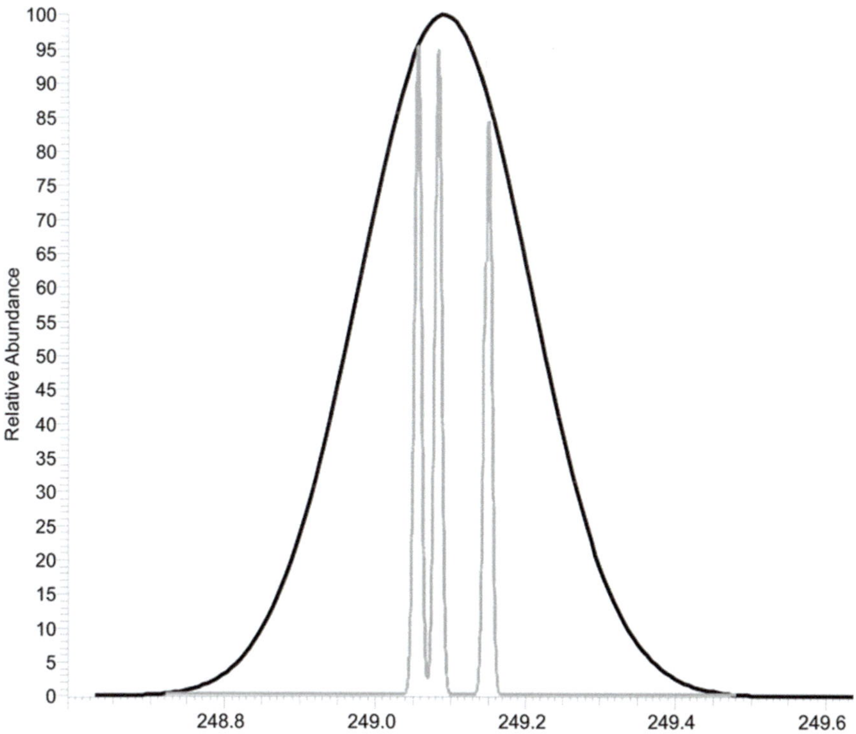

Fig. 5.3 Mass spectra recorded at different resolutions: mass spectrum of a mixture of three isobaric species $[C_{19}H_7N]^+$, $[C_{20}H_9]^+$, $[C_{13}H_{19}N_3O_2]^+$ obtained at low resolution (*black line*) and at resolving power 50,000 (*grey line*). It is noteworthy that at low resolution the three isobaric peaks are completely unresolved

Fig 5.4 Chromatographic peak profile and number of scans

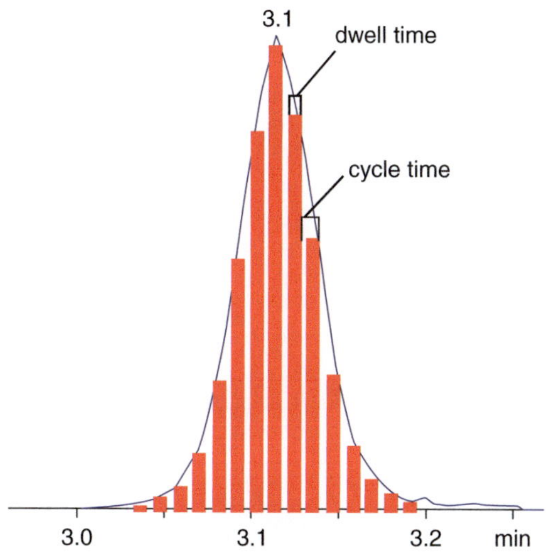

list, defined by the operator, of potential elements present in the ion with their ranges, *i.e.* C 5÷10, H 5÷30, N 1÷5, O 0÷10, etc. Starting from this list and from the accurate mass value, the software calculates potential chemical formulae. For each, it determines its own exact mass (the calculated mass considering the monoisotopic mass of a single isotope of each atom) which is compared to the accurate mass. In general, higher is the agreement between the two values, higher is the probability that the proposed chemical formula corresponds to the real formula of the ion. This is estimated by an error which can be expressed as mmu ((*accurate mass – exact mass)/m/z)* or ppm ((*(accurate mass – exact mass)/m/z)* $\times 10^6$) [20].

A reliable accurate mass value requires a correct mass calibration of the mass spectrometer, consisting of fitting the observed mass measurements to the exact masses of two or more different ions. Mass calibration can be external, *i.e.* reference masses from a mass spectrum of another analyte acquired under similar conditions, or internal, *i.e.* the reference masses are for ions of known elemental composition in the same mass spectrum as the analyte. Internal calibration is typically at least twice as accurate as external calibration.

Evaluation of repeatability and reproducibility of accurate mass measurements should be also carried out.

For an unambiguous identification, by increasing the *m/z* value, a higher mass accuracy is required. In fact increasing mass-to-charge ratios the number of potential formulae dramatically increases making identification more and more difficult.

Thus for the characterization of ions containing only C, H, O, N, with compositions $C_{0\div100}$, $H_{3\div74}$, $O_{0\div4}$ and $N_{0\div4}$, at *m/z* 118 only an error not exceeding 34 ppm is required for an unambiguous identification, but at *m/z* 750 an error better than 0.018 ppm is required to eliminate “all extraneous possibilities” [21]. This is graphically shown in Fig. 5.5. As an example, for ions containing only C, H, N, O, at *m/z*

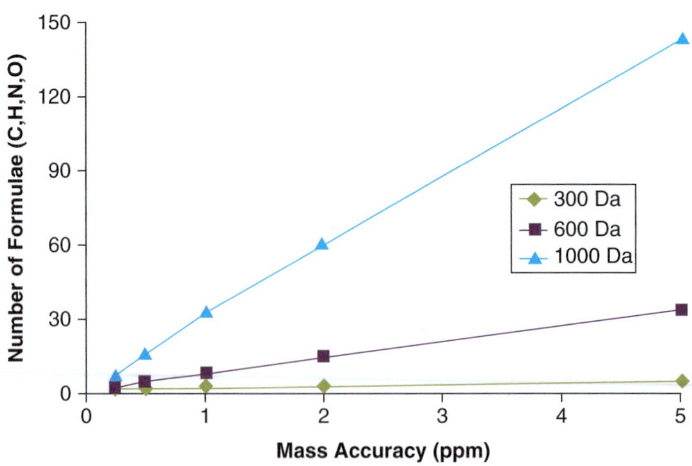

Fig. 5.5 The relationship between mass accuracy and m/z values on the number of potential elemental formulae [22]

1000 and mass accuracy of 5 ppm, about 150 chemical formulae can be proposed, at *m/z* 600 the number is reduced to about 30 and at *m/z* 300 the number of potential formulae is reduced to a very few [22].

Some constraints can be set up for reducing ambiguity and the number of potential chemical formulae. One is represented by the double bond/ring equivalents (RDB) defined in Eq. 5.1

$$RDB = 1 + \frac{\left[\sum_{i}^{i\max} Ni\left(Vi - 2\right)\right]}{2} \tag{5.1}$$

where N*i* is the number of atoms of element *i*, and V*i* is the valence of atom *i*. In addition to information on the presence of double bonds and/or rings in the ions, its decimal digit gives an indication on the nature of the ion. In fact for odd-electron ions, RDB value is x.0 while for even-electron ions this value is x.5. So if we are studying the molecular ion ($M^{+\bullet}$, a radical cation) produced by electron ionization, RDB values x.0 will be considered. It should be noted that when we are considering ions produced by ESI or MALDI, where in general only even-electron ions are formed, our attention will be focused on formulae having RDB x.5.

In addition to the accurate mass of the most abundant ion, also accurate masses of $[M+1]^+$ and $[M+2]^+$ should be determined together with the relative intensity of the isotopic cluster to ascertain the congruity of all the data (Fig. 5.6).

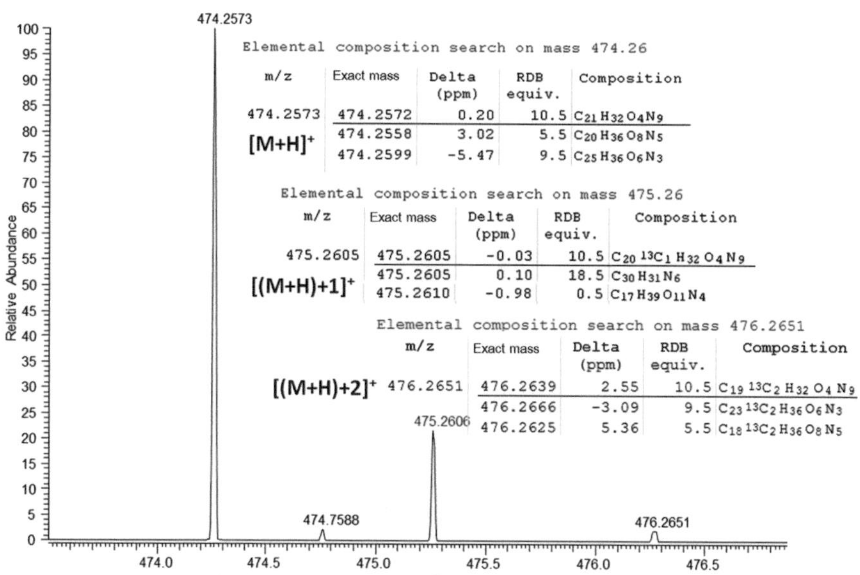

Fig. 5.6 High resolution mass spectrum of a cyclic peptide, accurate masses of the isotopic cluster of the protonated molecule, the first three proposed elemental formulae, their exact masses, errors and RDB calculated for each accurate mass

Table 5.2 Number of possible molecular formulae at different levels of mass accuracy and the impact of isotopic abundance accuracy [24]

	without isotope abundance information					2% isotopic abundance accuracy	5% isotopic abundance accuracy
molecular mass [Da]	10 ppm	5 ppm	3 ppm	1 ppm	0.1 ppm	3 ppm	5 ppm
150	2	1	1	1	1	1	1
200	3	2	2	1	1	1	1
300	24	11	7	2	1	1	6
400	78	37	23	7	1	2	13
500	266	115	64	21	2	3	33
600	505	257	155	50	5	4	36
700	1046	538	321	108	10	10	97
800	1964	973	599	200	20	13	111
900	3447	1712	1045	345	32	18	196

Another constrain for reducing ambiguity and the number of potential chemical formulae is the spectral ion abundance error which is an estimation on how well the experimental isotopic pattern from the detected compound matches to the isotopic pattern calculated for the proposed chemical formula [23]. In fact high mass accuracy (<1 ppm) alone is not enough to exclude enough candidates with complex elemental compositions (C, H, N, S, O, P, and potentially F, Cl, Br and Si). Use of isotopic abundance patterns as a single further constraint removes most of false candidates (Table 5.2).

Finally, it is possible performing accurate mass measurements on fragment/product ions, for example in EI spectra or in tandem mass spectra, and verify the agreement of all the data. For example $[C_{10}H_{20}N_2]^+$ cannot have $[C_9H_{18}O]^+$ as product ion. In addition, accurate mass measurements on fragment/product ions allows one to determine the formula of the species lost in the fragmentation.

High resolution mass spectrometry finds application in many fields: MS imaging, food, petroleomics, proteomics, forensic sciences, metabolomics, homeland security, etc.

5.5 Tandem Mass Spectrometry

In mass spectrometry the presence of fragment/product ions is crucial for elucidating the structure of an unknown. Until the ion remains intact, its structure remains unknown!

So the aim of tandem mass spectrometry is to induce selective fragmentation of an ion that is located outside of the ionization source, i.e. in the analyzer region. In tandem mass spectrometry experiments, two (MS/MS or MS^2) or more (MS^n) sequential stages of mass analysis (which can be spatially or temporally separated)

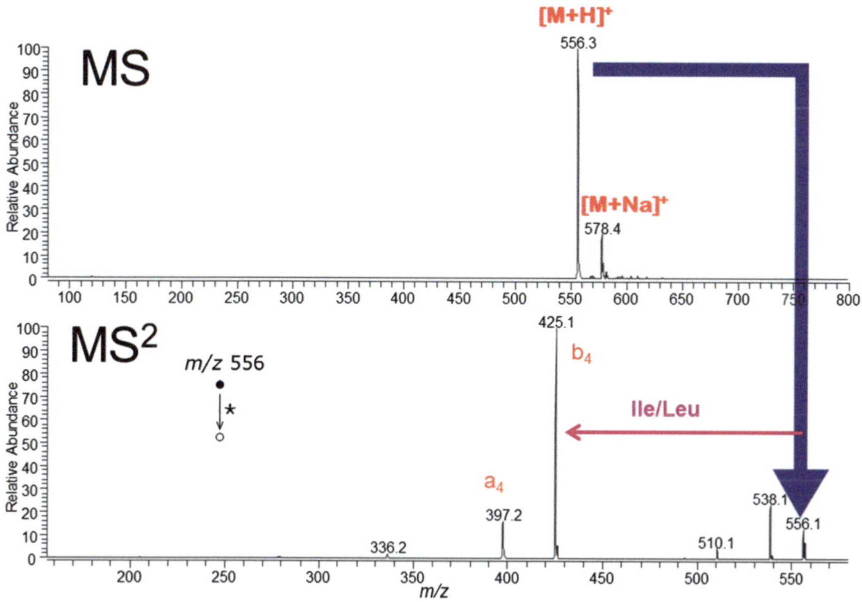

Fig. 5.7 ESI (+) MS (*top*) and MS/MS product ion scan (*bottom*) of a peptide

are used in order to examine selectively the decomposition of given ions in a mixture of ions [1–4].

Three main scan modes can be used in tandem mass spectrometry: (i) *product ion scan* which consists in selecting ions with a given *m/z* value, induce their dissociation and detecting their *product ions* (Fig. 5.7); (ii) *precursor ion scan*: in this case, all precursors of a given fragment ion are detected. This approach is particularly useful when a common fragment ion is produced from different compounds of a mixture.

The third kind of scan mode, i.e. *neutral loss scan*, consists of the detection of fragmentation pathways involving elimination of neutral species, such as H_2O, CO, CO_2, HCN.

Ion decomposition can be obtained by four main mechanisms: (a) collisions of ions with a gas, generally nitrogen, helium, or argon. This yields a conversion of part of the translational energy of the ion to internal energy with a consequent bond cleavage and dissociation so to produce *collision induced dissociation* (CID) spectra. CID experiments can be done in low energy regime (1–100 eV collision energy) in ion traps, triple quadrupoles, Q-TOF or in high energy regime (5–30 keV collision energy) in sectors and TOF-TOF instruments; (b) interactions of ions with electrons through electron capture dissociation (ECD) [6], occurring inside a FT-ICR analyzer, or through electron transfer dissociation (ETD) [7] by ion-ion reactions inside an ion trap; (c) interactions of ions with photons (infrared multiphoton dissociation (IRMPD)); (d) interactions of ions with surfaces (surface-induced dissociation (SID)) [8, 25].

5.6 Conclusion

Modern mass spectrometry offers very powerful tools for identification, structural characterization and quantitation of unknowns.

Its coupling with separative techniques, such as gas chromatography or liquid chromatography (HPLC, UPLC, UHPLC, etc.) allows to separate analytes based on their retention times and to study complex mixtures.

In addition to the molecular weight determination, high resolution mass spectrometry allows to have very high specificity, by separating isobaric ions, and to make accurate mass measurements from which it is possible to determine the chemical formula of the ion.

Different techniques of tandem mass spectrometry, through the formation of selective fragmentation and decomposition of ions, allow obtaining structural information or offering suitable tools for quantitative determinations.

Other tools, such as mass spectrometry imaging (see Chap. 7 of this book), native mass spectrometry, hydrogen-deuterium exchange mass spectrometry, isotope ratio mass spectrometry, etc) are nowadays available and yield wide amounts of information about different analytes. In addition to them, it is also possible to investigate size and shape (conformation) of ions by using ion mobility which is having more and more relevance and diffusion in modern mass spectrometry. It has a lot applications in different fields: with portable and standalone spectrometers in homeland security with detection of toxic industrial chemicals, chemical-warfare agents; with an ion mobility device inserted inside a mass spectrometer it can investigate size and shape of wide classes of molecules, from small to large and very large, such as carbohydrates, metabolites, proteins, etc.

References

1. De Hoffmann E, Stroobant V (2007) Mass spectrometry: principles and applications, 3rd edn. Wiley, New York
2. Gross JH (2011) Mass spectrometry. A textbook, 2nd edn. Springer, Berlin
3. Dass C (2007) Fundamentals of contemporary mass spectrometry. Wiley-Interscience, New York
4. Desiderio DM, Nibbering NM (2009) In: Ekman R, Silberring J, Westman-Brinkmalm AM (eds) Mass spectrometry: instrumentation, interpretation, and applications. Hoboken, Wiley
5. Gross ML, Caprioli RM (eds) (2003) The encyclopedia of mass spectrometry, vol 1–10. Elsevier Science, Oxford
6. Zubarev RA, Kelleher NL, Mc Lafferty FW (1998) Electron capture dissociation of multiply charged protein cations. A nonergodic process. J Am Chem Soc 120:3265–3266
7. Syka JE, Coon JJ, Schroeder MJ, Shabanowitz J, Hunt DF (2004) Peptide and protein sequence analysis by electron transfer dissociation mass spectrometry. Proc Natl Acad Sci USA 101:9528–9533
8. Mabud MA, Dekrey MJ, Cooks RG (1985) Surface-induced dissociation of molecular ions. Int J Mass Spectrom Ion Proc 67:285–294

9. Robb DB, Covey TR, Bruins AP (2000) Atmospheric pressure photoionization: an ionization method for liquid chromatography–mass spectrometry. Anal Chem 72:3653–3659
10. Revel'skii IA, Yashin YS, Kurochkin VK, Kostyanovskii RG (1991) Mass spectrometry with photoionization at atmospheric pressure and the analysis of multicomponent mixtures without separation. Chem Physicochem Methods Anal 243–248 (translated from Zavodskaya Laboratoriya 57:1–4)
11. Van Berkel GJ, Pasilis SP, Ovchinnikova O (2008) Established, emerging atmospheric pressure surface sampling/ionization techniques, for mass spectrometry. J Mass Spectrom 43:1161–1180
12. Chen H, Gamez G, Zenobi R (2009) What can we learn from ambient ionization techniques? J Am Soc Mass Spectrom 2009(20):1947–1963
13. Alberici RM, Simas RC, Sanvido GB, Romao W, Lalli P, Benassi M, Cunha IBS, Eberlin MN (2010) Ambient mass spectrometry: bringing MS into the "real world". Anal Bioanal Chem 398:265–294
14. Huang MZ, Yuan CH, Cheng SC, Cho YT, Shiea J (2010) Ambient ionization mass spectrometry. Annu Rev Anal Chem 3:43–65
15. Huang MZ, Cheng SC, Cho YT, Shiea J (2011) Ambient ionization mass spectrometry: a tutorial. Anal Chim Acta 702:1–15
16. Ifa DR, Wu C, Ouyang Z, Cooks RG (2010) Desorption electrospray ionization and other ambient ionization methods: current progress and preview. Analyst 135:669–681
17. Weston DJ (2010) Ambient ionization mass spectrometry: current understanding of mechanistic theory; analytical performance and application areas. Analyst 135:661–668
18. Harris GA, Galhena AS, Fernández FM (2011) Ambient sampling/ionization mass spectrometry: applications and current trends. Anal Chem 83:4508–4538
19. IUPAC Compendium of chemical terminology – the Gold Book, https://goldbook.iupac.org/
20. Brenton AG, Godfrey AR (2010) Accurate mass measurement: terminology and treatment of data. J Am Soc Mass Spectrom 21:1821–1835
21. Gross ML (1994) Accurate masses for structure confirmation. J Am Soc Mass Spectrom 5:57
22. Quenzer TL, Robinson JM, Bolanios B, Milgram E, Greig MJ (2002) Automated accurate mass analysis using FTICR mass spectrometry, Proceedings of the 50th ASMS conference, Orlando
23. Wang Y, Gu M (2010) The concept of spectral accuracy for MS. Anal Chem 82:7055–7062
24. Kind T, Fiehn O (2006) Metabolomic database annotations via query of elemental compositions: mass accuracy is insufficient even at less than 1 ppm. BMC Bioinformatics 7:234–243
25. Wysocki VH, Ding JM, Jones JL (1992) Surface-induced dissociation in tandem quadrupole mass spectrometers: a comparison of three designs. J Am Soc Mass Spectrom 3:27–32

Chapter 6
Primary Considerations in Quantitative Mass Spectrometry

Mark W. Duncan

Abstract Mass spectrometry is a powerful approach for generating precise, accurate and sensitive quantitative information on the components of a complex sample. This chapter addresses the importance of quantification and the key considerations in the process of developing and performing a quantitative analysis. Rather than providing specific details, the focus here is on introducing the most important considerations associated with developing a quantitative analysis.

6.1 Introduction

6.1.1 Overview

Chemical analysis allows us to improve our quality of life by offering safer, better, more cost efficient production of goods (e.g., food, clothing, industrial and household products) and services (e.g., diagnostics in healthcare and detection of hazardous substances). The analytical data we generate are used by our legal system as evidence of corruption, criminal activity, pollution, contamination, and of unfair, unsafe or illegal practices. For example, chemical testing might be used to detect potential terrorist activity, thereby saving hundreds, even thousands of lives. Alternatively, chemical analysis might also be used to establish that an individual possesses a banned substance (e.g., illicit drugs) and this finding could lead to fines, imprisonment or in some countries, execution.

Contrary to what many think, analytical chemistry is as much about carefully defining a problem, devising a suitable analytical strategy, and determining how the results will be used as it is about the process of instrumental analysis. What are the target analytes and their expected concentrations? What is the sophistication of the

M.W. Duncan (✉)
Biodesix Inc., Boulder, CO, USA
e-mail: Mark.Duncan@ucdenver.edu

J.H. Banoub, R.M. Caprioli (eds.), *Molecular Technologies for Detection of Chemical and Biological Agents*, NATO Science for Peace and Security Series A: Chemistry and Biology, DOI 10.1007/978-94-024-1113-3_6

"client",[1] how should the results (and related errors) be reported to them? What are acceptable error rates for detection (i.e., false positives, false negatives) and acceptable uncertainties for reported values (e.g., accuracy and precision requirements)? Only after addressing these issues can the analyst develop a method that meets the needs of the client and balances thoroughness against practical considerations such as time and cost. Throughout the whole process the analyst must document, not only the analytical process itself, but also all the critical assumptions (and the limitations) inherent in the method.

6.1.2 The Concept of Fitness for Purpose

As highlighted above, the results of a quantitative analysis are frequently used by others to make important decisions. However, because all analytical tasks are not equivalent, and because there is no universally applicable instrumental method or approach, compromises in the approach are always necessary. For example, it may be necessary to sacrifice speed and cost in some settings in the interests of delivering high accuracy; at other times, speed may be the overarching requirement and to achieve the necessary turn-around time some compromise in cost and accuracy is considered acceptable.

Typically, the client commissions a laboratory to perform analytical work, and in consultation with the laboratory, both parties agree on acceptable tolerances given how the data will be used. This process of tailoring the method and the reporting of the results to the intended purpose is referred to as *fitness for purpose*.* Thompson and Ramsey define fitness for purpose as "… the property of data produced by a measurement process that enables the user of the data to make technically correct decisions for a stated purpose. [1]"

6.1.3 Why Quantify?

Quantitative analysis acknowledges that simply saying that a component is either absent or present from a sample is essentially meaningless. For the analytical chemist, absent means "not detected" by the specific analytical strategy that was adopted, or more accurately, that the amount of the analyte present in the specified sample, and per the method adopted, is lower than a specific value defined by the limit of detection (LOD) for that method. Similarly, declaring that an analyte is "present" is almost always inadequate. Clients want to know the amount of analyte in the sample and the uncertainty associated with the stated amount.

[1] There is always a client or end-user of any data that the analytical scientist generates. It is essential for the analyst to be mindful of how the data they generate will be used.

6.1.4 Approaches to Quantification

Quantification typically adopts instruments and methods to separate /isolate specific components, then to identify and quantify them. Historically, wet chemistry methods were used, including isolation by approaches such as solvent extraction, precipitation or distillation; identification based on color, odor, boiling point, melting point, flame test, or specific chemical reactions; then, quantification of the isolated and identified analyte by either mass (i.e., gravimetric analysis) or volume (i.e., volumetric analysis). Integrated instrumental methods have supplanted classical wet chemistry approaches in almost all settings because they are safer, faster, less laborious, less costly, more sensitive, more accurate and can be automated. Examples of instrumental methods that integrate a separation, identification and quantification component include liquid chromatography (LC) combined with UV/visible detection, gas chromatography (GC) combined with single stage or tandem MS (GC-MS and GC-MS/MS) and LC combined with mass spectrometry (e.g., LC-MS or LC-MS/MS) [2].

6.1.5 The Unique Benefits of MS Analysis

MS is primarily adopted because it is sensitive (i.e., delivers a low limit of detection), allows multiplexed determinations and it delivers analytical specificity/selectivity superior to that of almost all other technologies. This means it is possible to detect and quantify a signal specific to a target analyte without interferences from other sample components.

Some downsides associated with the adoption of mass spectrometry include the instrument footprint, purchase and maintenance costs, technical complexity, downtime, and the requirement for highly skilled staff. Additionally, sample throughput is limiting for many applications because substantial pre-analytical sample manipulation is required prior to instrumental analysis.

6.1.6 Types of Quantitative Analysis

Quantitative analysis can deliver results expressed in absolute (e.g., mg/g or mmolar) or relative terms with respect to the components in that sample (e.g., there is 1 part X for every 10 parts of Y, without specifying the amounts of either X or Y), or across multiple samples (e.g., X is 10 times more abundant in one sample than another). Regardless of how the results are expressed, the process and primary considerations remain essentially the same. Although the discussion in this chapter is focused on the quantification of a single component within a sample, multiple components can be quantified by simply extending the same principles and practices to multiple target analytes.

6.2 The Practice of Quantitative Chemical Analysis

6.2.1 Introduction

Quantification is possible when a measured physical or chemical property (X) associated with an analyte is directly and reproducibly proportional to the concentration (C_A) of that analyte:

i.e., $C_A = kX$ where, k is a proportionality constant.

The process of determining k is an important step in quantitative chemical analysis and is called calibration. In the simplest manifestation of quantification by mass spectrometry, the signal response for a fixed *m/z* value is measured – an approach known as selected ion monitoring. Additional selectivity can, however, be achieved by measuring: (a) multiple *m/z* signals characteristic of the analyte, (b) mass differences (i.e., loss of specific structural elements), (c) ion abundance ratios, (d) isotopic ratios, (e) exact masses (e.g., corresponding to a specific molecular formula) or (f) MS/MS transitions [2].

To illustrate the general principle, Fig. 6.1 shows a typical electron ionization (EI) spectrum. If a single ion is used for quantification, selectivity is optimal when

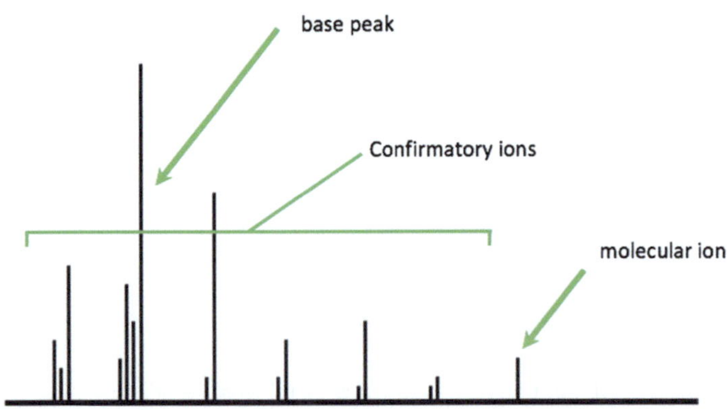

Fig. 6.1 A typical electron ionization (EI) spectrum. This spectrum shows the presence of a molecular ion and multiple fragment ions. Quantification can be based on measuring the signal response for one or more of these ions. When measuring one ion, optimal selectivity comes from quantification based on the molecular ion; optimal sensitivity is associated with quantifying the base peak. Increasing the number of ions monitored increases selectivity, but at the same time, there is a reduction in sensitivity (i.e., lower limit of quantification)

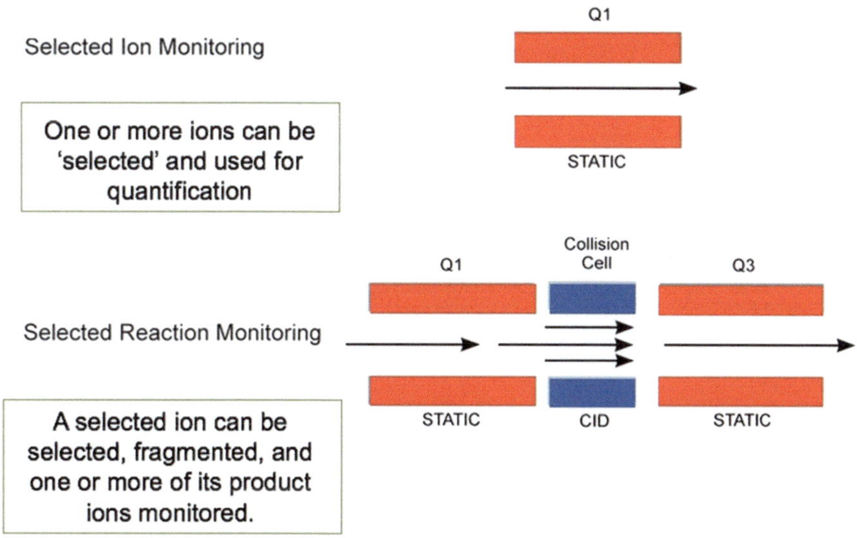

Fig. 6.2 Approach to quantification by mass spectrometry. The two most commonly used strategy involve selected ion monitoring (typically on a single stage instrument) and selected reaction monitoring (requiring multiple stages of mass separation)

that ion is the molecular ion; sensitivity is optimal, however, when that ion is the base peak (i.e., the most intense ion in the spectrum). Additional selectivity can be achieved by monitoring several ions (e.g., the molecular ion plus several confirmatory ions). Increasing selectivity comes, however, at the expense of sensitivity (i.e., the LOS and LLOQ are attenuated).

The most commonly adopted quantitative strategies are illustrated in Fig. 6.2. Selected ion monitoring (SIM) is used to quantify a target analyte based on one or more ions characteristic of that analyte. (If multiple ions are monitored, the approach is sometime referred to as multiple ion monitoring (MIM).) Selected reaction monitoring (SRM) is the process by which a precursor ion is fragmented to generate multiple product ions, typically by collision-induced dissociation in a triple quadrupole mass spectrometer. SRM provides for enhanced selectivity even in situations where the target analyte e.g., $[M + H]^+$ formed by electrospray ionization does not fragment directly [2].

6.2.2 General Strategies for Quantitative Analysis

The wealth of literature on approaches to quantitative analysis can be distilled into three basic strategies and these are outlined in the sections that follow.

The External Standard Method This is the most commonly adopted approach in analytical chemistry (e.g., quantification by spectrophotometry), but it requires that

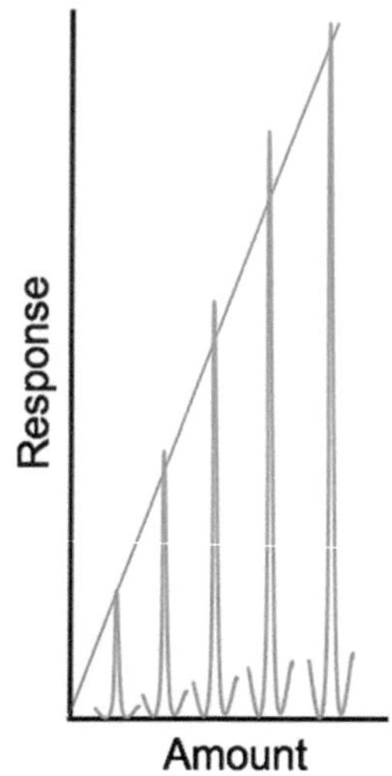

Fig. 6.3 Quantification based on the external standard method. Although commonly used in other areas of quantitative analysis (e.g., spectrophotometry), factors other than amount/concentration influence the measured signal intensity so that this approach is rarely used for quantification by mass spectrometry

there is direct, reproducible and rugged relationship between signal intensity and analyte amount independent of other components of the sample. With the external standard method, a series of standards is run under well-defined conditions and a plot is generated of response vs. concentration – the calibration curve [3, 4]. Thereafter, the amount of the target analyte in an unknown sample can be determined simply by measuring the signal response for that sample and then converting this to a concentration by reference to the calibration curve. The close to ideal relationship between measured response and analyte concentration that allows adoption of the external standard approach is illustrated in Fig. 6.3.

Internal Standard Method In almost all applications of mass spectrometry the measured ion current does not directly relate to the amount of sample introduced into the system. Instead, it fluctuates with time and from sample to sample because of changes in instrument pressure, temperature and voltages, and in response to other components in the sample matrix. To counter this problem an internal standard is typically incorporated into the sample prior to analysis [3, 4]. The internal standard approach is based on the principle that when two compounds of similar chemical and physical properties are introduced into the ion source at the same time, both will be influenced by these additional factors to the same extent. Consequently, the measured ion current ratio is proportional to the concentration ratio of those two components: i.e., if we know the concentration of one (i.e., the internal standard), we can calculate that of the second.

An ideal internal standard will be a near perfect chemical and physical mimic of the analyte; however, it must be distinguishable by mass (e.g., a stable isotope labeled form of the analyte or perhaps a minor structural variant). The internal standard is added to ALL samples and the standards that constituent the calibration curve at a known, fixed concentration. The signals for the target analyte and the internal standard are both measured, and the response (R) ratio $R_{analyte}/R_{internal\ std.}$ vs

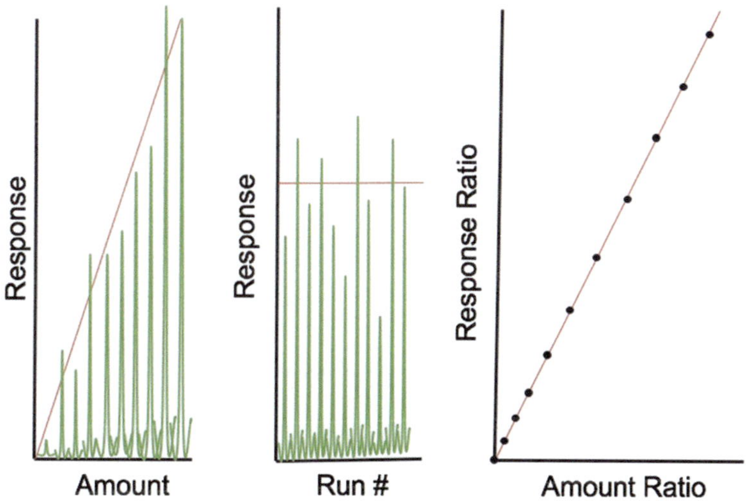

Fig. 6.4 Quantification based on the internal standard method. With the incorporation of an appropriate internal standard (IS) that closely mimics the behavior of the target analyte, and with recourse to a "gold standard" reference, it is possible to generate precise and accurate quantification by mass spectrometry (Note that the relationship between signal intensity and amount deviates from a direct linear relationship (panel A) because multiple factors others than amount influence response. The internal standard, however, is also subject to these fluctuations (panel B) and therefore the ratio of the responses, analyte/IS is linear. When plotted against analyte concentration (panel C))

concentration is plotted for all standards – i.e., the calibration curve. The ratio R/R_{is} is measured for all unknowns and converted to a concentration by reference to the calibration curve [3, 4]. Figure 6.4 shows the essential characteristics of the data generated by the internal standard method and illustrates how the internal standard "corrects" for fluctuations in response from sample to sample.

The Standard Addition Method Standard addition is commonly used in mass spectrometry when matrix affects are pronounced, target analyte levels are near the LOD, the sample set is unique or diverse in composition, or when analyte-free matrix is unavailable. With standard addition, each unknown sample is divided into two (or more) portions and a KNOWN amount(s) of analyte (spike) is/are added to one or more of these. The spiked sample(s) will show a larger analytical response and the difference in response between the spiked and unspiked samples is used to generate an intra-sample calibration curve from which the analyte in the original sample can be calculated [3, 4]. (A linear response is assumed.) The process of quantification by standard addition is illustrated in Fig. 6.5.

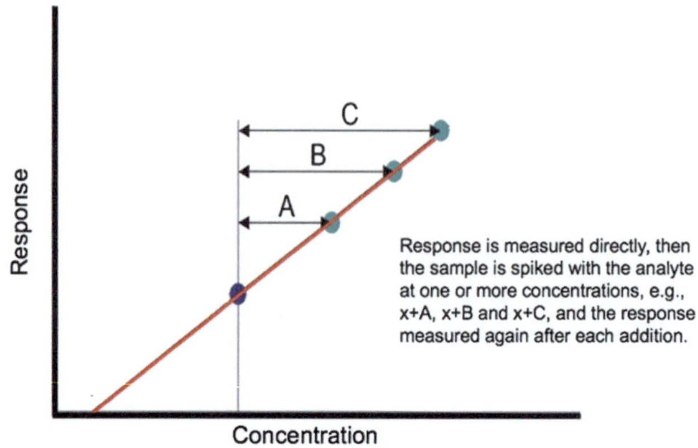

Fig. 6.5 Quantification based on the method of standard addition. Here the signal intensity of a sample solution is measured, then known amounts of the target analyte are added directly to that sample and the signal intensity is measured again after each addition. A plot of the signal intensities of each of these solutions vs. the added concentrations yields a straight line. The concentration of the target analyte is determined from the point at which the extrapolated line crosses the concentration axis at zero signal. Essentially, with standard addition a calibration curve is generated within each sample. This approach is particularly useful when the samples are diverse in composition

6.3 Key Terms

There is a precise language used to define the quality or performance characteristics of any quantitative analysis. The literature cited at the end of this chapter provide more rigorous definitions of many of these parameters, but for the purposes of this chapter, the following definitions are useful guides [3–5].

Accuracy: This is a measure of the "correctness" of any determination. Said differently, it is a measure of how close the experimental determined value the analyst's result is to the "true" value. While it is always possible to strive for greater accuracy, this battle comes at a significant price and is not always justified.

Precision: This is a measure of the repeatability of any determination and it is, of course, influenced by the irreproducibility or variance associated with each individual step in the analysis.

Specificity/selectivity: Selectivity refers to the ability to detect and quantify a specific analyte within a mixture without interference from other sample components.

Limit of Detection (LOD): The LOD is the lowest amount of analyte that can be *detected* at an acceptable signal-to-noise ratio (s/n) by the method. An acceptable s/n is often defined as 3:1.

Lower Limit of Quantification (LLOQ): The lowest amount of analyte that can be quantified at an acceptable precision and accuracy under the stated conditions.

Signal-to-Noise Ratio: The level of the specific signal attributable to the analyte relative to the level of background noise.

Robustness: The ability of an analytical method to be unaffected by small, deliberate changes in operating parameters.

6.4 Additional Considerations in Quantitative Analysis

6.4.1 Analyte, Sample and SampleCollection Issues

The nature of the sample, how it is collected, and the characteristics of the analyte itself are all critical factors that can profoundly influence the results. For example, most samples are heterogeneous, and therefore it is important to collect a representative sample. Clearly, accurate quantification on a *non-representative sample* is misleading, if not meaningless. In some settings it is best to take several random samples that are either analyzed individually, or in the interests of time and cost, mixed to generate a "representative" sample. If the sampling area is large - for example the soil in a field - it might be necessary to take many random samples. In some scenarios, the analyst may also have to consider the possibility that the target analyte may be present in multiple forms (e.g., free and bound) or that it is unstable, either in its isolated form or within the matrix [5, 6].

The sample or components of it may be toxic, explosive or highly flammable. Furthermore, the analytes of interest may be transformed during sample handling, transit or storage steps: e.g., evaporation of volatiles, reactions with air, photolytic decomposition, thermal decomposition, catalytic (e.g., enzymatic) activity or physicochemical reactions with other sample components (protein binding). Analyte stability should be assessed and accounted for in all analyses. Issues relating to the analyte, sample and sample selection are frequently the most troublesome and error-prone component of a quantitative analysis.

6.4.2 The Internal Standard

The nature of the internal standard and when it is added to the sample are additional considerations. Addition early in the process is necessary to account for losses and errors introduced during sample handling and processing. An appropriate internal standard should have chemical and physical properties as close as possible to those of the analyte of interest, but the practical issues of availability and cost also need to be taken into consideration [3–5].

6.4.3 Extraction and Sample Work Up

It may be necessary to extract or otherwise disrupt the sample to isolate the target analyte (e.g., mortar & pestle, blades/blenders, sonication or enzymes/chemicals). These processes need to be performed in a manner that minimizes analyte degradation while maximizing the extraction efficiency.

Minimal sample work-up reduces the workload, potential losses during extraction steps, and the potential for the chance introduction of contaminants. Contrastingly, although labor-intensive, extensive sample workup can markedly reduce the background and improve the achievable LOD and LLOQ. The demands on the hardware, the sensitivity requirements and multiple other factors come into play and determine the optimal balance between little versus extensive sample work up. For example, on-line chromatography reduces the need for off-line processing because it ensures that only one (or few) analytes simultaneously enter the source region. This minimizes interferences and maximizes the measured ion current for the analyte of interest [3–6].

6.4.4 Documentation

As highlighted earlier, detailed documentation is critical. Essential documentation for any quantitative analysis includes: (1) a written procedure (or protocol) that details all of the experimental steps; (2) a method validation report which includes both the experimental design and the data justifying the claim that the method, as written, performs as intended and (3) the system suitability criteria that define the minimum acceptable performance criteria prior to each analysis (i.e., criteria for the acceptance and/or rejection of the analytical data such as the minimum acceptable performance for standards and QCs for each subsequent application of method) [6, 7].

6.5 Quality Control Practices & Method Validation

Analytical chemists have developed a detailed prescription for defining the performance characteristics of any quantitative analysis so that the analyst and client can be sure that the data generated are suitable for their intended purpose. This process is referred to as validation. When performed properly and completely, validation: (1) demonstrates the ability of a specific assay to predict the concentration of unknowns and defines the associated accurately and reproducibly; (2) defines the sample storage conditions necessary to assure analyte stability and (3) defines the conditions for sample storage and work-up that maintain the integrity of the sample and the results. Importantly, the validation process defines criteria that can be used to accept or reject the analytical data in subsequent applications of the assay. Primary validation parameters include accuracy and precision; secondary parameters include a determination of the range (*i.e.*, upper and lower limits of quantification), selectivity (*i.e.*, assess interferences); recovery; stability; acceptance criteria and may also include a comparison with other existing methods [6, 7].

6.6 Conclusions

To provide analytical data that are correct and fit for their intended purpose should be the overarching consideration of the analytical scientist in any setting. Not only is the reporting of results important, but so too is an evaluation of the underlying uncertainties. Analysts must document, archive and share their methods, data and associated assumptions, errors and uncertainties. It is only with full disclosure of all this information that others can put the findings into context and draw sound conclusions. Analysts need to appreciate that their data will be scrutinized and that they will be held accountable for their findings. Frankly, in any given scenario, if the result of a test cannot be trusted, or are not fit for the intended purpose, then the analysis itself was a waste of time, and the finding are potentially misleading or damaging.

Acknowledgements The author thanks Dr. Alex Nickel, Biodesix Inc., for helpful discussions during the preparation of this manuscript.

References

1. Ramsey MH, Thompson M (2007) Uncertainty from sampling, in the context of fitness for purpose. Accred Qual Assur 12(10):503–513
2. Bethem R, Boison J, Gale PJ, Heller D, Lehotay S, Loo J, Musser S, Price P, Stein S (2003) Establishing the fitness for purpose of mass spectrometric methods. J Am Soc Mass Spectrom 14:528–541
3. Harris DC (2006) Quantitative chemical analysis, 7th edn. Freeman WH, pp 1008
4. Green MJ (1996) A practical guide to analytical method validation. Anal Chem 68(9):305A–309A
5. Little TA (2012) Assay development and method validation essentials. BioPharm Int 25(1):p8
6. Lee JW, Devanarayan V, Barrett YC et al (2006) Fit-for-purpose method development and validation for successful biomarker measurement. Pharm Res 23:312. doi:10.1007/s11095-005-9045-3
7. Tiwari G, Tiwari R (2010) Bioanalytical method validation: an updated review. Pharm Methods 1(1):25–38. doi:10.4103/2229-4708.72226

Chapter 7
Imaging Mass Spectrometry – Molecular Microscopy for Biological and Clinical Research

Richard M. Caprioli

Abstract Imaging mass spectrometry has seen tremendous advances over recent years and has become an important tool for molecular discovery in intact tissues with high molecular fidelity and spatial resolution. This chapter will focus on MALDI Imaging Mass Spectrometry (IMS) because it is the most useful technology for the mapping of metabolites, lipids, peptides and proteins at high sensitivity in intact tissue sections. Desorption of biomolecules is accomplished by direct laser irradiation of an array of spots (*i.e.*, pixels) on the tissue to map the location of specific molecules. Both fresh-frozen and formalin-fixed tissue sections can be imaged without the need for specific targeting reagents such as antibodies. Molecular images of this nature are produced based on specific *m/z* (mass-to-charge) values, or ranges of values. Thus, each specimen gives rise to many hundreds of specific molecular images from a single raster of the tissue. In a complementary approach, where only discrete areas within the tissue are of interest, such as in anatomic pathology, we have developed a histology-directed approach that integrates mass spectrometry and microscopy. Mass spectra are collected from selected discrete areas of cells within the tissue for laser ablation and analysis, and these are then correlated with microscopic images of the tissue section.

This chapter will illustrate the use of IMS in several biologically and medically relevant research projects. One area of interest is the mapping of molecular changes occurring in diabetes in both a mouse model and in the human disease. Major molecular alterations have been recorded in advanced diabetic nephropathy involving both proteins and lipids. Another application employs IMS to differentiate benign skin lesions from melanomas using in-house developed PIMS (Pathology Interface for Mass Spectrometry) software.

R.M. Caprioli (✉)
Department of Biochemistry, Mass Spectrometry Research Center, Vanderbilt University School of Medicine, Nashville, TN, USA
e-mail: Richard.M.Caprioli@Vanderbilt.edu

J.H. Banoub, R.M. Caprioli (eds.), *Molecular Technologies for Detection of Chemical and Biological Agents*, NATO Science for Peace and Security Series A: Chemistry and Biology, DOI 10.1007/978-94-024-1113-3_7

Although not primarily a technology review, this chapter briefly describes recent technological advances both in sample preparation and instrumental performance to achieve images at high spatial resolution (1–10 microns) and at high speeds (a typical sample tissue once prepared can be imaged in minutes). Applications will include the use of tandem mass spectrometry (MS/MS), ultra-high mass resolution, and ion accumulation technology for IMS. Finally, new biocomputational approaches will be described that address the high dimensionality of IMS data as well as our implementation of 'image fusion' in predictive integration applications of MS images with microscopy and other imaging modalities.

7.1 Introduction

The concept of a molecular microscope for producing images of biological samples goes back many decades and much effort has gone into experiments to establish the feasibility of such an instrument. However, only in the past decade has advanced instrumentation been developed that enables this concept to become a reality in biological and clinical laboratories. Major improvements in ionization and mass analyzer technologies have made possible instrumentation for the molecular mapping of tissues at reasonable spatial resolutions. While a brief history of imaging samples by mass spectrometry is given here that describes several approaches to imaging, the major portion of this chapter is devoted to MALDI imaging mass spectrometry. Technical aspects of this technology are described, such as instrumentation, sample preparation, and different imaging modes, as well as the applications of this technology to basic biological and clinical research.

7.2 Background/History

Mass spectrometry has a long history in the field of image analysis, dating back over 40 years to early studies with secondary ion mass spectrometry (SIMS) performed in microscope [1] and scanning microprobe modes [2]. This technology utilizes a primary ion beam (for example, Ga^+, In^+, O_2^+, Au^+, or Cs^+) to sputter ions and neutrals from a surface. The sputtered secondary ions are directed into a mass spectrometer analyzer, typically a time-of-flight (TOF), quadrupole, or magnetic sector instrument [3]. The high energy of the primary ion beam is best for generating elemental ions, atomic clusters, and organic fragments. Thus, many early SIMS imaging experiments involved elemental distributions of metal or contaminant compounds in alloys, ceramics, and semiconductors. One advantage of SIMS is the high lateral resolution attainable with the highly focused liquid metal ion guns that are typically used to generate the primary ion beams. This energetic beam can be focused onto the target with a diameter in the submicron range (100's of nm) that is much smaller than the diameter of an average human cell (~10–15 μm). Nevertheless, for biological tissue and cells, the limiting parameter is sensitivity, so that in practice spatial resolutions of 5–10 μm or more are typically used in imaging experiments.

Laser desorption/ionization (LDI) [4, 5] can utilize a number of different types of lasers (*e.g.*, excimer, solid-state, infrared [IR], visible, ultraviolet [UV], *etc.*), each having different desorption, focusing, energy, and pulse frequency characteristics. For imaging of biological samples, UV lasers are commonly used for imaging, although IR lasers have been used as well. The LAMMA-500 (laser microprobe mass analyzer) was a commercially available LDI instrument that coupled a Nd:YAG laser (256 nm) with a TOF mass analyzer [6, 7]. In general, LDI desorbs/ionizes molecular species with little molecular fragmentation, although the accessible mass range is low at about 1500 Da [8]. The spatial resolution attainable from a typical LDI source is in the 5–30 μm range. As with SIMS, there is a tradeoff between ablation spot size and sensitivity that ultimately limits the effective spatial resolution achievable with today's technology.

In the late 1980s, matrix-assisted laser desorption/ionization (MALDI) [9–11] and electrospray ionization (ESI) [12, 13] were developed, allowing intact high molecular weight gas-phase ions to be directly detected for the first time. MALDI is a laser desorption/ionization technology that requires the application of an energy-absorbing matrix to be applied to samples of interest. The absorption of laser energy and subsequent ablation of material allows intact compounds to be transferred and ionized into the gas-phase and subsequently detected as pseudo-molecular ions $[M + H]^+$ or $[M - H]^-$. The use of MALDI for molecular mapping was first reported in 1997, where the spatially resolved detection of protonated insulin (*m/z* 5802) in a rat pancreas tissue section was shown [14]. Each pixel was manually acquired every 25 μm across a 450 × 75 μm region, and the intensity of protonated insulin was plotted for each pixel in the image array. As expected, pancreatic islets were found to be rich in insulin, as well as other ions from molecules of molecular weights up to about 25,000 Da, presumably islet proteases. Since those initial experiments, improvements in sample preparation and instrument technology have come in areas such as automation [15], and a large number of diverse applications have demonstrated the use this technology in many areas of biological and clinical research [16].

In the ensuing years following the introduction of (MALDI) IMS, a number of liquid desorption and liquid/laser desorption combinations have been reported. Desorption electrospray ionization (DESI) [17] was the first of these, demonstrating that an electrospray jet of liquid droplets could desorb molecules from a tissue sample to generate a molecular map, although with limited spatial resolution and molecular weight range. A unique feature of DESI is that, since it is an ambient ionization technology, large or odd shaped samples could be easily interrogated since the sample is outside the vacuum system of the mass spectrometer. Since then, literally dozens of desorption/ionization techniques have been demonstrated. A review of these technologies is outside the scope of the present chapter and the reader is referred to several reviews of this topic [18, 19].

7.3 MALDI Imaging Mass Spectrometry: Technology and Protocols

The molecular mapping of biomolecules in tissues is best performed using MALDI MS because it is capable of high spatial resolution (~1 um) [20] at high sensitivity and is amenable to most classes of biomolecules, including proteins beyond 200,000 Da. For these reasons, this chapter will focus on the use of MALDI for the molecular mapping of biological tissues in both basic biological and clinical research.

IMS technology is generally used in two distinct but related modes: global imaging of a sample and targeted (*i.e.*, histology-directed) profiling (Fig. 7.1). The mode employed will depend primarily on the overall informational content desired from the sample. Global imaging is used for spatial molecular localization over the entire sample. Profiling is typically used for obtaining molecular information from specific small areas within the tissue. These areas are then mapped back onto digital images of the tissue to develop molecular signatures of diseased cells, either to differentiate between diseased and normal healthy tissue or to differentiate between levels or stages of disease. For clinical applications, this differentiation necessitates the analysis of a large number of patients followed by supervised statistical analysis using a training set of biopsies from patients having known outcomes. A validation set of analyses from a different group of patients then determines the level of accuracy of the classification. Both global imaging and histology directed profiling are frequently used together in the search of biomarkers that make up a signature for a disease or disease state.

The sample preparation for the two modes is similar although not identical [21–23]. Fresh frozen tissue samples are sectioned on a cryostat to ~5–20 μm thicknesses and then subsequently washed to remove salts that can interfere with matrix crystallization and diminish sensitivity. For protein analysis, alcohol washes are used to dehydrate and fix the proteins while maintaining their original spatial orientation. A typical procedure involves successive washes in graded ethanol (70%, 90%, 95% for 30 sec each). For validation and correlation with standard histological procedures, serial sections are cut and stained with hematoxylin and eosin (H&E) or other histological reagents.

IMS experiments require the application of matrix to the sample in nearly all cases. Typically, MALDI matrices are applied to tissues with the goals of maximizing analyte extraction, minimizing analyte migration, and achieving homogeneous matrix coverage. Commonly used matrices include sinapinic acid (SA) for protein analysis, α-cyano-4-hydroxycinnamic acid (CHCA) for peptides, and 2,5-dihydroxybenzoic acid (DHB) for lipids and small molecules. Other matrices reported for low molecular weight analytes include 1,5-diaminonapthalene (DAN), 2,5-dihydroxyacetophenone (DHA), and 5-chloro, 2-benzothiazolethiol (CMBT). The matrix may be applied manually or robotically. For tissues, manual application of ~250 nL to 1 μL of matrix solution (~5–50 mg/mL) produces the best sensitivity, but results in relatively large spots (~1 mm diameter or greater). Often times manual

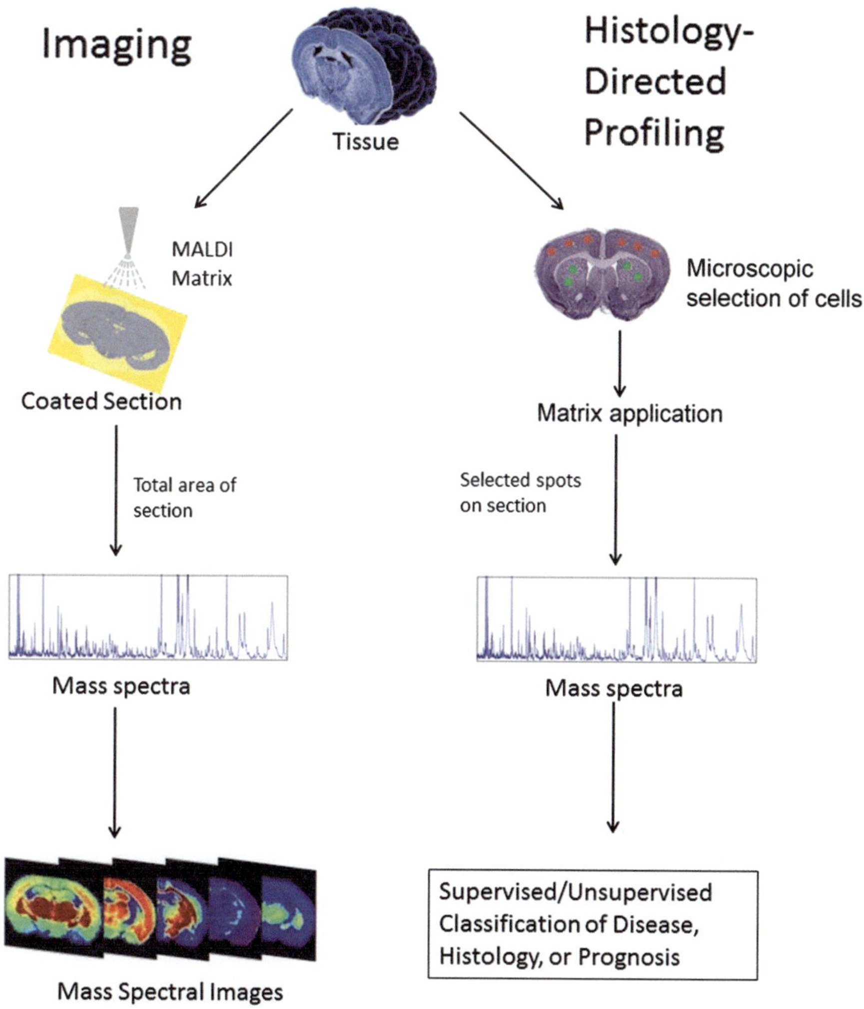

Fig. 7.1 Workflow for MALDI Imaging Mass Spectrometry showing both global imaging and histology directed profiling of a tissue sample

spotting is used to test and validate instrument performance on tissue prior to IMS data acquisition. Robotic matrix application devices are generally of the spotter or spray type. An example of a spotting robot is the Labcyte printer (Portrait 360), which is capable of depositing picoliter volumes at specified x, y coordinates on tissue. Typically, multiple passes are required for analyte extraction and crystal formation, resulting in matrix spots of ~100–200 μm on tissue. Arrays of spots have the advantage of limiting analyte migration to the diameter of the matrix spots, but suffer from having lower spatial resolution that is limited to the diameter of the deposited spot. A second robotic application method provides a homogeneous coating of matrix by spraying the matrix solution onto the tissue section (*e.g.*, using the HTX TM Sprayer). Finally, although not robotic in the same sense, homogeneous matrix deposition can also be accomplished using sublimation methods [23]. Of course, continuous matrix coatings can also be applied manually using an airbrush [24] or glass reagent sprayer. However, it can be difficult to obtain a uniform coating using manual techniques. Additionally, the spectral quality from array spots is typically better than that from a spray coated tissue because of enhanced analyte extraction, a larger analyte concentration per pixel, and a greater surface area from which to sample fresh matrix crystals. In general, the best results are obtained when the tissue is sufficiently wetted to allow efficient extraction of analytes without inducing delocalization and when matrix crystals are uniformly distributed across the surface of the tissue. Optimal matrix application often takes multiple cycles of spraying and drying to allow a crystal layer to slowly build up on the tissue. Examples of a spotted surface and a sprayed surface are shown in Fig. 7.2. In this example the homogeneous matrix application was accomplished by sublimation.

Data acquisition is performed similarly for both global imaging and histology-directed profiling. Today, commercial MALDI mass spectrometers come equipped

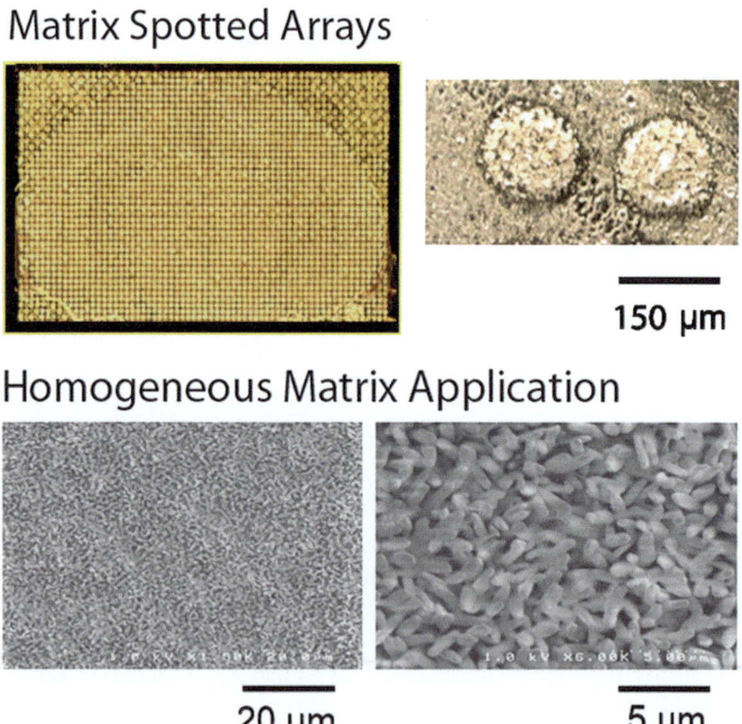

Fig. 7.2 Matrix deposition on a tissue demonstrating both the droplet spotted array protocol and the sublimation protocol

with software to facilitate ion image generation. Commonly, the sample stage is moved in x, y directions using precision stepper motors under software control while the laser is held in a stationary focus position. UV lasers (*e.g.*, N_2 at 335 nm, Nd:YAG at 355 nm, Nd:YLF at 349 nm) capable of firing at repetition rates of ~60 Hz to over 5 kHz are generally used in imaging experiments. Each pixel or ablation spot on the target is typically the result of ~50–400 laser shots that are summed into a single spectrum at a given location. Of course, there is no limit on the number of laser shots used, and some published works use thousands or more shots per pixel. However, coherent signal accumulation can diminish so that excessive shots add more to the noise of the spectrum than to the coherent signal from the tissue (*e.g.*, with MALDI-TOF instruments). In selecting a mass analyzer, tradeoffs exist between sensitivity, mass range, mass resolution, acquisition speed, and data file size. Protein analyses are generally performed on TOF instruments because they have a large mass range (~2000–50,000 Da). Higher molecular weight proteins over 200 kDa have been measured from tissue using TOF instruments, however this type of analysis is not routine, requiring optimized instrumental parameters and additional acquisition time. Other analyzers that have been used for IMS include quadrupole time-of-flight (Q-TOF), ion trap, and Fourier transform ion cyclotron resonance (FT-ICR) platforms. However, these instruments are usually employed for the analysis of peptides from enzymatic digestions on tissues or for lower molecular weight compounds such as lipids and metabolites.

Following acquisition, mass spectral data are preprocessed using baseline subtraction, noise reduction, and, in many cases, some type of normalization (*e.g.*, to total ion current [TIC]). In profiling experiments, cellular areas that meet certain threshold criteria (*i.e.*, minimal S/N, prevalence of a marker in a certain percentage of sampled spectra, *etc.*) are exported for biostatistical analysis. Although there are several tools available for the statistical analysis of the data, one effective software package, ClinProTools (Bruker), enables spectral preprocessing as well as statistical evaluation (*e.g.*, average values, standard deviation, t-test) and classification (*e.g.*, via hierarchical clustering, genetic algorithm, support vector machine) directly from the MS data files. General image processing may be performed using vendor-specific software or by exporting data into formats compatible with freeware (*e.g.*, BioMap). The major goals of an imaging experiment are to identify unique localizations, to identify signals that may co-localize or exhibit mirror-image localizations, and to identify analytes that correlate with known histology.

7.4 Molecular Identification

The unambiguous molecular identification of ion signals is a key step in an IMS experiment. For low MW analytes such as metabolites and lipids, accurate mass measurements coupled with on-tissue MS/MS analysis is sufficient to identify these compounds. For protein identification, experiments are conducted following an IMS experiment in which well-vetted analytical proteomics methods are used to fractionate and identify proteins of interest. Briefly, tissue samples are homogenized in a buffer solution and then treated with protease inhibitors and prepared for high-performance liquid chromatography (HPLC) separation [25]. In one method, protein fractionation is performed on a reverse phase LC column using an acetonitrile-water gradient containing 0.1% trifluoroacetic acid. Fractions are collected at 1 minute intervals and subsequently assayed by MALDI MS. Fractions containing the proteins of interest are further separated using one dimensional sodium dodecyl sulfate polyacrylamide gel electrophoresis (SDS-PAGE). Following electrophoresis, the gels are fixed, stained with Colloidal Blue, and then de-stained using water. Gel bands containing the molecular weight region of interest are excised from the gel and the proteins reduced and alkylated. Trypsin is added and the samples are enzymatically digested overnight. The resulting peptides are then extracted from the gel, reconstituted in 0.1% TFA, desalted, and kept at 4 °C until analysis. LC-MS/MS analyses are generally performed on an ion trap mass spectrometer equipped with a microelectrospray (nanospray) ion source. Reversed-phased separation of tryptic peptides is performed using a data dependent scanning mode, with the acquisition of one full precursor MS scan (*m/z* 400–2000) followed by the acquisition of five sequential MS/MS scans. Following acquisition, the MS/MS data are then searched against a mass spectral database using commercially available software (*e.g.*, Sequest [Thermo Electron], Maxquest [freeware], Spectrum Mill [Agilent]). The resulting protein assignments are filtered using several metrics to ensure accurate protein identifications (*e.g.*, a false discovery rate of 5%, a requirement of two or more peptide identifications per protein).

7.5 Applications to Diabetes Research

While there are many emerging examples of IMS applied to basic biological research, perhaps one of the most important areas of current biomedical investigation focuses on the study of diabetes. Diabetes is a chronic disease that often results in the progressive development of several debilitating complications including nephropathy, retinopathy, neurodegenerative diseases, and atherosclerosis. Diabetic nephropathy (DN) is a major cause of glomerular disease both in the United States and around the world, and often leads to end-stage renal disease (ESRD), accounting for 44% of ESRD cases [26]. ESRD is now a serious health care problem given the requirement for dialysis or kidney transplantation in order to sustain the patient's life. The main site of damage in diabetic kidney diseases is the glomerulus, the functional filtration unit of the kidney. A single glomerulus consists of a small bundle of capillaries surrounded by specialized cells and basement membranes to filter the blood. Humans have ~1 million glomeruli per kidney. Over prolonged periods of diabetes, damage builds in the glomeruli, disrupting the filtration process and impairing kidney function.

IMS has been used to study molecular changes in glomeruli in order to determine pathogenic mechanisms of DN, identify molecular markers of the disease, and characterize the effects of prospective treatments. Since progression of glomerular pathology can be followed histologically, high spatial resolution IMS at different stages of disease progression can be used to isolate and characterize specific subsets of renal glomeruli. Spatially-directed proteomic analysis of these glomerular populations provides unique information about pathogenic molecular events that cannot be obtained by an analysis of the proteome at the level of the whole-organ. There are three major cell types present in the glomerulus surrounded by a specialized glomerular basement membrane (GBM), each component contributing uniquely to pathologies that lead to DN. Understanding of the molecular events that specifically contribute to glomerulosclerosis from each cell type is critical to deciphering pathogenic pathways.

Recent studies have demonstrated that both protein glycation and lipid glycation may be involved in disease pathogenesis. Increased levels of glycated phosphatidylethanolamines (PEs) have been reported in plasma, erythrocytes, and atherosclerotic plaques of diabetic subjects [27–30]. Glycated PEs have also been shown to alter the structure and stability of cell membrane proteins [31] and promote lipid peroxidation [32] *in vitro* and are now being considered an important factor in diabetic complications. The application of high spatial resolution IMS to the analysis of single cells is unique in its capacity to study the proteomes of specific glomerular cells within the renal tissue environment while avoiding the artifacts of cell isolation methods. Moreover, with the capability for analysis of glomeruli at high spatial resolutions, it may be possible to look into molecular events at the very early stages of glomerular disease, when only a small subset of cells is affected. Such experiments require advanced instrumentation to ensure IMS spatial resolutions of at least 1–5 μm. These technologies include a high spatial resolution MALDI source capable of subcellular imaging, high sensitivity instruments, and advanced sample preparation techniques.

Recent work in our lab has utilized a gas-phase ion enrichment technique termed continuous accumulation of selected ions (CASI) for the analysis of low abundance analytes [33]. Briefly, CASI allows the dynamic range of a selected mass window to be expanded by selective ion accumulation. Ions generated by MALDI are isolated using a mass selective quadrupole prior to storage in a linear hexapole ion trap. The enriched ion population is then transferred to the ICR cell for detection. The use of CASI can increase signal intensities over 100-fold, representing a significant dynamic range improvement over full-scan mass spectral profiling methods. This allows for the detection of many more biological molecules than was previously possible. We have utilized CASI FTICR IMS to analyze PE glycation in the kidney directly from a thin tissue section (Fig. 7.3) [34, 35]. The identities of the modified PE species of interest were determined using MS/MS (Fig. 7.3a). We readily observed lipids that were specific to different morphological regions of the kidney. For example, the signal at *m/z* 1151.7055 corresponding to ganglioside GM3 was detected only in glomeruli whereas the sulfatide SM3 signal at *m/z* 1042.6712 was localized to the tubulointerstitial region of the kidney. Similarly, the signal at *m/z*

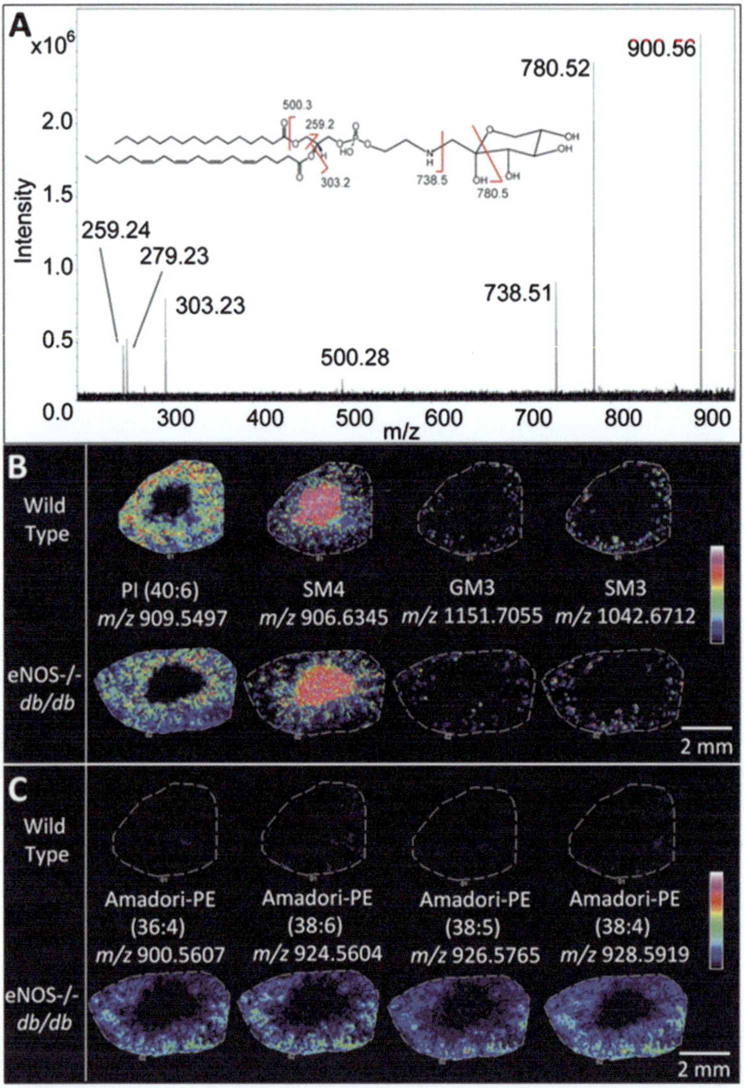

Fig. 7.3 (**a**) Lipid identification using Amadori-PE (36:4) as an example. The FTICR MS/MS spectrum with molecular ion (m/z 900.56) and product ions and Amadori-PE (36:4) fragmentation assignments (inset) are shown. (**b**) Localization patterns of different lipids. Signals at m/z 909.5497 (PI(40:6)), m/z 906.6345 (SM4), m/z 1151.7055 (GM3), and m/z 1042.6712 (SM3) were localized to the cortex, medulla, glomeruli, and tubulointerstitial area of the kidney, respectively. These signals showed minimal or no difference between the wild type and eNOS −/− db/db kidney. (**c**) Localization and relative abundance of four different Amadori-PE species. These glucose-modified species were detected in the cortex of eNOS −/− db/db kidney but not in wild type kidney

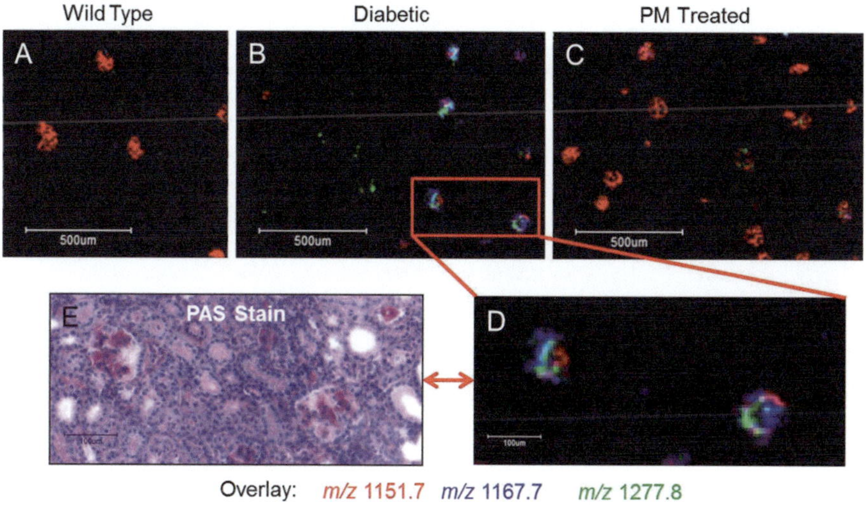

Fig. 7.4 IMS of lipids in individual glomeruli at 10 μm spatial resolution. Localization of m/z 1151.7, m/z 1167.7, and m/z 1277.7 in (**a**) wild type (**b**) diabetic and (**c**) PM treated diabetic mouse glomeruli. Scale bars =500 μm. (**d**) Zoom in of region boxed in *red* in *panel B* from diabetic kidney highlighting two diseased glomeruli. (**e**) PAS stain of same kidney section and region shown in *panel D*. Scale bars =100 μm

909.5497 (phosphatidylinositol (PI) 40:6) was localized to the cortex whereas the signal at *m/z* 906.6345 (sulfatide SM4) was localized to the medulla (Fig. 7.3b). A number of lipid species were differentially expressed in the wild-type versus the eNOS−/− *db/db* (DN mouse model) kidneys, most notably a series of PEs containing an Amadori moiety, which is a glucose adduct to the lipid amino group. Amadori-PEs were localized to the cortex of the eNOS−/− *db/db* kidney but were not detected in the wild-type kidneys (Fig. 7.3c).

Another area of investigation involves studies of molecular changes within the glomerulus after exposure to pyridoxamine (PM), a drug that has been shown to help amerliorate complications of DN [36]. As the main sites of damage in nephrotic diseases are the glomeruli, we have focused on achieving high spatial resolution MALDI images of the glomerulus [35]. Routine MALDI protein imaging analysis is performed at 25 μm and MALDI-MS lipid imaging analysis is performed at 10 μm with commercial instrumentation. At these resolutions the substructures of the glomeruli can be differentiated within the kidney cortex. In order to correlate the ion images of individual glomeruli with known histology, the same tissue section that is imaged with IMS is then washed to remove the MALDI matrix and stained with periodic acid Schiff (PAS). Using this approach, we are able to detect signals specific to the glomerulus as seen in Fig. 7.4. The signal at *m/z* 1151.7 is a GM3 ganglioside and is located in the glomeruli of wild type, untreated diabetic, and PM treated diabetic mouse kidneys. There are two additional lipids at *m/z* 1167.7 and *m/z* 1277.7 that localize to the glomeruli of the DN mouse and seem to be much

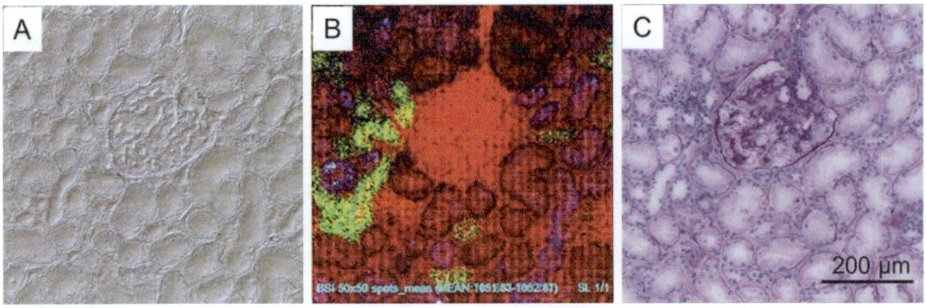

Fig. 7.5 IMS at 2 µm spatial resolution. (**a**) Differential interference contrast microscopy image of human kidney section imaged with MALDI IMS in *panel B*. The glomerulus is seen in the center of the image. (**b**) Overlay of three ions detected differentiating the glomerulus (*red*) and two distinct tubular areas (*green* and *blue*). (**c**) PAS stain of the serial section to section in *panel A* and *B*. Scale bar =200 µm

reduced upon PM treatment. These approaches have allowed us to correlate regio-specific biochemical changes to histological features at unprecedented scales, revealing disease processes which have not been previously described.

The next step in high spatial resolution IMS involves the analysis of subcellular compartments. Using an instrument modified in our laboratory specifically for ultra-high spatial resolution IMS, we have achieved 1–2 µm spatial resolution imaging of a human glomerulus. Discrete ion images of glomeruli (red) and two distinct tubular regions (blue and green) are seen in Fig. 7.5.

7.6 Applications to the Diagnosis of Melanoma

Another important area of IMS research focuses on the application of this technology to cancer classification and tumor prognosis. Malignant melanomas are the 5th and 6th most common cancers in the adult population in the U.S. The rate of incidence has been increasing at 3% per year since 2004 and represents a substantial public health problem. Early and accurate detection of melanoma malignancy is the most important factor in increasing survival rates. The current protocol for the diagnosis of melanoma is microscopic examination of H&E-stained tissue sections. However, classifying a tumor as benign or malignant remains challenging in a large number of cases, due to the fact that many melanocytic lesions have unclear histopathology, that is often times interpreted differently by different dermatopathologists. Accurate diagnoses are imperative for effective treatment and failures in diagnosis have important medical and legal consequences. Currently there is no tissue-based assay routinely used by pathologists that reliably discriminates between benign and malignant melanocytic tumors, or that can be used for prognostic classification [37]. New technology is needed to fill this gap and it is clear that a molecular approach would most accurately describe the phenotype of the lesion. IMS represents an

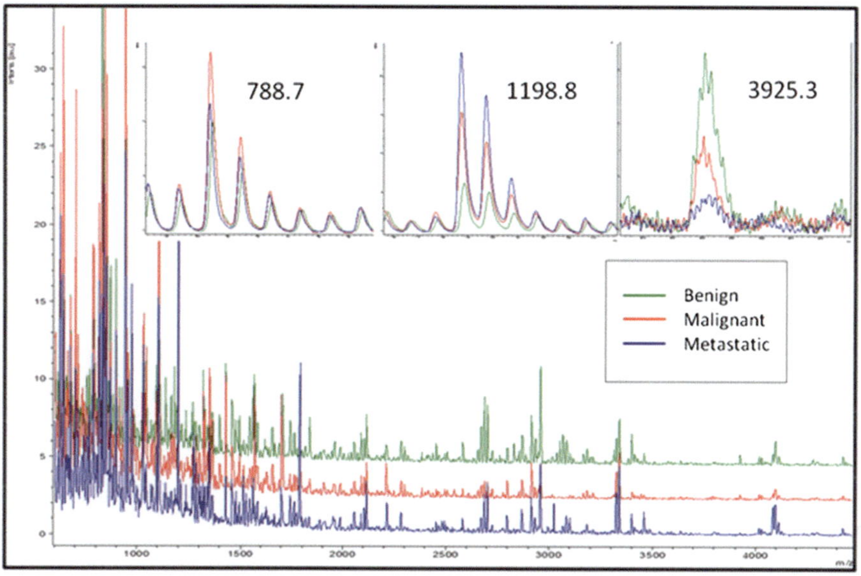

Fig. 7.6 Mass spectra comparing a primary and metastatic malignant melanoma with a benign dermal nevus. Selected regions of the spectra are expanded in intensity

exciting new technology with high sensitivity and molecular specificity for discriminating between benign and malignant melanocytic tumors and it has been shown that this technology is amenable for the analysis of the formalin-fixed, paraffin-embedded (FFPE) tissue types common to these types of studies.

Initially, several different samples of melanocytic skin tumors were investigated for differences in molecular profiles. These samples were controlled for age, gender, and anatomical location variation and included benign dermal nevi, primary malignant melanomas, and metastatic malignant melanomas. Several peaks at *m/z* 788.7, 1198.8, and 1349.9 were preferentially over-expressed in both cases of primary and metastatic melanomas, but not in the dermal nevus (Figs. 7.6 and 7.7). Several peptides were also identified at *m/z* of 3069.7 and 3925.3 that are highly expressed in the benign dermal nevus but not in the malignant lesions [38]. IMS data from over 150 melanocytic FFPE skin tumors were collected using MALDI- TOF mass spectrometry in order to generate IMS data to successfully discriminate benign from malignant conditions. Specifically, Spitz nevi (SN) and Spitzoid malignant melanomas (SMM) were classified using this technology. IMS classifications were ultimately compared to final diagnoses based on histopathologic and clinical data as well as patient follow-up data. Classification models were generated from a training set of 26 SN and 25 SMM samples. The classification algorithms were then validated on a separate set of 30 SN and 29 SMM samples. A set of five specific peptides produced by the protease digestion of biopsy material was sufficient to differentiate between the two lesions, with 29 of 30 SN and 26 of 29 SMM correctly

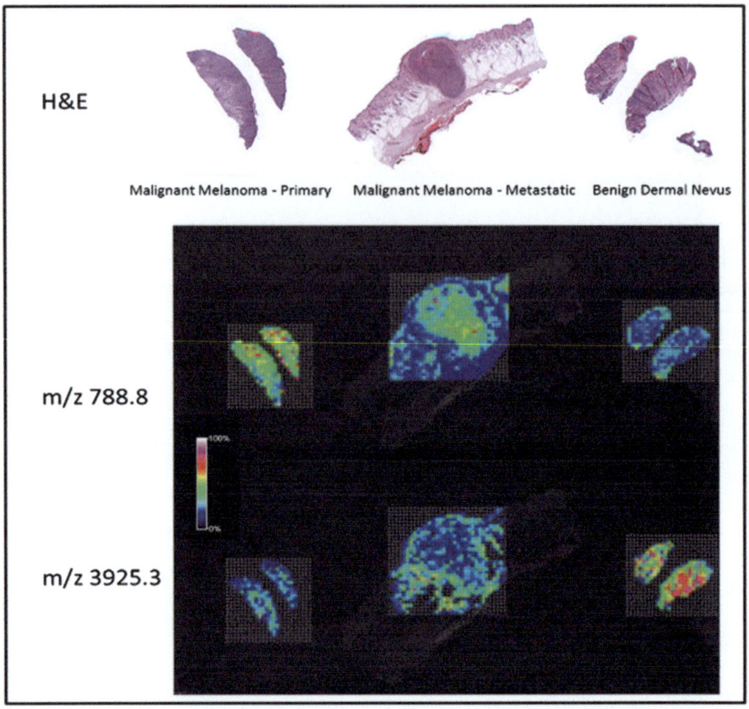

Fig. 7.7 Comparison of IMS of malignant melanomas and benign nevi. The ions at m/z 788.7 are highly expressed in melanomas, and that at m/z 3925.3 is highly expressed in benign dermal nevi. The *upper panel* shows tissue stained with H&E, and the *middle* and *lower panels* showing matching IMS profiles, with the color bar indicating relative intensity

identified. From this initial set of patient samples, IMS correctly classified SN with 97% sensitivity and 90% specificity.

7.7 Fusion of IMS and Microscopy Images

Important to the advancement of IMS technology is the development of new computational tools. Image fusion, as it relates to IMS, is a bio-computational process that integrates two distinct technologies: IMS, which is rich in chemical information but coarse in spatial resolution, and microscopy, which has low chemical specificity but is rich in spatial information. The resulting 'predicted' ion images combine the advantages of both technologies: the high molecular specificity afforded by mass spectrometry and the high spatial resolution of conventional microscopy. This provides for the prediction of the spatial distribution of specific ions at higher spatial resolution than would normally be achievable via IMS, in some cases increasing the spatial resolution tenfold. An image fusion algorithm has been developed that can

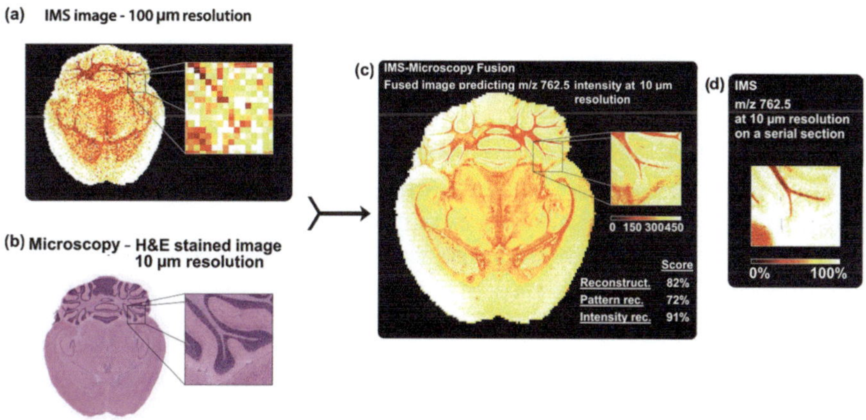

Fig. 7.8 Sharpening of ion spatial distributions through Image Fusion of IMS and microscopy data. This example in mouse brain fuses a measured ion image for m/z 762.5 at 100 µm spatial resolution (**a**) with a measured H&E-stained microscopy image at 10 µm resolution (**b**). The resulting image (**c**) predicts the ion distribution of m/z 762.5 at 10 µm spatial resolution with an overall reconstruction score of 82%. A neighboring section imaged at 10 µm spatial resolution (**d**) compares a physical measurement to the fusion approach for m/z 762.5 (Reprinted with permission from Ref. [39])

capture integrative connections between IMS and microscopy using multivariate regression that models variables in one technology using variables from the other technology [39]. Moreover, this method enriches biological signals and attenuates instrumental artifacts, revealing insights that cannot be extracted from either microscopy or IMS alone. Figure 7.8 illustrates this fusion process using a higher spatial resolution microscopy image and a lower spatial resolution mass spectrometry image taken across a transverse section of a mouse brain. Here, we are demonstrating only one ion of the many hundreds that are measured in a typical IMS analysis. In this example, the ion image for *m/z* 762.5 from a section that was analyzed at 100 µm spatial resolution using IMS is fused with a microscopy image and can now displayed at a predicted 10 µm spatial resolution (Fig. 7.8c). Figure 7.8d shows a serial section of tissue actually analyzed at 10 µm spatial resolution by IMS, clearly demonstrating the reconstruction accuracy of the prediction process shown in Fig. 7.8c. Applications of this methodology include spatial sharpening, out-of-sample predictions of ion images, and enrichments of biological signals while attenuating instrumental artifacts. The IMS process thus gains in sensitivity, speed, and a smaller data file size while still generating a 10 µm resolution image. The reconstruction scores shown in the figure are measures of how effective the algorithm is at fitting the two data sets. Generally, scores above 70% show good correlation and high confidence in a predicted ion image. This fusion process can be performed between any imaging modalities to reveal relationships that yield images that surpass what can be obtained from the original source technologies alone.

7.8 Conclusion and Perspective

The initial MALDI imaging publication in 1997 [14] stated, "The work presented here is a beginning, clearly demonstrating that ion images can be produced by MALDI MS. Much work needs to yet be done in areas of obtaining higher resolution images, higher sensitivity, better sample preparation, and faster data acquisition and analysis." Indeed, over the last decade, this technology has undergone tremendous growth and development, with significant progress made in all of these areas. The current state of the art is capable of routinely generating images at a spatial resolution of ~5–20 μm with 1–10 ppm mass accuracy using commercial instruments. In the research lab, spatial resolutions of 1 μm have been demonstrated [20]. Software enabling the facile generation of ion images has been developed by most of the leading mass spectrometer manufacturers, including the implementation of impressive algorithms to facilitate high throughput analyses and data processing. High-frequency lasers and new-age electronics have dramatically improved the speed with which mass spectrometers can operate, reducing the acquisition rate from ~1 pixel/minute in 1997 to up to ~50–100 pixels/second today. Finally, robotic devices are commercially available for automatic matrix deposition which has led to more robust and reproducible sample preparation.

Nonetheless, there are still significant challenges to overcome in various aspects of IMS workflows. For example, higher spatial resolutions are required in order to effectively image inside a single cell. More effective sample preparation protocols are needed to image higher molecular weight proteins as well as hard-to-detect species such as membrane-bound proteins. As spatial resolution is improved, the amount of data generated from the same area of tissue will greatly increase, providing a need for advances in data storage and data processing as well as increasing analysis speed.

In conclusion, IMS has demonstrated its effectiveness for mapping biological molecules in complex tissue samples in both basic biological research and in helping to solve clinical diagnostic dilemmas. The technology possesses a molecular specificity second to none, with sufficient sensitivity and spatial resolution that enables the discovery of new biology *in situ*. There is no need for specific targeting reagents such as antibodies, making this technology ideal for discovery-based imaging of many analytes at the same time. IMS offers a unique molecular perspective that, in combination with other imaging modalities, promises to bring new multimodal imaging capabilities beyond what each technology can deliver alone.

Acknowledgements The author acknowledges funding from the National Institute of Health (NIH/NIGMS 8P41 GM103391).

References

1. Castaing R, Slodzian G (1962) Microanalysis by secondary ionic emission. J Microsc 1(6):395–410
2. Liebl H (1967) Ion microprobe mass analyzer. J Appl Phys 38(13):5277–5283
3. Pacholski ML, Winograd N (1999) Imaging with mass spectrometry. Chem Rev 99(10):2977–3006
4. Van Vaeck L, Struyf H, Wim Van R, Fred A (1994) Organic and inorganic analysis with laser microprobe mass spectrometry. Part I: instrumentation and methodology. Mass Spectrom Rev 13(3):189–208
5. Denoyer E, Van Grieken R, Adams F, Natusch (1982) DFS laser microprobe mass spectrometry. 1. Basic principles and performance characteristics. Anal Chem 54 (1): 26A-41A
6. Hercules DM, Day RJ, Balasanmugam K, Dang TA, Li CP (1982) Laser microprobe mass spectrometry. 2. Applications to structural analysis. Anal Chem 54(2):280A–305A
7. Wechsung R et al (1978) LAMMA – a new laser – microprobe – mass – analyzer. Microsc Acta Suppl (2):281–296
8. Karas M et al (1990) Principles and applications of matrix-assisted UV-laser desorption/ionization mass spectrometry. Anal Chim Acta 241(2):175–185
9. Karas M, Bachmann D, Bahr U, Hillenkamp F (1987) Matrix-assisted ultraviolet laser desorption of non-volatile compounds. Int J Mass Spectrom Ion Processes 78:53–68
10. Karas M, Hillenkamp F (1988) Laser desorption ionization of proteins with molecular masses exceeding 10,000 daltons. Anal Chem 60(20):2299–2301
11. Tanaka K et al (1988) Protein and polymer analyses up to *m/z* 100 000 by laser ionization time-of-flight mass spectrometry. Rapid Commun Mass Spectrom 2(8):151–153
12. Whitehouse CM, Dreyer RN, Yamashita M, Fenn JB (1985) Electrospray interface for liquid-chromatographs and mass spectrometers. Anal Chem 57(3):675–679
13. Fenn JB, Mann M, Meng CK, Wong SF, Whitehouse CM (1989) Electrospray ionization for mass spectrometry of large biomolecules. Science 246(4926):64–71
14. Caprioli RM, Farmer TB, Gile J (1997) Molecular imaging of biological samples: localization of peptides and proteins using MALDI-TOF MS. Anal Chem 69(23):4751–4760
15. Stoeckli M, Farmer TB, Caprioli RM (1999) Automated mass spectrometry imaging with a matrix-assisted laser desorption ionization time-of-flight instrument. J Am Soc Mass Spectrom 10(1):67–71
16. Stoeckli M, Chaurand P, Hallahan DE, Caprioli RM (2001) Imaging mass spectrometry: a new technology for the analysis of protein expression in mammalian tissues. Nat Med 7(4):493–496
17. Wiseman JM, Puolitaival SM, Takats Z, Cooks RG, Caprioli RM (2005) Mass spectrometric profiling of intact biological tissue by using desorption electrospray ionization. Agnew Chem Int Ed Engl 44(43):7094–7097
18. McDonnell LA, Heeren RMA (2007) Imaging mass spectrometry. Mass Spectrom Rev 26(4):606–643
19. Norris JM, Caprioli RM (2013) Analysis of tissue specimens by matrix-assisted laser desorption/ionization imaging mass spectrometry in biological and clinical research. Chem Rev 113:2309–2342
20. Zavalin A, Yang J, Hayden K, Vestal M, Caprioli RM (2015) Tissue protein imaging at 1 μm laser spot diameter for high spatial resolution and high imaging speed using transmission geometry MALDI TOF MS. Anal Bioanal Chem 8(407):2337–2342
21. Cornett DS et al (2006) A novel histology-directed strategy for MALDI-MS tissue profiling that improves throughput and cellular specificity in human breast cancer. Mol Cell Proteomics 5(10):1975–1983
22. Schwartz SA, Reyzer ML, Caprioli RM (2003) Direct tissue analysis using matrix-assisted laser desorption ionization mass spectrometry: practical aspects of sample preparation. J Mass Spectrom 38(7):699–708

23. Hankin JA, Barkley RM, Murphy RC (2007) Sublimation as a method of matrix application for mass spectrometric imaging. J Am Soc Mass Spectrom 18:1646–1652
24. Todd PJ, Schaaff TG, Chaurand P, Caprioli RM (2001) Organic ion imaging of biological tissue with secondary ion mass spectrometry and matrix-assisted laser desorption/ionization. J Mass Spectrom 36(4):355–369
25. Reyzer ML et al (2004) Early changes in protein expression detected by mass spectrometry predict tumor response to molecular therapeutics. Cancer Res 64(24):9093–9100
26. U.S. Renal Data System, USRDS (2009) Annual data report: atlas of end-stage renal disease in the United States, 2009. Bethesda, National Institutes of Health, National Institute of Diabetes and Digestive and Kidney Diseases
27. Sookwong P et al (2011) Amadori-Glycated Phosphatidylethanolamine, a potential marker for hyperglycemia, in Streptozotocin-induced diabetic rats. Lipids 46(10):943–952
28. Shoji N et al (2010) LC-MS/MS analysis of carboxymethylated and carboxyethylated phosphatidylethanolamines in human erythrocytes and blood plasma. J Lipid Res 51(8):2445–2453
29. Breitling-Utzmann CM et al (2001) Identification and quantification of Phosphatidylethanolamine-derived Glucosylamines and Aminoketoses from human erythrocytes—influence of Glycation products on lipid peroxidation. Arch Biochem Biophys 391(2):245–254
30. Ravandi A, Kuksis A, Shaikh AN (2002) Glucosylated Glycerophosphoethanolamines are the major LDL Glycation products and increase LDL susceptibility to oxidation. Arterioscler Thromb Vasc Biol 20(2):467–477
31. Levi V et al (2008) Effects of phosphatidylethanolamine glycation on lipid–protein interactions and membrane protein thermal stability. Biochem J 416(1):145–152
32. Oak J-H, Nakagawa K, Miyazawa T (2000) Synthetically prepared Amadori-glycated phosphatidylethanolamine can trigger lipid peroxidation via free radical reactions. FEBS Lett 481(1):26–30
33. Spraggins JM, Grove KJ, Cornett DS, Caprioli RM (2012) Massively enhanced sensitivity and dynamic range for molecular imaging by high dynamic range (HDR) MALDI FTICR MS, 60th ASMS conference on mass spectrometry and allied topics. Vancouver, Canada
34. Grove KJ, Voziyan PA, Spraggins JM, Wang S, Paueksakon P, Harris RC, Hudson BG, Caprioli RM (2014) Diabetic nephropathy induces alterations in the glomerular and tubule lipid profiles. J Lipid Res 55(7):1375–1385. doi:10.1194/jlr.M049189
35. Grove K (2014) Imaging mass spectrometry for the elucidation of lipid and protein changes in diabetic nephropathy and assessment of drug efficacy, Ph.D. thesis, Vanderbilt University
36. Voziyan PA, Metz TO, Baynes JW, Hudson BG (2002) A post-Amadori inhibitor pyridoxamine also inhibits chemical modification of proteins by scavenging carbonyl intermediates of carbohydrate and lipid degradation. J Biol Chem 277:3397–3403
37. Sabel MS, Liu Y, Lubman DM (2011) Proteomics in melanoma biomarker discovery: great potential, many obstacles. Int J Proteomics 2011:181890
38. Lazova R et al (2012) Imaging mass spectrometry-a new and promising method to differentiate Spitz nevi from Spitzoid malignant melanomas. Am J Dermatopathol 34(1):82–90
39. Van de Plas R, Yang J, Spraggins J, Raprioli RM (2015) Fusion of mass spectrometry and microscopy: a multi-modality paradigm for molecular tissue mapping. Nat Methods 12(4):366–372

Chapter 8
MALDI Mass Spectrometry and Infectious Diseases

Jessica L. Moore and Eric P. Skaar

Abstract MALDI mass spectrometry is an emerging technology that has revolutionized the field of diagnostic microbiology. MALDI mass spectrometry has the potential to significantly impact the field of infectious diseases as new technology emerges. In particular, the use of histology-directed MALDI-MS profiling of biofluids and tissues presents a way to potentially perform bacterial typing and monitor host response without an additional culture step to culture the microorganism. MALDI Imaging Mass Spectrometry is another attractive technology to study the pathogen-host interaction within infected tissues. This allows an unprecedented view of the molecular changes associated with infection. Pairing advanced Imaging Mass Spectrometry with emerging analyte identification strategies, the identity of these molecular changes can be determined. Utilizing such technologies could revolutionize the study of microbial pathogenesis and may reveal a number of novel targets for future antimicrobial intervention.

8.1 Introduction

According to the Centers for Disease Control, an estimated two million people each year become infected with antibiotic-resistant bacteria. Of these two million, there are a reported 23,000 deaths directly attributed to the microbes [1]. Compounding this threat is the tremendous rate at which pathogens are gaining resistance to antimicrobial strategies. This is partially because antibiotics are overly prescribed, with suboptimal prescriptions occurring in as often as 50% of all cases. Between 1935 and 2003, there were fourteen emerging classes of antibiotics [2]. The decrease in emerging therapeutics is convoluted; novel antibiotic development is an expensive

J.L. Moore
Department of Chemistry, Vanderbilt University, Nashville, TN, USA

E.P. Skaar (✉)
Department of Pathology, Immunology, and Microbiology, Vanderbilt University Medical Center, Nashville, TN, USA
e-mail: eric.skaar@vanderbilt.edu

J.H. Banoub, R.M. Caprioli (eds.), *Molecular Technologies for Detection of Chemical and Biological Agents*, NATO Science for Peace and Security Series A: Chemistry and Biology, DOI 10.1007/978-94-024-1113-3_8

endeavor that is not often profitable for pharmaceutical companies [2]. In addition, the time from antibiotic deployment to emergence of clinically relevant resistance has dramatically decreased when new antimicrobials do become available [3]. This led the Infectious Disease Society of America to release a 2009 call to action to study microbial resistance and to target seven bacteria of particular concern [4]. These pathogens were deemed the "ESKAPE" pathogens and include *Enterococcus faecium, Staphylococcus aureus, Klebsiella pneumonia, Acinetobacter baumannii, Pseudomonas aeruginosa*, and *Enterobacter* species. In order to continue treating infections caused by these bacteria, it is important to study their pathogenesis using novel and emerging analytical technologies. The application of such technologies might isolate novel targets for antimicrobial intervention, and further aid the treatment of antimicrobial resistance at the cusp of the post-antibiotic era.

8.2 MALDI Mass Spectrometry and Infectious Diseases

Mass spectrometry has greatly contributed to the scientific body of knowledge in the field of infectious diseases. From large scale untargeted proteomics experiments to drug metabolism and pharmacokinetic studies, the literature contains numerous examples showcasing how mass spectrometry helps further our understanding of antimicrobial resistance and microbial pathogenesis.

Matrix-assisted laser desorption/ionization mass spectrometry (MALDI-MS) is an analytical approach that is suited for the analysis of complex systems due to its high sensitivity, wide dynamic range, and applicability to numerous analyte classes. MALDI-MS utilizes a matrix, typically a small organic acid having strong ultraviolet absorbance, to desorb the analytes into the gas phase and aid their ionization [5]. Analytes are detected as a mass-to-charge ratio, or *m/z*, which yields specific molecular signatures within complex biological samples. Matrix can be applied directly to biological samples to obtain spectra from a variety of analytes, including small molecules, metabolites, lipids, peptides, and proteins. This workflow has been successfully used to profile a range of systems, including tissue sections [6], plants [7], insects [8], whole animals [9], and microbial colonies [10]. The ability to analyze such a wide range of complex systems has led MALDI MS to be used as a diagnostic tool, particularly in diagnostic microbiology.

8.3 Diagnostic Microbiology

MALDI-MS has been used in diagnostic microbiology to rapidly identify bacteria based on molecular profiles of small cell populations. Microbial colonies from culture are moved onto MALDI targets, mixed with matrix, and analyzed using a mass spectrometer [Fig. 8.1]. This approach has resulted in new FDA-approved instrumentation and methods for the identification of bacteria in clinical laboratories.

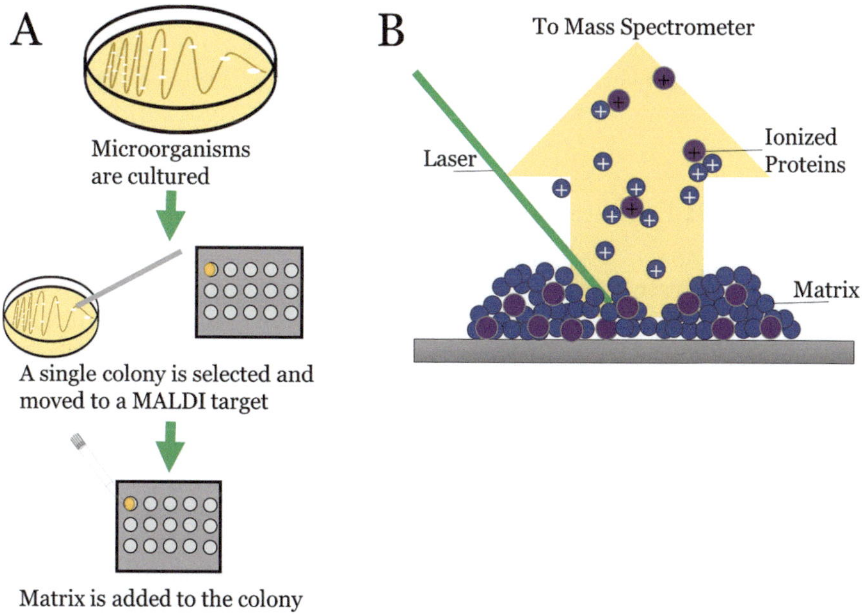

Fig. 8.1 A typical workflow for MALDI MS-based spectral matching for microbial identification. (**a**) The unknown pathogen is cultured using standard techniques on agar plates. A single colony is selected and moved onto a MALDI target using a sterile device. Care is taken to not move agar with the colony, as it introduces interferences into the sample. Finally, cells are lysed on target when they are mixed with a MALDI matrix in an organic solvent, typically α-cyanocinnamic acid in 50% Acetonitrile. (**b**) Once the cells are lysed, the proteins extracted into the organic solvent will co-crystalize with the MALDI matrix. Upon interrogation in the mass spectrometer, a laser passes energy to the matrix, ionizing the small proteins. The ionized proteins are measured using a mass spectrometer and result in a characteristic mass spectrum that can be used to classify bacteria

MALDI-MS presents a way to interrogate intact cells from agar plates or liquid media with robust signatures independent of culture conditions [11]. The ability of MALDI-MS to distinguish molecular fingerprints associated with specific microorganisms has improved rapid diagnostics of infectious agents and led to the inclusion of mass spectrometry in the diagnostic microbiology laboratory [10, 12, 13]. This has been expanded to include identification of fungi, anaerobic bacteria, highly infectious bacteria, and microbes that are difficult to culture [14–17].

Initial work using MALDI-MS microbial fingerprinting created databases that can be queried to classify clinical isolates quickly and accurately [10]. Instrumentation, automated analysis, and database searching are available from commercial instrument manufacturers, like the MALDI BioTyper system (Bruker Daltonics) and Vitek microbial identification system (Biomérieux) [18]. Databases are perpetually evolving to meet biological needs and to include emerging strains, bacterial subtypes, and to differentiate pathogenic and non-pathogenic strains [19–22]. Clinical analyses have also included profiles from complex mixtures, including blood cultures and complex polymicrobial infections [23]. These continued advancements decrease the time needed to identify a pathogen, shortening the amount of time until antibiotic intervention and therefore decreasing the cost of care [24]. This has revolutionized clinical care of bacterial infections, in particular in cases of sepsis [25].

8.4 MALDI Profiling of Infected Human Tissues

The analytical power to rapidly identify microorganisms from culture has been a major advancement. However, a culture step is still required, removing the microorganism from the complex environment of the host organism, and severely delaying the time required to positively identify the causative agent of infection. This approach makes the study of microbial pathogenesis and virulence difficult because microorganisms in culture are not experiencing the same conditions as microorganisms growing within the host. However, there are significant analytical challenges associated with the study of bacteria actively causing diseases within the host. Bacteria and their products are generally of limited abundance when compared to host markers and require high sensitivity in analytical technology. However, MALDI-MS is emerging to fill this gap and to perform analyses from complex mixtures. One study attempted to determine host-specific markers for sepsis directly from serum [26]. Another study utilized MALDI-MS profiling to study host response to *Staphylococcus aureus* skin wounds over time. At set time points post infection, a wound was swabbed into sterile water and analyzed. Signals were detected from the wound bed that were thought to belong to mouse defensins and blood, highlighting the ability of MALDI profiling to monitor host response to infection and wound healing over time [27]. In addition to analyzing spotted homogenates or serum, MALDI MS provides the distinct advantage of *in situ* tissue profiling [28, 29]. This method allows spectra to be obtained directly from tissue sections in a targeted approach, allowing pathologists to focus analyses on histological regions of disease [5, 28]. Approaching large sample sets in a histologically-directed way presents a reasonable method to differentiate signals associated with disease in a high-throughput manner. This approach also presents an alternative way to characterize bacterial signatures *in vivo* at a point when these organisms are actively causing disease within host tissues.

8.5 Imaging Mass Spectrometry

8.5.1 MALDI-IMS

MALDI profiling and spectral fingerprint analyses have no doubt provided great analytical power in the previous examples. Very similarly, MALDI-MS can be used as an imaging modality by systematically interrogating samples at defined x/y

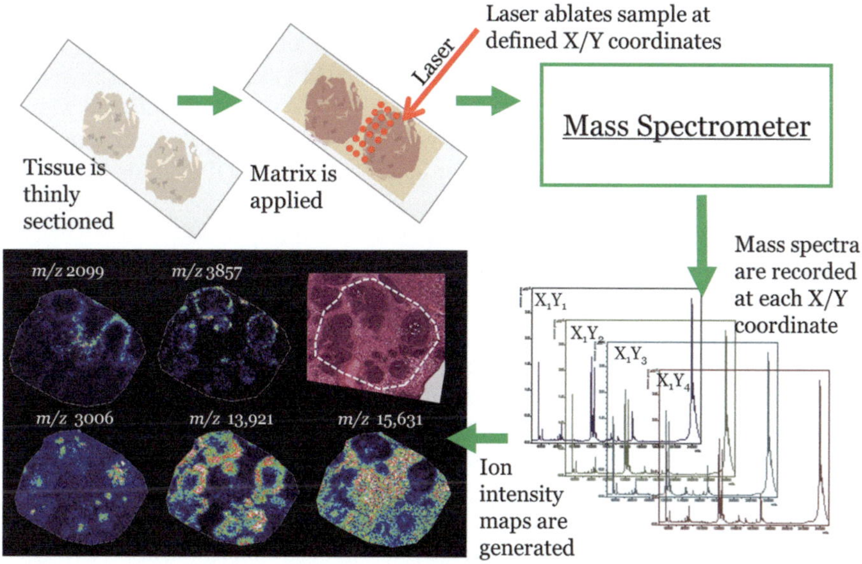

Fig. 8.2 The MALDI Imaging Mass Spectrometry workflow begins with tissues sectioned very thin and mounted to a slide. Tissues can be washed to remove interfering lipids and salts or not treated at all. MALDI matrix, typically small organic acids, are then applied to the surface in a homogeneous way. A laser is used to interrogate the sample at defined *x/y* coordinates and a mass spectrum is collected at each location. Heat maps can be generated by setting mass windows and integrating the area under the peak in the selected windows. These heat maps reveal thc spatial distribution of analytes within thin sections

coordinates (Fig. 8.2). The spectral data can be correlated and heat maps can be generated for each ion of interest. This reveals not only the specific *m/z* of the analyte but also the spatial origin of that analyte. Using this modality, molecular species of interest can be tracked regiospecifically to biological foci to help draw conclusions. In cases of infection, this has been used to help define biological systems in many ways- from tracking antibiotics in tuberculosis lesions [30] to tracking metals in bacterial biofilms [31].

8.5.2 **Staphylococcus aureus** ***Infections***

One noted ESKAPE pathogen that has been extensively studied by IMS is *Staphylococcus aureus*. *S. aureus* colonizes the anterior nares of approximately 30% of the population, with approximately 2% of those strains being antibiotic resistant [32]. It is a cause of significant morbidity, accounting for half of all reported deaths from antibiotic resistant bacteria reported in the CDC's 2013 report [1].

S. aureus infection causes the formation of purulent inflammatory foci, called abscesses, which are characterized by the recruitment of host immune cells, in particular neutrophils [33]. Previous studies have used IMS to analyze the staphylococcal abscess. For these studies, 6 week old mice were systemically infected with *S. aureus* and were sacrificed at varying stages of abscess development. Organs were aseptically removed and flash-frozen for downstream MALDI analysis. An initial study revealed accumulation of the host protein calprotectin at sites of neutrophil aggregation [34]. This protein has been reported to make up to 40% of the cytosolic protein content of a neutrophil [35]. Calprotectin consists of two subunits, S100A8 and S100A9. Together the subunits have the ability to chelate metals at sites of infection [36]. This contributes to a host immune strategy called nutritional immunity- a process by which host cells sequester nutrient metals from invading pathogens [37]. Calprotectin has been shown via IMS to be a temporally transient protein- it aggregates at sites of active infection and is cleared when the immune stimulant is no longer a threat [38]. Accumulation at infectious foci is not without risk to the protein; these regions are under significant oxidative stress from other neutrophils attempting to kill pathogens [39, 40]. Utilizing high resolving power mass spectrometers, such as Fourier transform ion cyclotron resonance mass spectrometers (FT-ICR), the protein can be studied in even greater detail. High mass resolving power IMS allows researchers to study not only the protein, but the protein in different modified forms. Recent work has revealed that calprotectin becomes oxidatively modified in a regiospecific way (Fig. 8.3) [41]. Most interestingly, this work revealed a modified form of S100A8, with a cysteine residue experiencing a trioxidation, to be localized to a staphylococcal colony. Such studies allow an unprecedented view of the pathogen-host interaction by revealing how host proteins are modified in their attempt to destroy the pathogen.

These previous studies have provided significant insight into the host response to bacterial infection within the staphylococcal abscess. However, studying the pathogen has remained analytically challenging. Within the center of these abscesses resides a small colony of bacteria, called the nidus. This biofilm is situated in a unique environment of host-induced stresses and represents the physiologically relevant form of bacteria causing disease within the host. The nidus of a staphylococcal abscess can grow quite large in cases of severe sepsis, making this model system ideal for applying new IMS technologies. Utilizing high mass and spatial resolution MALDI-IMS, researchers can directly analyze bacteria actively causing disease within host tissues for the first time. This provides novel molecular insights at the pathogen-host interface.

One of the first key analyses of study is the determination of the spatial resolution needed to resolve biological features. Several key experimental parameters affect spatial resolution. They include the laser spot size and the step spacing at which spectra are collected. Combined these parameters yield the effective spatial resolution [42].

Figure 8.4 illustrates four serial sections of a mouse kidney infected with *S. aureus*. Each section was interrogated at a different pitch with constant

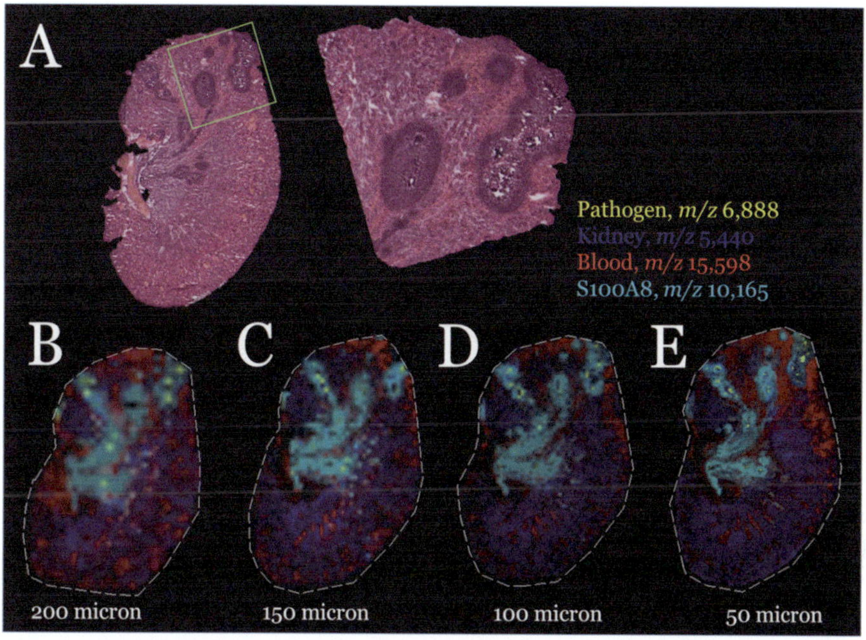

Fig. 8.3 Spatial resolution is important in MALDI IMS to define anatomical features of interest. (**a**) An H & E stain of a staphylococcal abscess in a mouse kidney. The inset shows the anatomy of the abscess, with staphylococcal colonies residing in the center. (**b**–**e**) Serial sections of the mouse kidney analyzed at increasing spatial resolutions. **b**, **c**, **d**, and **e** were interrogated at 200, 150, 100, and 50 μm spatial resolution. A signal co-localizing with the bacterial colony is displayed in yellow. At increasing spatial resolution, it is much easier to distinguish this signal from surrounding host tissues

instrumental conditions. As the spatial resolution of the analysis increases from 200 to 50 microns, the features differentiating the staphylococcal colony within the lesion can be determined. A signal co-localizing with bacterial colonies appears at *m/z* 6,888 and is displayed in yellow in Fig. 8.4. As spatial resolution increases, co-localization with bacterial colonies becomes much more apparent.

MALDI-IMS instrumentation has also experienced great advancements in the last 5 years, further enabling such analyses. For example, laser optics can be refined to decrease the footprint of the beam, allowing high spatial resolution experiments to be possible [43]. Next generation instrumentation is being designed to rapidly and accurately interrogate tissue at higher spatial resolution [44, 45]. These technological advancements have allowed tissues to be studied at extremely high spatial resolution and are important innovations to support the continued study of the pathogen-host interface [43, 46].

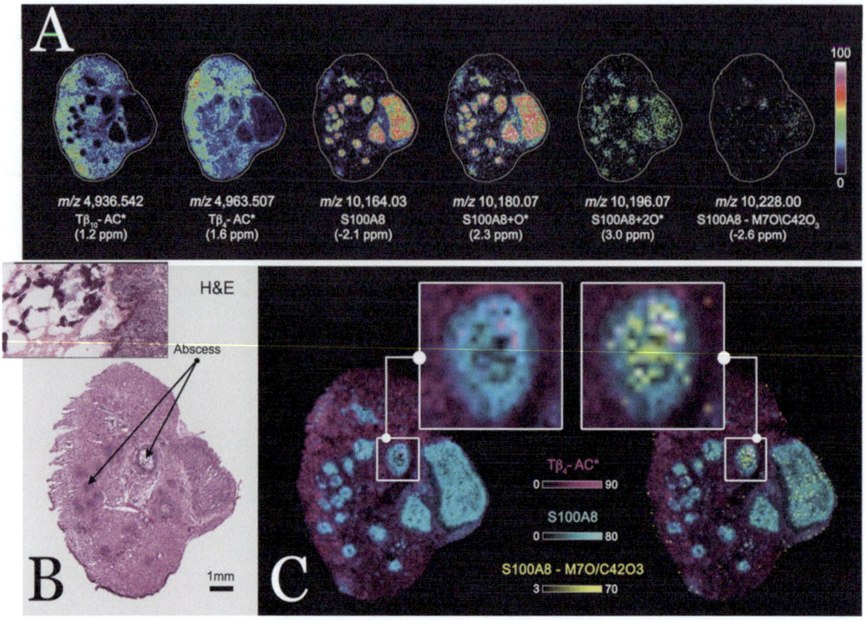

Fig. 8.4 "Selected ion images of intact proteins from kidney tissue from a mouse infected with *S. aureus* collected using MALDI FTICR MS (**a**). Ions were identified using mass accuracy to correlate imaging results with separate top-down proteomics experiments. For comparison, a serial tissue section was H&E stained (**b**). Ion image overlays show the advance oxidation product S100A8 – M7O/C42O_3 is localized specifically to the center of infectious foci (**c**). Italicized ions were tentatively identified by mass accuracy only" (Reprinted with permission of Springer, Ref. [41])

8.5.3 Analyte Identification Strategies

Both MALDI profiling and MALDI imaging approaches yield a wealth of data from tissue. A number of analyses have been able to successfully perform diagnostic assays simply by fingerprint matching. For example, it is hypothesized that most of the signatures in microbial fingerprinting from agar are actually small ribosomal proteins [47–50]. These analytical strategies have made a large impact on clinical microbiology and the time to diagnosis and antibiotic intervention even without knowing the identities of the peaks in the classifier. Recently, the field has begun the arduous task of identifying the proteins yielding the fingerprint patterns used in classification [47, 49]. Identification of analytes that denote antibiotic resistance or increased pathogenicity has utility not only in clinical settings but also in the primary research laboratory.

MALDI profiling and imaging experiments performed in the research laboratory have the potential to reveal unprecedented information about microbial pathogenesis and the pathogen-host interaction by allowing for the study of bacteria within host tissues actively causing disease. However, in order to drive biological conclusions, identification of these analytes is necessary. IMS experiments yield an enormous amount of data, making identification a cumbersome process. Recently there have been many advancements in the field that facilitate analyte identification. These strategies are depicted in Fig. 8.5.

8.5.4 Protein Identification Strategies

A majority of microbial typing and histology-directed experiments focus on small proteins, typically under 30 kDas. Protein identification strategies typically begin with analyte extraction. This can be performed in bulk, by homogenization of the sample, or in more spatially directed approaches, including laser capture microdissection or punch biopsies. Another recent method, called microextraction, allows researchers to extract directly from tissue sections in a spatially refined manner [51]. When trying to identify species from IMS experiments, enriching the sample for the species of interest by spatially directed extraction can greatly enhance the chance at analyte identification.

Bottom-up proteomics approaches have been used with success for protein identification. In these experiments, an enzyme, typically trypsin or LysC, is used to reduce the protein of interest to peptides for traditional LC-MS/MS analysis. Since the MS/MS analysis is performed at the peptide level, it is sometimes difficult to isolate which intact protein mass corresponds to the MALDI-MS target. In solution digestions of bulk extracts are very rich and will yield a large number of protein identifications. While this presents a wealth of knowledge related to the biology of the sample, it is not the best approach for matching protein identities to MALDI-MS experiments.

A large majority of identification of small proteins from MALDI-MS experiments have utilized in-gel digestions for identification. In this strategy, the protein of interest is loaded into a high percentage SDS-PAGE gel. The gel band or region of interest is excised and subjected to in-gel tryptic digestion and analyzed by LC-MS/MS analysis. This method is relatively straightforward and has been used with success [31, 34]. However, there is a plethora of proteins that are successfully identified from a single gel band or a region of a gel. This requires the researcher to make an educated association for identification. This includes the consideration of peptide counts for each identified species as well as the theoretical mass of the protein. The analyte of interest could be a small fragment of a much larger protein, introducing uncertainty in the identification. Additionally, this method has difficulties in assessing post-translational modifications or truncations that cause deviation from the anticipated intact mass. Additional certainty can be gained by

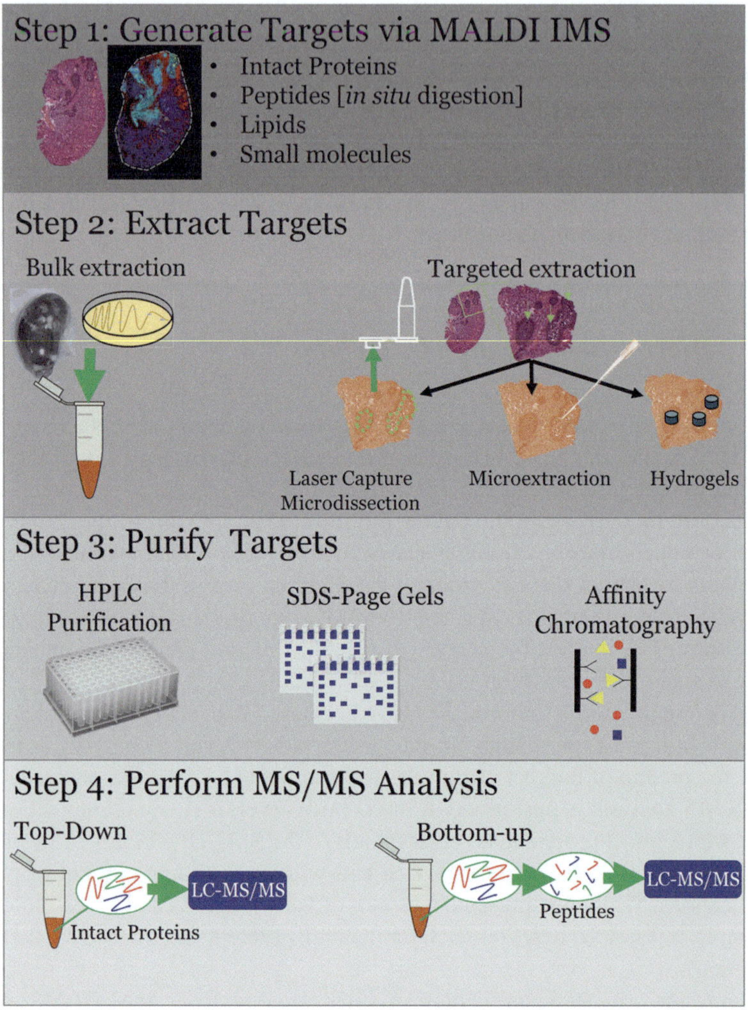

Fig. 8.5 An overview of analyte identification strategies. Step 1 includes generating targets from MALDI IMS experiments. These can be a range of analytes from proteins to lipids and small molecules. An ideal analyte is robust, appearing in technical replicates of the same tissue sections and reproducible, appearing in biological replicates of the same biological model. Step 2 involves physically extracting the targets from the tissue. This can be done from bulk homogenates or spatially from a tissue section. Care should be taken to match the sample preparation used in IMS experiments when using the tissue section methods. This includes washing strategies to remove interfering analytes. Step 3 is to further purify samples if needed. A broad range of strategies can be used, including offline HPLC fractionation, SDS-PAGE gels, or affinity chromatography. Step 4 is used to actually fragment the analyte of interest. Two main proteomics approaches are used: top-down sequencing which fragments intact proteins and bottom-up sequencing, which utilizes an enzymatic digestion to create peptides which are framented

adding a separation before the gel step. Here, a protein extract is fractionated by another analytical technique, traditionally HPLC or affinity chromatography, with fractions collected and stored. When MALDI-MS is used to interrogate the fractions, the analyte of interest can be isolated to a more purified fraction. Using this approach will decrease the amount of protein identified from the gel band and can therefore give the researcher more confidence in the identification [31].

Top-down sequencing presents an ideal method for protein identification. In this workflow, the intact protein is fragmented using electron transfer dissociation. In this approach, the intact mass of the protein is recorded prior to fragmentation. This removes any ambiguity about which parent ion the fragments derived from. When trying to assign identities to MALDI-MS experiments, this approach is superior because the mass of the parent can be matched back to MALDI data [41].

8.6 Peptide Identification Strategies

MALDI-MS profiling and imaging experiments are not limited to intact proteins. Performing a digestion step either in solution or *in situ* prior to MALDI analyses has great benefit. It allows for the analysis of proteins that are very large or for the analysis of biopsies that have been fixed. Many mechanisms of antibiotic resistance require large protein machinery to confer that resistance. Therefore, implication of a digestion step prior to MALDI analysis could yield increased sensitivity for microbial fingerprinting assays. Traditional enzymatic digestion protocols require an incubation step to allow the enzyme to act upon the proteins in the sample. This could prove prohibitive in the clinical identification workflow, when time to pathogen identification and antibiotic intervention is paramount for successful treatment [52]. Recent advancements in microwave-assisted enzymatic digestion are promising approaches to reduce the time needed for enzymatic digestion [53, 54].

In situ enzymatic digestion has been performed on both fresh-frozen and formalin-fixed, paraffin embedded (FFPE) tissue for peptide profiling and classification [55, 56]. Though these methods have not yet been applied to infectious diseases, it is an attractive solution for handling clinical biopsies of unknown origin or those of significant biosafety risk. Additionally, since the analytes measured in this approach are peptides, it is much easier to match the analyte mass to the species sequenced in a parallel LC-MS/MS analysis, circumventing the problems discussed above.

Another emerging technology that can be used for peptide identification is the hydrogel. In this approach, a small punch of acrylamide is loaded with enzyme and placed on the tissue for digestion [57]. This allows for spatially defined proteomics to be performed on the surface of tissue samples. Hydrogel technologies have been used to successfully interrogate both fresh-frozen and formalin-fixed biopsies [58]. Additionally, hydrogel technologies have been coupled to microwave-assisted enzymatic digestion to decrease the time needed to perform a digest [54]. Rapid tryptic digestion allows the analysis of peptides to be within the time range needed for practical implementation into the clinical workflow of microbial biotyping.

8.7 Conclusions

MALDI mass spectrometry has had broad positive impacts on the field of infectious diseases. Through microbial biotyping by matching an intact bacterial colony's spectral fingerprint to known databases, the ability to rapidly identify pathogens from culture has enabled many clinicians to rapidly provide appropriate antimicrobial intervention for patients. This can be further extrapolated to histology-directed MALDI mass spectrometry, where histological features within tissue sections can be targeted for interrogation with a mass spectrometer. Future work in this field could involve targeting areas within biopsies that were suspect for microbial infection and obtaining a spectral fingerprint to identify the pathogen causing disease from directly within host tissues. Imaging Mass Spectrometry takes these experiments one step further by systematically interrogating a tissue surface to reconstruct the molecular pathology of a biopsy. This allows researchers to target not only the pathogen signals but also the changing host environment surrounding the threat, allowing for unprecedented molecular study of the pathogen-host interface. Analyte identification strategies are another crucial area of research so that researchers can learn the identity of these differentially localized molecular species. Utilizing a mixture of the technologies discussed here has the potential to revolutionize the study of microbial pathogenesis and could present a number of novel targets for antimicrobial intervention.

Acknowledgements This work was supported by 5P41 GM103391-06

References

1. (CDC), C. f. D. C. a. P. Threat Report (2013) | Antimicrobial resistance | CDC. http://www.cdc.gov/drugresistance/threat-report-2013/. Accessed 05 Dec 2014
2. Doron S, Davidson LE (2011) Antimicrobial stewardship. Mayo Clin Proc 86(11):1113–1123
3. Clatworthy AE, Pierson E, Hung DT (2011) Targeting virulence: a new paradigm for antimicrobial therapy. Nat Chem Biol 3(9):541–548
4. Boucher HW, Talbot GH, Bradley JS, Edwards JE, Gilbert RLB, Scheld M, Spellberg B, Bartlett J (2008) Bad bugs, no drugs: no ESKAPE! An update from the infectious diseases society of America. Clin Infect Dis 48(1):1–12
5. Norris JL, Caprioli RM (2013) Analysis of tissue specimens by matrix-assisted laser desorption/ionization imaging mass spectrometry in biological and clinical research. Chem Rev 113(4):2309–2342
6. Caprioli RM, Farmer TB, Gile J (1997) Molecular imaging of biological samples: localization of peptides and proteins using MALDI-TOF MS. Anal Chem 69(23):4751–4760
7. Kueger S, Steinhauser D, Willmitzer L, Giavalisco P (2012) High-resolution plant metabolomics: from mass spectral features to metabolites and from whole-cell analysis to subcellular metabolite distributions. Plant J 70(1):39–50
8. Schoenian I, Spiteller M, Ghaste M, Wirth R, Herz H, Spiteller D (2011) Chemical basis of the synergism and antagonism in microbial communities in the nests of leaf-cutting ants. Proc Natl Acad Sci U S A 108(5):1955–1960

9. Rath CM, Alexandrov T, Higginbottom SK, Song J, Milla ME, Fischbach MA, Sonnenburg JL, Dorrestein PC (2012) Molecular analysis of model gut microbiotas by imaging mass spectrometry and nanodesorption electrospray ionization reveals dietary metabolite transformations. Anal Chem 84(21):9259–9267
10. Sogawa K, Watanabe M, Sato K, Segawa S, Ishii C, Miyabe A, Murata S, Saito T, Nomura F (2011) Use of the MALDI bioTyper system with MALDI-TOF mass spectrometry for rapid identification of microorganisms. Anal Bioanal Chem 400(7):1905–1911
11. Valentine N, Wunschel S, Wunschel D, Petersen C, Wahl K (2005) Effect of culture conditions on microorganism identification by matrix-assisted laser desorption ionization mass spectrometry. Appl Environ Microbiol 71(1):58–64
12. Jackson KA, Edwards-Jones V, Sutton CW, Fox A (2005) Optimisation of intact cell MALDI method for fingerprinting of methicillin-resistant *Staphylococcus aureus*. J Microbiol Methods 62(3):273–284
13. Toh-Boyo GM, Wulff SS, Basile F (2012) Comparison of sample preparation methods and evaluation of intra- and intersample reproducibility in bacteria MALDI-MS profiling. Anal Chem 84(22):9971–9980
14. Coltella L, Mancinelli L, Onori M, Lucignano B, Menichella D, Sorge R, Raponi M, Mancini R, Russo C (2013) Advancement in the routine identification of anaerobic bacteria by MALDI-TOF mass spectrometry. Eur J Clin Microbiol Infect Dis 32(9):1183–1192
15. Biswas S, Rolain J-M (2013) Use of MALDI-TOF mass spectrometry for identification of bacteria that are difficult to culture. J Microbiol Methods 92(1):14–24
16. Chalupova J, Martin Raus MS, Marek S (2014) Identification of fungal microorganisms by MALDI-TOF mass spectrometry. Biotechnol Adv 32(1):230–241
17. Drevinek M, Dresler J, Klimentova J, Pisa L, Hubalek M (2012) Evaluation of sample preparation methods for MALDI-TOF MS identification of highly dangerous bacteria. Lett Appl Microbiol 55(1):40–46
18. Mather CA, Rivera SF, Butler-Wu SM (2014) Comparison of the Bruker Biotyper and Vitek MS matrix-assisted laser desorption ionization–time of flight mass spectrometry systems for identification of mycobacteria using simplified protein extraction protocols. J Clin Microbiol 52(1):130–138
19. Boehme K, Fernandez-No IC, Barros-Velazquez J, Gallardo JM, Calo-Mata P, Canas B (2010) Species differentiation of seafood spoilage and pathogenic gram-negative bacteria by MALDI-TOF mass fingerprinting. J Proteome Res 9(6):3169–3183
20. Marko DC, Saffert RT, Cunningham SA, Hyman J, Walsh J, Arbefeville S, Howard W, Pruessner J, Safwat N, Cockerill FR, Bossler AD, Patel R, Richter SS (2012) Evaluation of the Bruker Biotyper and Vitek MS matrix-assisted laser desorption ionization-time of flight mass spectrometry systems for identification of nonfermenting Gram-negative bacilli isolated from cultures from cystic fibrosis patients. J Clin Microbiol 50(6):2034–2039
21. Rizzardi K, Wahab T, Jernberg C (2013) Rapid subtyping of *Yersinia enterocolitica* by matrix-assisted laser desorption ionization-time of flight mass spectrometry (MALDI-TOF MS) for diagnostics and surveillance. J Clin Microbiol 51(12):4200–4203
22. Boggs SR, Cazares LH, Drake R (2012) Characterization of a *Staphylococcus aureus* USA300 protein signature using matrix-assisted laser desorption/ionization time-of-flight mass spectrometry. J Med Microbiol 61(5):640–644
23. Vlek ALM, Bonten MJM, Boel CHE (2012) Direct matrix-assisted laser desorption ionization time-of-flight mass spectrometry improves appropriateness of antibiotic treatment of Bacteremia. PLoS One 7(3):e32589
24. Miller JM (2013) Cost-saving strategies for diagnostic microbiology laboratories. Clin Microbiol Newsl 35(24):195–204
25. Zilahi G, Artigas A, Martin-Loeches I (2016) What's new in multidrug-resistant pathogens in the ICU? Ann Intensive Care 6(1):96
26. Ng PC, Ang IL, Chiu RWK, Li K, Lam HS, Wong RPO, Chui KM, Cheung HM, Ng EWY, Fok TF, Sung JJY, Lo YMD, Poon TCW (2010) Host-response biomarkers for diagnosis of late-onset septicemia and necrotizing enterocolitis in preterm infants. J Clin Investig 120(8):2989–3000

27. Narayana JL, Gopal J, Wu H-F (2012) Wound infection kinetics probed by MALDI-MS: rapid profiling of *Staphylococcus aureus* in mice. Analyst 137(14):3372–3380
28. Norris JL, Caprioli RM (2013) Imaging mass spectrometry: a new tool for pathology in a molecular age. Proteomics Clin Appl 7(11-12):733–738
29. Norris JL, Tsui T, Gutierrez DB, Caprioli RM (2016) Pathology interface for the molecular analysis of tissue by mass spectrometry. J Pathol Inform 7(13):PMC4837791
30. Prideaux B, Dartois V, Staab D, Weiner DM, Goh A, Via LE, Barry CE III, Stoeckli M (2011) High-sensitivity MALDI-MRM-MS imaging of Moxifloxacin distribution in Tuberculosis-infected rabbit lungs and granulomatous lesions. Anal Chem 83(6):2112–2118
31. Wakeman CA, Moore JL, Noto MJ, Zhang Y, Singleton MD, Prentice BM, Gilston BA, Doster RS, Gaddy JA, Chazin WJ, Capriol RM, Skaar EP (2016) The innate immune protein calprotectin promotes *Pseudomonas aeruginosa* and *Staphylococcus aureus* interaction. Nat Commun 7:11951
32. Zipperer A, Konnerth MC, Laux C, Berschei A, Janek D, Weidenmaier C, Burian M, Schilling NA, Slavetinsky C, Marschal M, Willmann M, Kalbacher H, Schittek B, Brötz-Oesterhelt H, Grond S, Peschel A, Krismer B (2016) Human commensals producing a novel antibiotic impair pathogen colonization. Nature 535:511–516
33. Cheng AG, DeDent AC, Schneewind O, Missiakas DA (2011) A play in four acts: *Staphylococcus aureus* abscess formation A play in four acts: *Staphylococcus aureus* abscess formation. Trends Microbiol 19(5):225–232
34. Corbin BD, Seeley EH, Raab A, Feldmann J, Miller MR, Torres VJ, Anderson KL, Dattilo BM, Dunman PM, Gerads R, Caprioli RM, Nacken W, Chazin WJ, Skaar EP (2008) Metal chelation and inhibition of bacterial growth in tissue abscesses. Science 319(5865):962–965
35. Clohessy PA, Golden BE (1995) Calprotectin-mediated zinc chelation as a biostatic mechanism in host defence. Scand J Immunol 42(5):551–556
36. Damo SM, Kehl-Fie TE, Sugitani N, Holt ME, Rathi S, Murphy WJ, Zhang Y, Betz C, Hench L, Fritz G, Skaar EP, Chazin WJ (2013) Molecular basis for manganese sequestration by calprotectin and roles in the innate immune response to invading bacterial pathogens. Proc Natl Acad Sci 110(10):3841–3846
37. Hood MI, Skaar EP (2012) Nutritional immunity: transition metals at the pathogen-host interface. Nat Rev Microbiol 10(8):525–537
38. Moore JL, Becker KW, Nicklay JJ, Boyd KL, Skaar EP, Caprioli RM (2013) Imaging mass spectrometry for assessing temporal proteomics: analysis of calprotectin in *Acinetobacter baumannii* pulmonary infection. Proteomics 14(7-8):820–828
39. Kehl-Fie TE, Chitayat S, Hood MI, Damo S, Restrepo N, Garcia C, Munro KA, Chazin WJ, Skaar EP (2011) Nutrient metal sequestration by calprotectin inhibits bacterial superoxide defense, enhancing neutrophil killing of Staphylococcus aureus. Cell Host Microbe 10(2):158–164
40. Damo S, Chazin WJ, Skaa EP, Kehl-Fie TE (2012) Inhibition of bacterial superoxide defense: A new front in the struggle between host and pathogen. Virulence 3(3):325–328
41. Spraggins JM, Rizzo DJ, Moore JL, Rose KL, Hammer ND, Skaar EP, Caprioli RM (2015) MALDI FTICR IMS of intact proteins: using mass accuracy to link protein images with proteomics data. JASMS 26(6):974–985
42. Zubair F, Prentice DM, Norris JL, Laibinis PE, Caprioli RM (2016) Standard reticle slide to objectively evaluate spatial resolution and instrument performance in imaging mass spectrometry. Analy Chem 88(14):7302–7311
43. Zavalin A, Yang J, Rm C (2013) Laser beam filtration for high spatial resolution MALDI imaging mass spectrometry. J Am Soc Mass Spectrom 24(7):1153–1156
44. Spraggins JM, Rizzo DG, Moore JL, Noto MJ, Skaar EP, Caprioli RM (2016) Next-generation technologies for spatial proteomics: integrating ultra-high speed MALDI-TOF and high mass resolution MALDI FTICR imaging mass spectrometry for protein analysis. Proteomics 16(11-12):1678–1689

45. Spraggins JM, Capriol RM (2011) High-speed MALDI-TOF imaging mass spectrometry: rapid ion image acquisition and considerations for next generation instrumentation. J Am Soc Mass Spectrom 22(6):1022–1031
46. Zavalin A, Todd EM, Rawhouser PD, Yang J, Norris JJL, Caprioli RM (2012) Direct imaging of single cells and tissue at sub-cellular spatial resolution using transmission geometry MALDI MS. J Mass Spectrom 47(11):1473–1481
47. Lasch P, Jacob D, Grunow R, Schwecke T, Doellinger J (2016) Matrix-assisted laser desorption/ionization time-of-flight (MALDI-TOF) mass spectrometry (MS) for the identification of highly pathogenic bacteria. Trends Analy Chem 85:103–111
48. Debois D, Hamze K, Guerineau V, Le Caer JP, Holland B, Lopez P, Ouazzani J, Seror SJ, Brunelle A, Laprevote O (2008) In situ localisation and quantification of surfactins in a *Bacillus subtilis* swarming community by imaging mass spectrometry. Proteomics 8(18):3682–3691
49. Holland RD, Duffy CR, Rafii F, Sutherland JB, Heinze TM, Holder CL, Voorhees KJ, Lay JO Jr (1999) Identification of bacterial proteins observed in MALDI TOF mass spectra from whole cells. Anal Chem 71(15):3226–3230
50. Fenselau C, Demirev P (2001) Characterization of intact microorganisms by MALDI mass spectrometry. Mass Spectrom Rev 20(4):157–171
51. Schey KL, Anderson DM, Rose KL (2013) Spatially-directed protein identification from tissue sections by top-down LC-MS/MS with electron transfer dissociation. Anal Chem 85(14):6767–6774
52. Perez KK, Olsen RJ, Musick WL, Cernoch PL, Davis JR, Land GA, Peterson LE, Musser JM (2013) Integrating rapid pathogen identification and antimicrobial stewardship significantly decreases hospital costs. Arch Pathol Lab Med 137(9):1247–1254
53. Sun W, Sun J, Zou L, Shen K, Zhong D, Zhou D, Sun W, Li J (2016) The successful diagnosis and typing of systemic amyloidosis using a microwave-assisted filter-aided fast sample preparation method and LC/MS/MS analysis. PLoS ONE 10(5):e0127180
54. Taverna D, Norris JL, Caprioli RM (2015) Histology-directed microwave assisted enzymatic protein digestion for MALDI MS analysis of mammalian tissue. Anal Chem 87(1):670–676
55. Lazova R, Seeley EH, Keenan M, Gueorguieva R, Caprioli RM (2012) Imaging mass spectrometry – a new and promising method to differentiate Spitz nevi from Spitzoid malignant melanomas. Am J Dermatopathol 34(1):82–90
56. Seeley EH, Washington MK, Caprioli RM (2013) Proteomic patterns of colonic mucosal tissues delineate Crohn's colitis and ulcerative colitis. PROTEOMICS Clin Appl 7(7-8):541–549
57. Harris GA, Nicklay JJ, Caprioli RM (2013) A localized in-situ hydrogel-mediated protein digestion and extraction technique for on-tissue analysis. Anal Chem 85(5):2717–2723
58. Taverna D, Pollins AC, Nanney LB, Caprioli RM (2016) Histology-guided protein digestion/extraction from formalin-fixed and paraffin-embedded pressure ulcer biopsies. Exp Dermatol 25(2):143–146

Chapter 9
Soft X-ray Radiation Applied in the Analysis of Intact Viruses and Antibodies by Means of Nano Electrospray Differential Mobility Analysis

Guenter Allmaier, Victor U. Weiss, Nicole Y. Engel, Martina Marchetti-Deschmann, and Wladyslaw W. Szymanski

Abstract The analysis of an intact infectious virus and a therapeutic monoclonal antibody with a nano electrospray differential mobility analyzer, a kind of ion mobility device, (nES-DMA aka nES GEMMA, macroIMS or SMPS) incorporating a commercial soft X-ray (SXR) source for charge reduction was successfully demonstrated representing a clear alternative to a ^{210}Po-based charge reducing source for bio nanoparticles. Not only bio nanoparticle detection and size determination will be shown with the SXR device but also particle number-based quantification demonstrated with an antibody. These achievements will open up new avenues in life sciences in general as well as in biomonitoring.

9.1 Introduction

Nearly every technique for the analysis and handling of aerosolized nanoparticles (NPs) applied electrostatic principles which make the well-defined charging of NPs an essential requirement for the application of such techniques. The most popular application of a charging device in the field of bio aerosol science is the use in combination with a differential mobility analyzer (DMA). For this purpose a bipolar diffusion charger based on radioactive sources as ^{85}Krypton, ^{210}Polonium or ^{241}Americum is in practice to generate bipolar ions from air inside the charge

G. Allmaier (✉) • V.U. Weiss • N.Y. Engel • M. Marchetti-Deschmann
Institute of Chemical Technology and Analytics, Vienna University of Technology (TU Wien), Getreidemarkt 9, A-1060 Vienna, Austria
e-mail: Guenter.allmaier@tuwien.ac.at

W.W. Szymanski
Faculty of Physics, University of Vienna, Boltzmanngasse 5, A-1090 Vienna, Austria

J.H. Banoub, R.M. Caprioli (eds.), *Molecular Technologies for Detection of Chemical and Biological Agents*, NATO Science for Peace and Security Series A: Chemistry and Biology, DOI 10.1007/978-94-024-1113-3_9

manipulation device. The big advantages of this approach are the ease of use, low cost and the well-defined charge distribution which is achieved. However, the increasing legal regulations in a huge number of countries concerning any use of radioactive material as well as potential safety risks are making this approach increasingly difficult in everyday life in the lab (particular in biochemical and chemical laboratories) and have stimulated the search for suitable alternatives. The alternatives for the mentioned radioactive source to generate the air ions for the charge manipulation process are electrical discharge as corona discharge (but generating a monopolar atmosphere) or soft X-ray (SXR) irradiation. X-ray radiation has been applied to charge aerosols already at the beginning of the twentieth century [1]. To a certain extent it was reinvented or better reintroduced in aerosol physics by Shimada et al. [2] during the last years. This type of charge manipulation device in combination with nano electrospray (nES) still generates some critical radiation, but this can be easily shielded and the system can be switched on and off. In a number of papers it was shown that a similar charging performance of SXR devices could be achieved [3–5] (see Fig. 9.1a containing two soft X-ray tubes in one device [3]). Nevertheless there is no consistent demonstration that a commercial SXR charger (Fig. 9.1b) generate similar data for bio nanoparticles (bio NPs), i.e. intact viruses and labile proteins from the type of an antibody, which might be more sensitive to the high intensity radiation. Furthermore we investigated the quantitative response of a commercial integrated SXR device with a nano DMA and water-based condensation particle counter (CPC) for an antibody.

Separation of single-charged (by means of radioactive sources) bio NPs in the gas-phase by application of a constant sheath flow of air and a tunable electric field after a nES process and charge conditioning – a concept first presented by Kaufman and colleagues in 1996 [6] for globular proteins – finds increasing attention in bio NP analysis. This system is called either nES-DMA, nES GEMMA (gas-phase electrophoretic mobility molecular analyzer), LiquiScan-ES, macro ion mobility spectrometer (macroIMS) or scanning mobility particle sizer (SMPS), i.e. a lot of acronyms describing all the same concept. What is of great importance in contrast to hydrodynamic particle size values as obtained for liquid-phase separations, nES-DMA, i.e. gas-phase electrophoresis, yields dry NP diameters. In case of spherical NPs, detected diameters corresponds to the NP size. Number-concentration based data evaluation allows NP characterization as recommended by the European Commission (2011/696/EU from October 18th, 2011). In the field of bio NPs, gas-phase electrophoresis (nES-DMA) is especially interesting for the analysis of viruses, virus-like particles and large proteins as shown by Bacher et al. [7] in the year 2001 and reviewed by Pease a decade later [8] as these analytes often exhibit spherical or near-spherical shapes. nES-DMA data can be employed for bio NP size determination and investigation of virus formation, studies of virus binding to receptor molecules, antibodies and antibody fragments [9, 10] or particle molecular weight (MW) determination via comparison to standard molecules of known MW [6–7, 11].

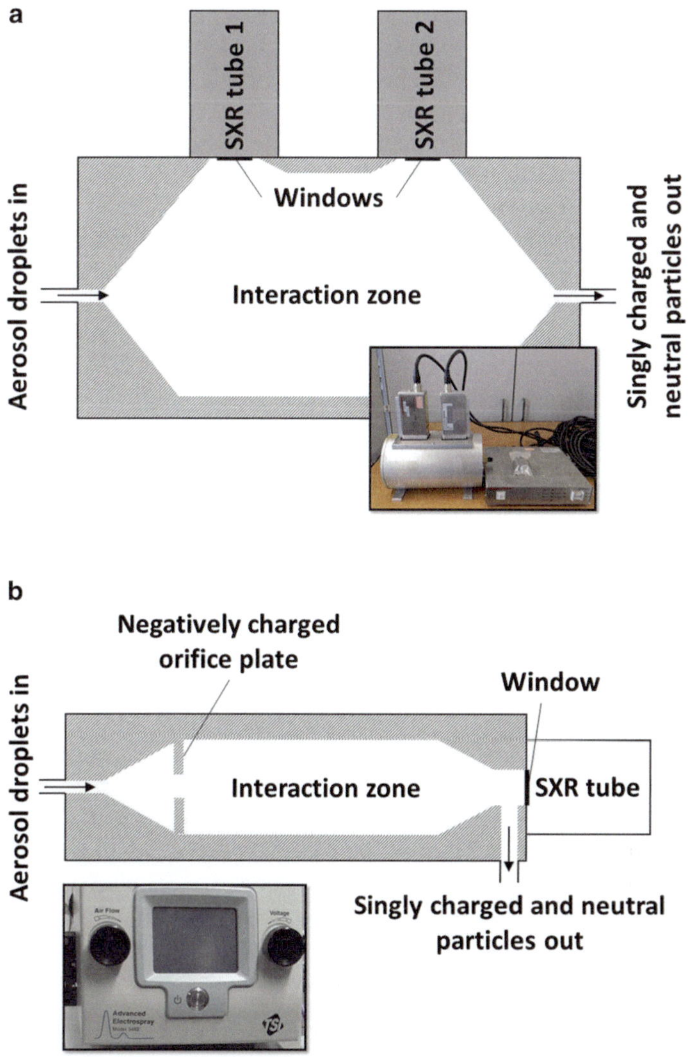

Fig. 9.1 (**a**) Scheme and picture of a home-built SXR source containing two photo ionizers (9.5 kV tube voltage) covering a large charge manipulation volume [3]. Everything is on-axis from the aerosol droplet entering and the leaving of the singly charged and neutral nanoparticles. (**b**) Scheme and picture of a commercial SXR containing device (advanced aerosol neutralizer model 3482) consisting of one photo ionizer axially mounted (to show a maximum of interaction volume) at the charge manipulation zone (with an added orifice plate after the spray capillary tip). The singly charged and neutral particles leave the device perpendicular to the axis of photo ionizer tube

In a series of papers we concentrated on the analysis of human Rhinovirus (HRV) serotype 2 (HRV). HRV, a non-enveloped, icosahedral bio NP of approx. 30 nm diameter, is comprised of four viral proteins, sixty copies each, and a single stranded RNA genome bound to one copy of an additional viral protein. The MW of such a macromolecular assembly was calculated to be 8.09 MDa [12]. With the current investigation we aimed to analyze HRV with a commercial SXR device incorporated in a nES-DMA system for charge conditioning [13].

Monoclonal antibodies have become major therapeutic and diagnostic agents and have been used for the treatment of many diseases as well as for biomedical research. Thus, it is of importance to find highly sensitive methods to monitor the antibodies' production, purity, and the formation of aggregates or degradation products. Due to working at atmospheric pressure and under relatively soft conditions, the nES-DMA system containing a commercial SXR source [13] allows the analysis of bio macromolecules as proteins in native-like conditions keeping non-covalent interactions intact [7]. A 148 kDa therapeutic monoclonal antibody was evaluated in terms of the analytical feasibility with the new commercial SXR device.

9.2 Experimental

9.2.1 Instrumentation

nES-DMA, i.e. gas-phase electrophoresis was carried out on a macroIMS system (TSI, Shoreview, MN, USA) consisting of a so-called advanced electrospray charge conditioning device with one SXR tube (model 3482, concept and picture shown in Fig. 9.1b), an electrostatic classifier (model 3082) including a nano DMA and a water-based CPC (model 3788). The virus samples were electro sprayed from ammonium acetate (20 mM, pH 8.4, filtered through a 0.2 μm pore size syringe filter) by application of a NE-1000 syringe pump (New Era Pump Systems, Farmingdale, NY, USA) set to 5–6 μL/min flow. Application of a 40 μm inner diameter, tipped fused silica capillary (tip prepared as described [14]) was found beneficial for obtained results. Compressed air for the nES process was additionally dried (Variodry Membrane Dryer Superplus, Donaldson, Bloomington, MN, USA). Parameters for the nES process to obtain a stable Taylor cone typically were 1.4–1.5 standard liters per minute (slpm) air, 0.1 slpm CO_2 and 1.4–1.7 kV (spray voltage). Analytes were separated in the gas-phase by application of 30 Lpm laminar air flow. Spectra were recorded in the EMD size range of 2–45 nm by variation of the applied voltage (56 s scan time in case of virus samples and 11 or 56 s scan time in case of the antibody samples). Reset of the applied voltage to starting values (3 s) and a 1 s lag lead to the recording of 20 individual spectra within the measurement time of 20 min. All data result from triplicate measurements.

9.2.2 Samples

A HRV serotype A2 sample prepared as described (including an additional lipase digestion step) [15] was subjected to buffer exchange applying 10 kDa MW cut-off spin filters (polyethersulfone membrane, VWR, Vienna, Austria) [16] and dilution to yield samples of 4 nM virus concentration for nES-DMA experiments.

376 nM monoclonal antibody (therapeutic grade) dissolved in 5 mM histidine and 60 mM trehalose buffer (pH 6.0) was used. Sample concentration and buffer exchange to 40 mM ammonium acetate (pH 8.0 or 5.0) was carried out in 10 kDa centrifugal filters (VWR, Vienna, Austria) and dilutions ranging from 5 to 7527 nM were prepared for nES-DMA investigations.

9.3 Results and Discussion

As demonstrated in Fig. 9.2 nES-DMA analysis with the SXR device for charge reducing (generating only singly charged bio NP species) of the intact HRV (including the complete genetic information in the form of the RNA) is possible. A well-defined peak ($[M]^+$) at 30.4 nm was obtained, representing the virus particle. Besides the peak for the human Rhinovirus at 30.4 nm EMD also two additional sample components at approx. 14 nm EMD (probably a complex protein) and a broad peak with an apex at approx. 25 nm EMD (cell membrane fragments or exosomes residues derived from the cell culture) are detected. The latter two peaks originate from contaminating material from the virus preparation process, which were not completely removed during virus preparation. What has to be pointed out is the fact that the inner diameter of the spray capillary has to be 40 μm. The use of 25 μm inner

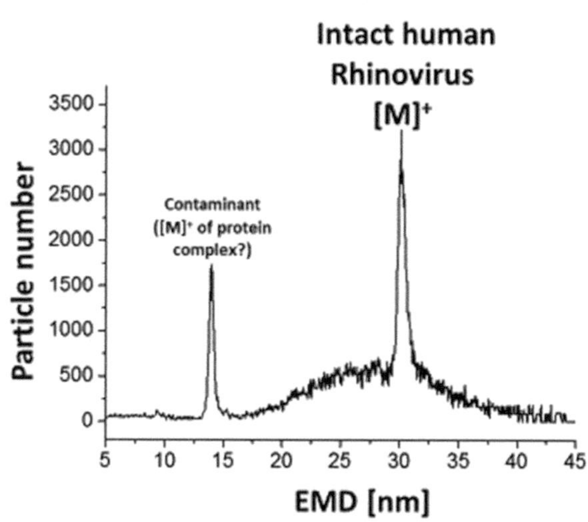

Fig. 9.2 nES-DMA results of the analysis of an intact HRV sample (4 nM virus concentration in 20 mM ammonium acetate at pH 8.4). The most abundant peak ($[M]^+$) is detected with an EMD of 30.4 nm corresponding to the intact virus including the RNA and the other abundant peak corresponds to contaminating material from the virus production

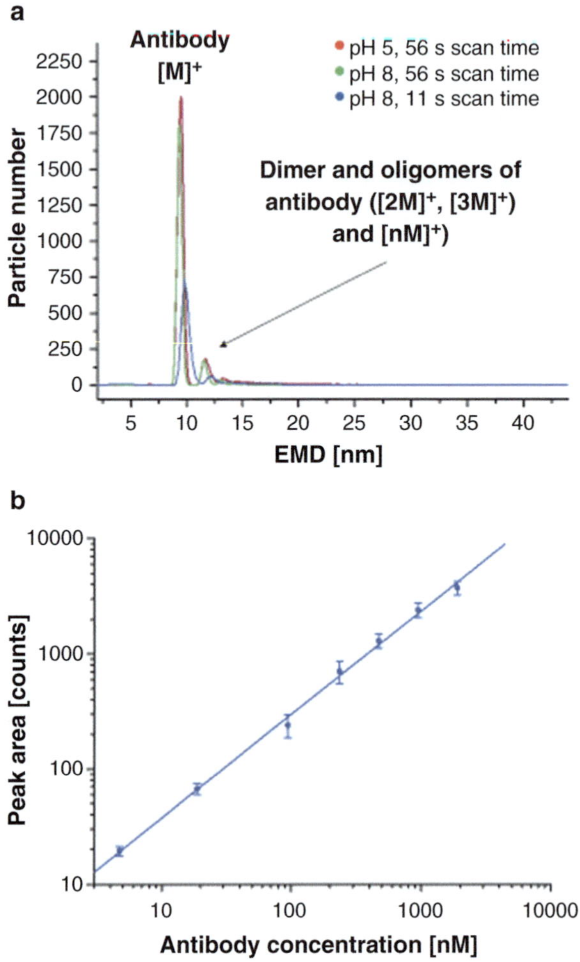

Fig. 9.3 (**a**) nES-DMA results of the analysis of a monoclonal antibody at different pH values (5 and 8) of the spray solution and at two different DMA scan times (11 and 56 s). The most abundant peak ($[M]^+$) is detected with an EMD of 9.46 ± 0.00 nm and 9.46 ± 0.13 nm (with a DMA scan time of 56 s at pH 8 and pH 5) corresponding to the singly charged antibody. The most abundant peak at pH 8 and only 11 s DMA scan time was 9.90 ± 0.05 nm. Peaks corresponding to higher EMD values are corresponding to antibody dimers ($[2M]^+$, 11.71 ± 0.00 nm pH 5 and 5 s DMA scan time) and oligomers. (**b**) Plot of peak area of $[M]^+$ ion derived from the well-resolved antibody monomer versus antibody concentration of the spray solution (at pH 8 and 56 s DMA scan time) across three decades of nM concentration range

diameter fused silica capillaries was not successfully so far and maybe related to clogging (adsorption of the virus NPs on the inner surface) effects. The obtained data are in excellent agreement with data obtained by nES device combined with a commercial ^{210}Po source. The device has to be evaluated further in terms of sensitivity and application to other type of virus species.

Monoclonal antibody dilutions were analyzed by the nES-DMA with the SXR device in combination with a water CPC and in all cases successfully singly charged molecules ($[M]^+$) could be obtained with high abundance shown in Fig 9.3a. An average size of 9.46 nm (at pH 5 and 8 at DMA scan time of 56 s) based on the EM diameter was obtained corresponding to values obtained by a ^{210}Po-based source and an identical nDMA. Furthermore low abundant peaks representing dimers

($[2M]^+$) and trimers ($[3M]^+$) and oligomers ($[nM]^+$). The latter are not well-resolved anymore. Different DMA scan times, namely 11 and 56 s at two different pHs, 5.0 and 8.0 were evaluated further (see Fig. 9.3a). The shorter scan time of 11 s was chosen for fast analysis times needed for *on-line* coupling with high performance liquid chromatography as size exclusion chromatography or capillary zone electrophoresis systems. 56 s was chosen for better DMA resolution. Results were compared in regard to EMD values, peak form, and peak height or area. Limit of detection (LOD) and limit of quantitation (LOQ) were determined for analysis. Inter-day and intra-day variability was evaluated for measurements with 56 s scan time. Good repeatability for (i) EMDs, (ii) peak widths, and (iii) peak area/heights were observed. The latter parameter, however, exhibits the highest standard deviations (SDs), especially in regard to measurements on different days. Inter-day variation was 28%, but intra-day variation only below 6%. Reproducible measurement set-up using similar parameters, the same capillary, and the formation of a stable cone-jet mode are of importance for such results. SDs are very likely to be further reduced by replacing the syringe pump for direct infusion with a liquid phase separation system. Directly comparing different DMA scan times at different pH shows the influence (Fig. 9.3a). Switching from 56 to 11 s drastically changes the signal intensity, peak width, and EMD. EMDs were detected at significantly higher values (9.46 nm versa 9.90 nm EMD) at lower scan speed. However, a mere variation of the pH at 56 s scan time did not affect the EMDs. In all cases, the determined EMD stayed very reproducible for each condition with SDs below 1.3%. The lower scan time of 11 s also caused peak broadening of about 30%. In contrast, the corresponding FWHM (full-width half maximum) values for each scan time stayed quite constant over the concentration ranges with a slight increase towards lower concentrations. Therefore, the SDs of the mean values were higher than for the EMDs (below 8% for the monomer signals and below 15% for dimers and trimers). Figure 9.3b displays a dilution experiment for the monoclonal antibody covering a wide concentration range from single digit beyond 10^3 nM applying the SXR tube and a water CPC (solution pH 8.0 and DMA scan time 11 s). The corresponding peak areas for the monomeric peak ($[M]^+$) were plotted against the antibody concentration. As expected, signal intensity increased with higher concentrations and the calibration showed a good linearity ($r^2 = 0.9968$). The upper limit of the linear range for the antibody monomer was at approx. 2000 nM at 11 s DMA scan speed. The particular reason for that is the increased tendency to form gas-phase dimers, trimers and unresolved oligomers during the nES process. In order to define LOD and LOQ of the analysis, pure background electrolyte was analyzed and its mean value as well as the SD in the range of 9–11 nm calculated. LOD was then determined as the mean value plus 3× SD, LOQ as the mean value plus 10× SD, which resulted in a LOD and LOQ of 5 and 10 nM, respectively. These values were in good accordance with the measurements of 4.7 and 18.8 nM having a signal-to-noise ratio over 5 and 20, respectively.

9.4 Conclusions

The analysis of an intact human virus and a therapeutic monoclonal antibody with a nES-DMA incorporating a commercial SXR source (Fig. 9.1b) was successfully demonstrated representing a clear alternative to a ^{210}Po-based charge reducing source for bio NPs. Not only sizing will be feasible with the commercial SXR device but also particle number-based quantification will be possible across a considerable linear range as shown for the monoclonal antibody. These achievements will openup new avenues in life sciences in general as well as in bio NP monitoring e.g. in airplanes, space crafts or indoor and QC (PAT) in biotechnology as well as vaccine production.

Acknowledgements The authors acknowledge funding by the Austrian Science Foundation (FWF) TRP29. Furthermore we thank D. Blaas (Vienna, Austria) for supplying the virus samples and F. Foret (Brno, Czech Republic) for the help of designing the spray capillary tip grinding device.

References

1. Millikan RA (1913) On the elementary electrical charge and the Avogadro constant. Phys Rev 2:109–143
2. Shimada M, Han B, Okuyama K, Otani Y (2002) Bipolar charging of aerosol nanoparticles by a soft X-ray photoionizer. J Chem Eng Jpn 35:78–793
3. Kallinger P, Steiner G, Szymanski WW (2012) Characterization of four different bipolar charging devices for nanoparticle charge conditioning. J Nanopart Res 14:944–951
4. Kallinger P, Szymanski WW (2015) Experimental determination of the steady-state charging probabilities and particle size conservation in non-radioactive and radioactive bipolar aerosol chargers in the size range of 5–40 nm. J Nanopart Res 17:171–182
5. Yoon YH, Bong C, Kim DS (2015) Evaluation of the performance of a soft X-ray charger for the bipolar charging of nanoparticles. Particuology 18:165–169
6. Kaufman SL, Skogen JW, Dorman FD, Zarrin F, Lewis KC (1996) Macromolecule analysis based on electrophoretic mobility in air: globular proteins. Anal Chem 68:1895–1904
7. Bacher G, Szymanski WW, Kaufman SL, Zoellner P, Blaas D, Allmaier G (2001) Nano ESI with charge reduction combined with differential mobility analysis of peptides, proteins, glycoproteins, noncovalent protein complexes and viruses. J Mass Spectrom 36:1038–1052
8. Pease LF 3rd (2012) Physical analysis of virus particles using electrospray differential mobility analysis. Trends Biotechnol 30:216–224
9. Laschober C, Wruss J, Blaas D, Szymanski WW, Allmaier G (2008) Gas Phase Electrophoretic Molecular Mobility Analysis (GEMMA) of size and stoichiometry of complexes of a common cold virus with antibody and soluble receptor molecules. Anal Chem 80:2261–2264
10. Bereszczak JZ, Havlik M, Weiss VU, Marchetti-Deschmann M, van Duijn E, Watts NR, Wingfield PT, Allmaier G, Steven AC, Heck AJ (2014) Sizing up large protein complexes by electrospray ionization-based electrophoretic mobility and native mass spectrometry: morphology selective binding of Fabs to hepatitis B virus capsids. Anal Bioanal Chem 406:1437–1446
11. Kaddis CS, Lomeli SH, Yin S, Berhane B, Apostol MI, Kickhoefer VA, Rome LH, Loo JA (2007) Sizing large proteins and protein complexes by electrospray ionization mass spectrometry and ion mobility. J Am Soc Mass Spectrom 18:1206–1216

12. Weiss VU, Bereszcazk JZ, Havlik M, Kallinger P, Goessler I, Kumar M, Blaas D, Marchetti-Deschmann M, Heck AJ, Szymanski WW, Allmaier G (2015) Analysis of a common cold virus and its subviral particles by gas-phase electrophoretic mobility molecular analysis and native mass spectrometry. Anal Chem 87:8709–8717
13. http://www.tsi.com/macroIMS-Macroion-Mobility-Spectrometer-3982/ from 2016-03-24
14. Tycova A, Prikryl J, Foret F (2015) Reproducible preparation of nanospray tips for capillary electrophoresis coupled to mass spectrometry using 3D printed grinding device. Electrophoresis 37:924–930
15. Weiss VU, Subirats X, Kumar M, Harutyunyan S, Goessler I, Kowalski H, Blaas D (2015) Capillary electrophoresis, gas-phase electrophoretic mobility molecular analysis, and electron microscopy: effective tools for quality assessment and basic rhinovirus research. Methods Mol Biol 1221:101–128
16. Weiss VU, Lehner A, Kerul L, Grombe R, Kratzmeier M, Marchetti-Deschmann M, Allmaier G (2013) Characterization of cross-linked gelatin nanoparticles by electrophoretic techniques in the liquid and the gas phase. Electrophoresis 34:3267–3276

Chapter 10
Mass Spectrometry in Environmental Chemistry and Toxicology

Ksenia J. Groh and Marc J.-F. Suter

Abstract The aquatic environment has long been a sink for diverse chemicals released as a result of human activities. In order to assess the associated risks for the aquatic life, both the chemicals' occurrences (exposure) as well as their effects in the living organisms (hazard) need to be known. These two aspects are studied by environmental chemistry and toxicology, respectively. Mass spectrometry has been successfully applied in both fields, as it can be used not only for measuring the pollutants in different environmental compartments, but also for gaining novel insights into the molecular mechanisms of toxicity. We here discuss the application of mass spectrometry in environmental chemistry and toxicology, illustrated by several case studies drawn from research carried out at our institute.

10.1 Introduction

Environmental sciences have evolved from being able to detect a limited set of environmental contaminants found at relatively high concentrations to where thousands of targets can be detected in a data-independent way. Examples for the first case are accidents that were catastrophic for the people involved, but raised awareness of the imminent risk of certain industrial procedures. One example is the Seveso disaster of 1976, where a runaway reaction, caused by overheated steam, led to the accidental release of the highly toxic dioxin 2,3,7,8-tetrachlorodibenzodioxin (TCDD) in kg amounts [1]. Another industrial accident, considered the world's worst, happened in 1984 in Bhopal, India, when water leaking into a tank containing methyl

K.J. Groh
Food Packaging Forum Foundation, 8045 Zürich, Switzerland

M.J.-F. Suter (✉)
Eawag, Swiss Federal Institute of Aquatic Science and Technology, 8600 Dübendorf, Switzerland

Department of Environmental Systems Science, ETH Zürich, Swiss Federal Institute of Technology, 8092 Zürich, Switzerland
e-mail: suter@eawag.ch

J.H. Banoub, R.M. Caprioli (eds.), *Molecular Technologies for Detection of Chemical and Biological Agents*, NATO Science for Peace and Security Series A: Chemistry and Biology, DOI 10.1007/978-94-024-1113-3_10

isocyanate initiated an exothermic reaction that led to the release of 40 tons of methyl isocyanate, killing over 3000 people within a day [2]. Finally, in 1986 a fire in a storehouse for chemicals at Schweizerhalle in Northern Switzerland and the subsequent run-off of firefighting water contaminated the River Rhine from Basel to its estuary near Rotterdam in The Netherlands, wiping the ecosystem clean of life along long stretches [3]. All three disasters produced high concentrations of pollutants in a very short time and were caused by neglected safety features or design errors, such as a missing retention basin. The Schweizerhalle case was the only one involving a multitude of initially unknown chemicals potentially responsible for the observed toxic effects. Both, the Seveso and the Schweizerhalle incident kick-started the development of highly specific and sensitive mass spectrometers. For instance, in order to accurately quantify the dioxin congeners of differing toxicity in the presence of other polychlorinated compounds (such as polychlorinated biphenyls – PCBs), mass spectrometers providing a resolution $m/\Delta m$ of better than 10,000 were required. At that time double-focusing sector field instruments were the only commercially available mass spectrometers capable of such a feat.

Another limitation was given by the fact that target analytes had to be ionized by electron ionization (EI), the most commonly used technique in the 1970s and 1980s. This limited the range of chemicals amenable for mass spectrometry (MS) analysis to those being volatile, i.e. having a mass typically below 1000 Dalton, and being thermally stable. Only in the 1990s, with the advent of electrospray ionization (ESI) and matrix-assisted laser desorption ionization (MALDI), the field of MS applications expanded towards polar and ionic target analytes. In addition to that, ESI extended the accessible mass range by orders of magnitude, due to the fact that it produces multiply charged species. This in turn provided the possibility to analyze biological macromolecules such as enzymes, allowing for instance to investigate stress response of organisms on the molecular level [4, 5].

Figure 10.1 illustrates the fields of application of gas chromatography (GC)-MS and liquid chromatography (LC)-MS in a biased personal representation. Of course the polar part can be partially addressed with GC-MS after derivatization.

10.2 The Universe of Chemicals

The "Universe of Chemicals" shown schematically in Fig. 10.1 consists of millions of substances. According to the Chemical Abstracts Service (CAS – an operating division of the American Chemical Society) the 100 millionth CAS registry number was assigned on June 29, 2015 [6]. Over 300,000 of the 100 million registered chemicals are inventoried/regulated substances. Of course, not all of these will be necessarily present in the aquatic environment. However, even if it is only a fraction of these 300,000 plus regulated chemicals that pose a risk, the number is likely still beyond what can routinely be monitored simultaneously. In addition to the sheer size of the chemical universe, it also covers a wide range of physical-chemical properties, from apolar with $\log K_{ow} > 5$, to polar and ionic, from volatile to non-volatile.

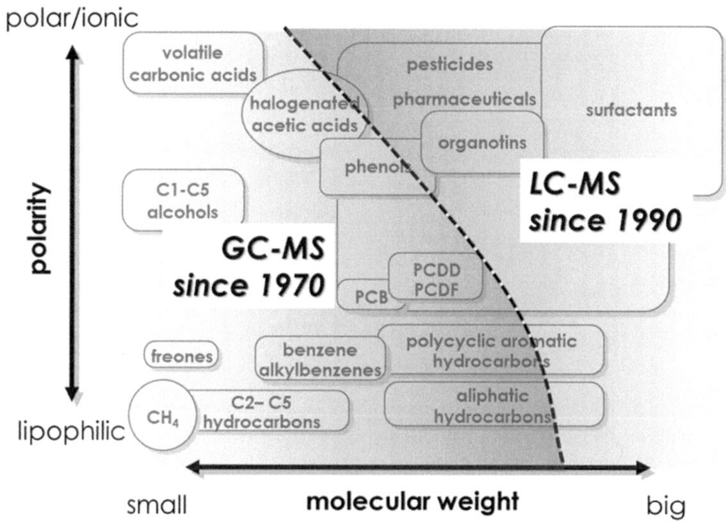

Fig. 10.1 The "Universe of Chemicals" as seen from the point of view of a mass spectrometry-practicing environmental scientist (author's biased personal representation). The *lower left* part of the diagram can be analyzed using GC-MS, while the polar and ionic high molecular weight compounds have only become accessible with soft desorption ionization techniques such as electrospray, typically used in LC-MS

Furthermore, the speciation of a polar or ionic species depends on pH and potentially other (charged) chemicals present in the matrix, thus strongly influencing sorption onto sediments, interaction with natural organic matter and uptake into aquatic organisms [7]. To simultaneously capture this multitude of chemicals, mixed-mode enrichment [8, 9] and separation [10] are required. A combination of four different separation mechanisms has been achieved by coupling two commercial mixed-mode columns, containing C18 that also acts like hydrophilic interaction chromatography (HILIC) down to 70% organic, and either weak anion (WAX) or weak cation exchanger phases (WCX – see Fig. 10.2). An environmental extract is injected onto a C18 trap, which captures hydrophobic compounds, while anionic or cationic chemicals pass through to the WAX-WCX columns.

They are then separated by a simultaneous gradient for HILIC, anion and cation exchange chromatography [10], caused by the increasing ionic strength in the eluents (see Fig. 10.2c).

After switching of the divert valve, this is then followed by a classical reversed-phase gradient. This unique setup allows the simultaneous separation of hydrophobic and ionic species (see Fig. 10.3).

For the environmental toxicologist the question however remains how best to identify the chemicals present in a very complex mixture that cause adverse effects in an ecosystem. One approach is to develop a hypothesis of what causes the effect and then target the chemicals known to be linked to it. Hence, when it was observed that brown trout catches had declined by 60% over a period of 20 years in Switzerland

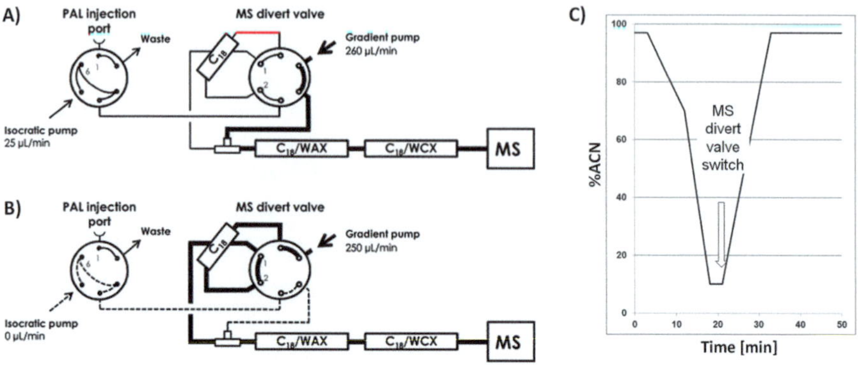

Fig. 10.2 Schematic of the multimode chromatographic setup. (**a**) Sample loading followed by a HILIC (hydrophilic interaction chromatography) and IEX (ion exchange) run; 3–21 min. (**b**) Reversed phase (RP) run after switching of the divert valve; 21–36 min. The flow rate of the isocratic delivery pump was reduced to 0 after 3 min. (**c**) Gradient program – Eluent A: Acetonitrile (ACN) with 3% H_2O, 3 mM NH_4HCO_3 (pH 7.3), Eluent B: H_2O with 10% ACN, 30 mM NH_4HCO_3 (pH 7.3) ([10], reproduced with permission)

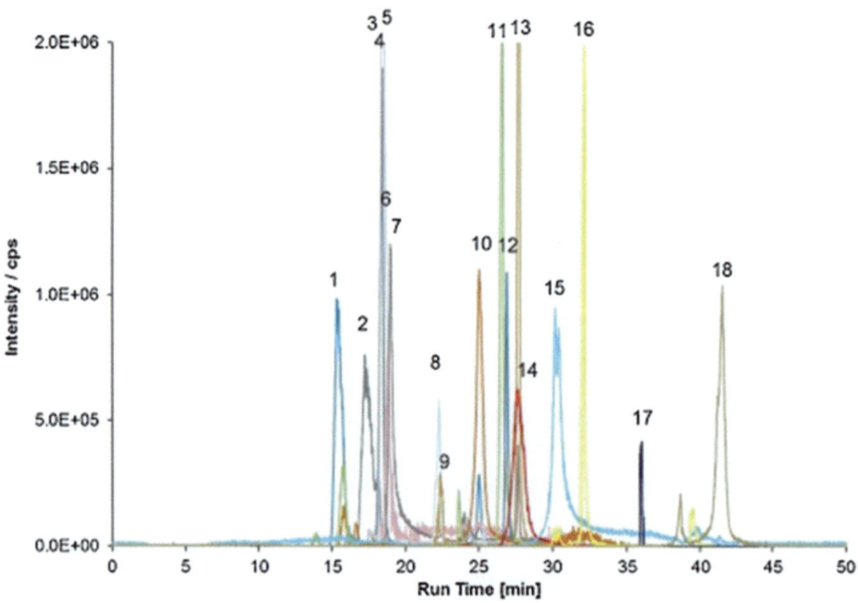

Fig. 10.3 Typical multimode separation chromatogram obtained from 18 standard compounds; *1* phenylalanine, *2* ascorbate, *3* galacturonic acid, *4* glutamate, *5* cystine, *6* hexanoic acid, *7* glutathione, *8* glucose-1-phosphate, *9* glutathione disulfide, *10* lysine, *11* tryptophan, *12* sucralose, *13* fluconazole, *14* arginine, *15* cysteine, *16* clotrimazole, *17* tocopherol, *18* dodecyl sulfate ([10], reproduced with permission)

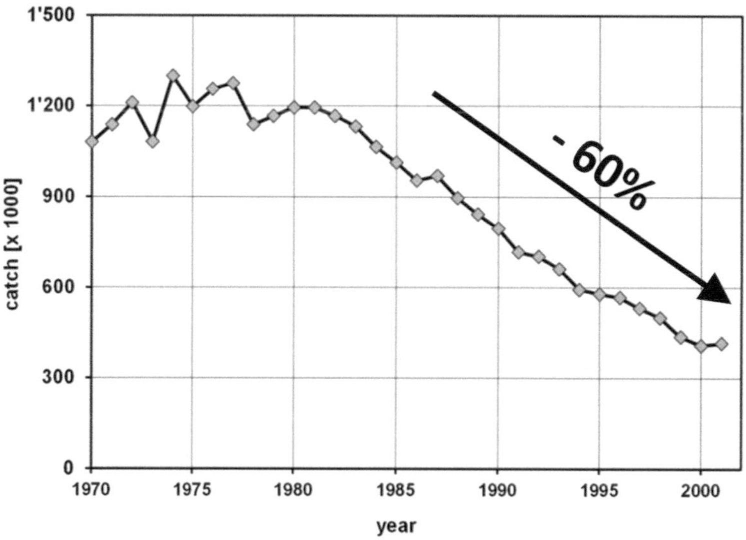

Fig. 10.4 Observed decline in brown trout catches in Switzerland [12]

(see Fig. 10.4), the most popular hypothesis was that endocrine-disrupting compounds, specifically estrogens, were affecting reproductive success, thus lowering the population levels of brown trout as a consequence.

However, the following investigation lasting from 1998 to 2003 showed that (xeno)estrogens could be made potentially responsible for reduced fish fertility and hatching success only at hotspots, e.g. downstream of wastewater treatment plants (WWTP), where they indeed could be detected at effect-causing levels by chemical analysis. In the end, it turned out that one of the main reasons for the observed declining catches was deteriorated habitat quality [11].

10.3 Effect-Directed Analysis

Another popular strategy, called Effect-Directed Analysis (EDA), is to prioritize for chemical analysis only the samples or fractions thereof that show an effect in an *in vitro* or *in vivo* assay (see Fig. 10.5). This allows greatly reducing the number of samples that need to be chemically analyzed. In the ideal case the measured activity can be explained by the chemicals found to be present in the sample. For example, in a Swiss survey of rivers, estradiol equivalents determined in the yeast estrogen screen (YES) were shown to correlate with estradiol equivalents calculated based on determined concentrations and relative potencies of 17ß-estrogen (E2), estrone (E1) and 17α-ethinlyestradiol (EE2), both in grab samples and extracts from passive samplers [13].

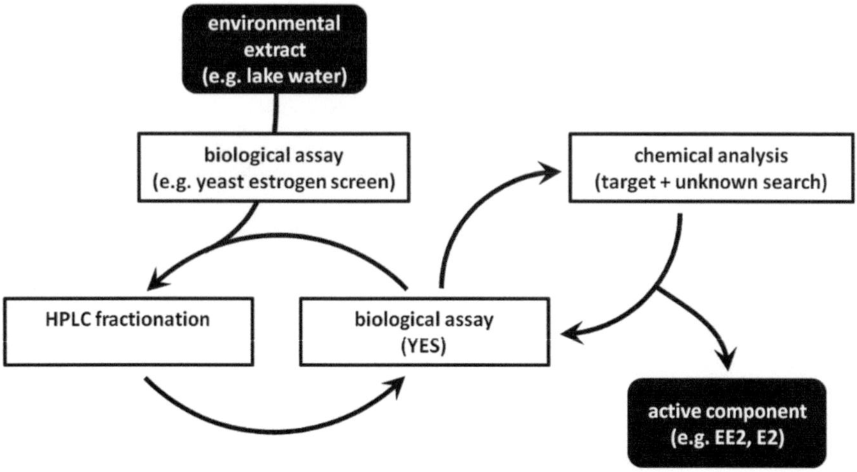

Fig. 10.5 Schematic representation of the Effect-Directed Analysis (Adapted from W. Brack [14])

The same is true for tissues and body fluids as shown in the case of muscle and bile from white fish (*Corregonus lavaratus)* in an alpine lake in Switzerland. Here too, estrogenicity determined in the YES could be explained by the natural estrogens E2 and E1 [15]. Similarly, the main cause for estrogenicity measured in the River Thames could be attributed to 17ß-estradiol [16], and in another project, triclosan was identified as being responsible for growth inhibition observed in green algae [17].

Although this method is straightforward when established targeted analysis is used as in the case of the estrogens mentioned above, the investigation becomes much more challenging as soon as targeted analysis returns a blank and the identity of the effect-causing chemical remains unknown. In this case a differential analysis becomes necessary. The following case study will illustrate the procedure.

At the turn of the millennium, fishermen in the Bernese Oberland, Switzerland, observed an increased incidence of gonad malformations in white fish [18]. Potential causes could have been biological, e.g. genetic factors or infectious diseases, or related to environmental conditions such as temperature, habitat quality, food availability or of course water quality. In the search for causes, sediment, water and plankton samples were collected and analyzed for estrogenic activity. Figure 10.6 shows the estrogenic activity in plankton extracts, determined with two different assays, the YES, a reporter gene assay that measures estrogen receptor activation, and the E-Screen that measures estrogen-dependent cell proliferation. Both endpoints indicate estrogenic potential of a sample, which means that e.g. the plankton collected in Lake Brienz, August 2005 (BRI 08/05), is estrogenic. When comparing this sample to any sample that is not estrogenic, features found in the estrogenic but absent in non-estrogenic samples could potentially be the active compound. Identifying features of interest is done using LC-MS/MS and data-dependent analysis [19, 20]. With this technique, routinely used in global proteomics [5], everything eluting off the column is collisionally activated and the fragments generated detected.

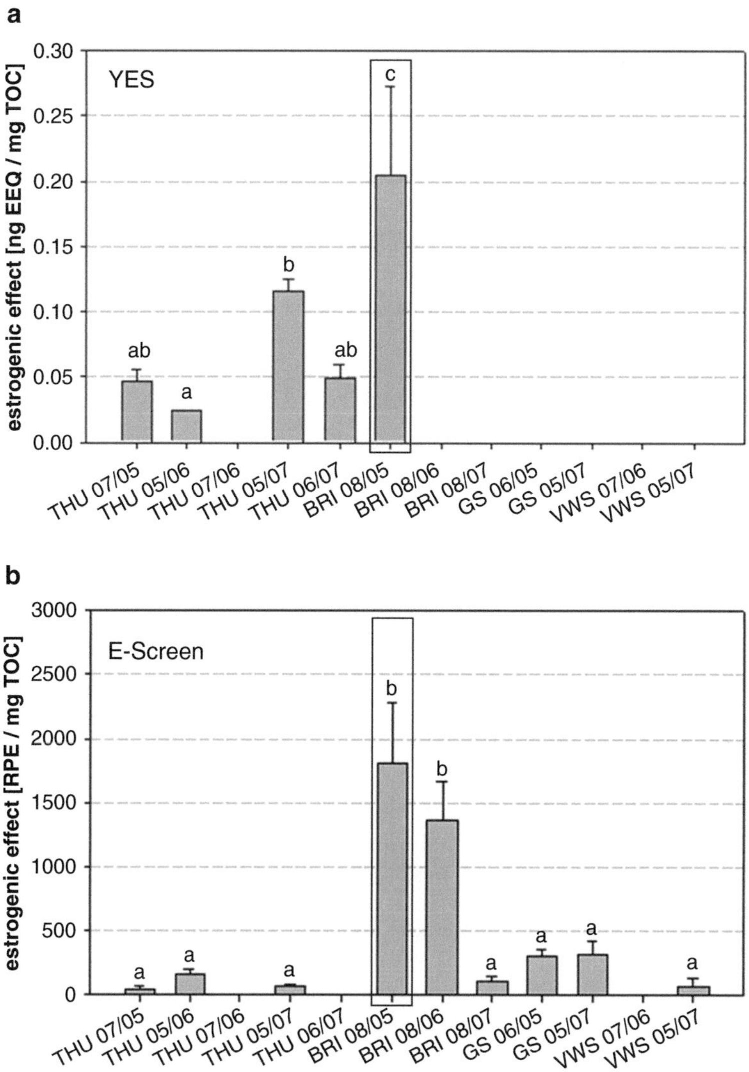

Fig. 10.6 Estrogenic activity determined in plankton extracts collected over 3 years (2005–2007) from four different lakes in Switzerland: Lakes Thun (THU), Brienz (BRI), Greifen (GS) and Lucerne (VWS) [21]. The samples were analyzed using the YES and the E-Screen assays [22]

In a first step the most intense signals, determined in a full scan spectrum, are selected and their MS/MS spectra acquired a few times. Following this, these signals are then put on an exclusion list to allow less intense features to be analyzed as well. Depending on the settings, this then also gives information related to signals of low intensity. Another technique, becoming increasingly popular in proteomics and other fields is scan-independent analysis or SWATH [23]. Instead of selecting a single precursor ion as in data-dependent acquisition mentioned above, an entire window of masses is collisionally activated and the corresponding fragments acquired. Assignment of the fragments to precursor ions is then done based on overlapping chromatographic profiles. By repeatedly shifting the window to higher masses the whole mass spectral information can be collected from a sample. This allows reanalyzing the data at a later stage, when new evidence leads to new hypotheses.

A differential analysis identifies compounds that have significantly different concentrations in an active versus an inactive sample. It provides accurate mass and fragment ion information in addition to the retention time. All this helps narrowing down the elemental composition of an unknown substance, identifying functional groups through MS/MS spectra, and the interaction with the separation phase, reflected in the retention time, informs on the physical-chemical properties of the chemical.

The seven golden rules established by the Fiehn group [24] tell us for instance that the mass 180.1020 singled out in Fig. 10.7 can only have a maximum of 15 carbon, 178 hydrogens, 11 oxygens atoms and 12 nitrogens, if it is composed of only this one element, which of course is chemically unlikely or impossible (Rule 1). The maximal number of elements can further be restricted based on the empirical information available from libraries (e.g. Wiley mass spectral database [25], or the dictionary of natural products [http://dnp.chemnetbase.com]). Another requirement for an elemental sum formula is that chemical structures can be drawn that fulfill the Lewis and Senior rules, or in other words that "molecules consisting of main group elements, especially carbon, nitrogen and oxygen, share electrons in a way that all atoms have completely filled s, p-valence shells" (Rule 2) [24]. The experimental data also contains isotopic information which helps filtering potential sum formulas even further (Rule 3). Additionally, from basic MS interpretation courses it is well known that when the monoisotopic ion M ($M^{+\cdot}$ in EI or $[M + H]^+$ in ESI) shows a signal with high

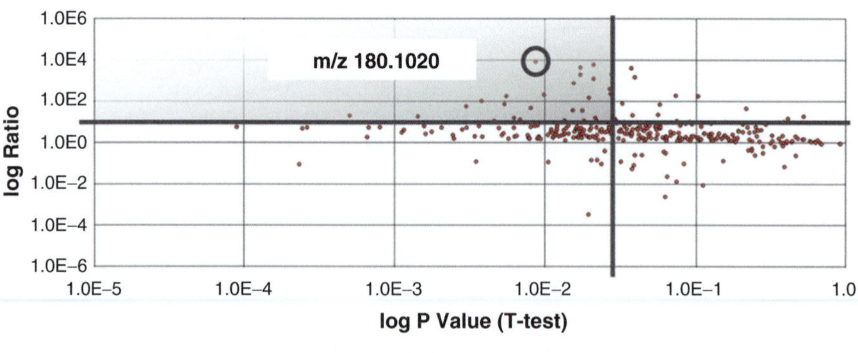

Fig. 10.7 Differential analysis of an estrogenic plankton extract from Lake Brienz (BRI 08/05 in Fig. 10.6) versus an inactive sample using the software SIEVE (ThermoScientific, USA). The x-axis gives the significance of the difference, the y-axis the log ratio of the areas (estrogenic/inactive). The shaded area corresponds to a p-value < 0.02 and a ratio >10

intensity for the M + 2 isotope, at least one chlorine (>32%) or one bromine (>97%) is likely to be present. Similarly, the natural relative abundance of the other elements allows clearly assigning the natural isotopes to the higher isotope signals (see Fig. 10.8). For instance, when looking at the number of potential candidate sum formulas at mass 500, with a mass accuracy of 10 ppm, 266 candidates are possible. With an accuracy of 1 ppm, easily achievable with an Orbitrap, only 21 remain possible. When taking into account the isotopic distribution, the candidates are further narrowed down to 3 at 3 ppm and 2% isotope accuracy. An additional filter to reduce the number of potential sum formulas is the H/C ratio (Rule 4). As shown by Kind and Fiehn [24], the H/C ratio for most chemical compounds (99.7%) lies between 0.2 and 3.1, based on the Wiley mass spectral database (45′000 chemicals with a mass range of 30 to 1,500 Da, containing C, H, N, S, O, P, F, Cl, Br, I, Si). Similarly, other element ratios again based on the Wiley library help filter the data further (Rule 5): F/C 0–1.5, Cl/C and Br/C 0–0.8, N/C 0–1.3, O/C 0–1.2, P/C 0–0.3, S/C 0–0.8, Si/C 0–0.5; these ranges cover 99.7% of the library entries. Rule 6 performs an element probability check and Rule 7 only applies to GC/MS data, since it does a trimethylsilylation (TMS) check for derivatized compounds.

Not all of these rules are needed to find a sum formula that fits the ion at m/z 180.1020, since its mass is quite low (see Table 10.1). Table 10.2 lists possible candidate sum formulas for the protonated molecular ion $[M + H]^+$ of 180.1020 Da. With a mass accuracy of 5 ppm, routinely achieved on time-of-flight instruments and Orbitraps, only seven candidates are possible, using the elements C, H, N, S, O, P, F, Cl, Br, I, and Si. $C_2H_{14}ON_7Si$ and $C_3H_{14}N_7S$ (marked with an X in Fig. 10.9) can be ruled out because of their high H/C and N/C ratios. Figure 10.9 shows the intensities of the M + 1 and M + 2 isotope peaks relative to the monoisotopic ion, of the ten candidates for ion m/z 180.1020 listed in Table 10.2. If the unknown were carrying a chlorine, the isotope pattern would be telling (see Fig. 10.9, the two points at M + 2 > 30%). What Fig. 10.9 shows is that the ten candidates can easily be distinguished based on their isotopic pattern (Rule 3), if the relative intensities of the isotopes are reproducible, which is a given with current instrumentations. But candidate 1 in Table 10.2 is not only the prime candidate based on isotopic distribution, but also because this investigation targeted estrogenic compounds. Most estrogens have a steroid structure with ring A being phenolic. This means that the double bond equivalents (DBE) must be four or higher (three for the double bonds of the aromatic system, and one for the ring).

This then identifies candidate 1 from Table 10.2 as the only possible candidate. A potential structure, which fits the sum formula and MS/MS data could be proposed and synthesized, but turned out to be non-estrogenic in the YES.

Even though the sum formula of the compound responsible for the estrogenic activity found in the plankton extract from Lake Brienz could unequivocally be determined, the structure remains unknown. Nonetheless, the case study presented here nicely shows the power of the EDA approach.

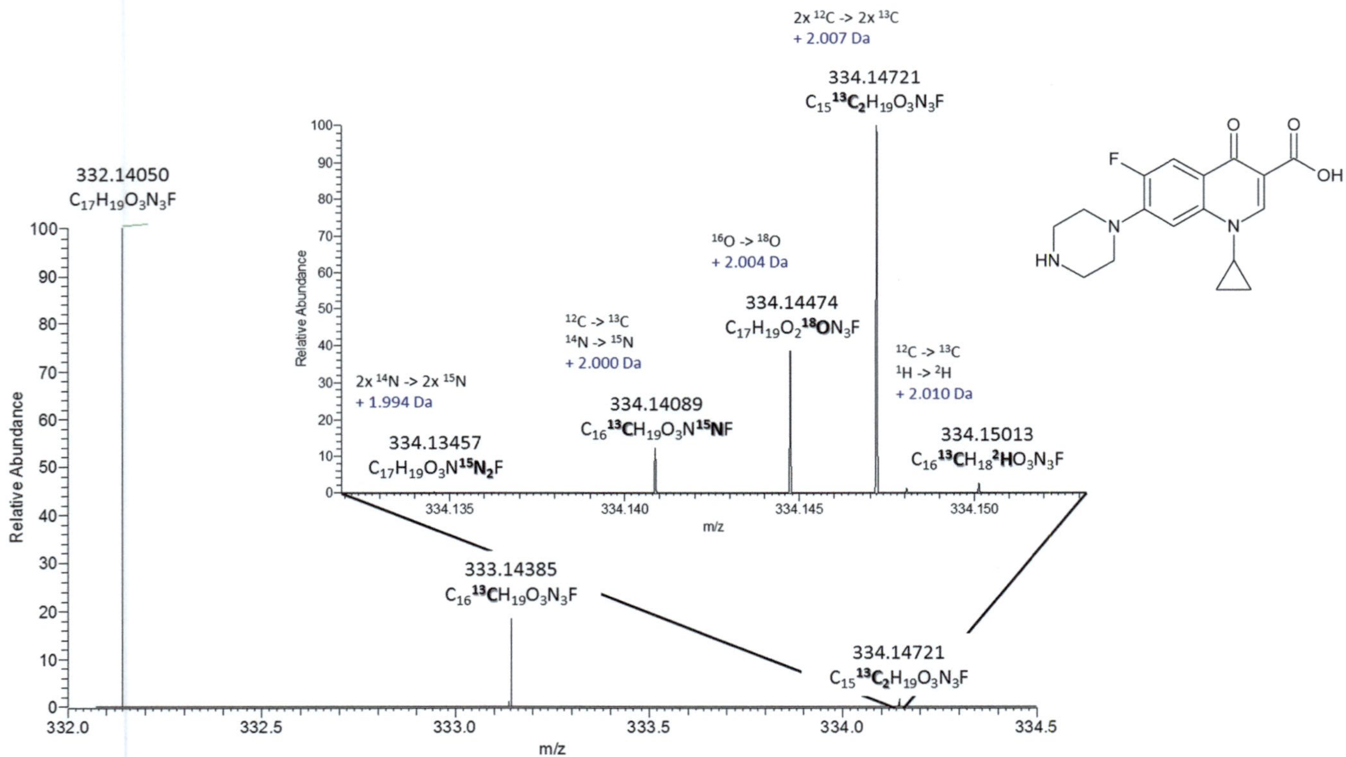

Fig. 10.8 Isotopic fine structure of the M + 2 isotope of ciprofloxacin ($[M + H]^+ = C_{17}H_{19}O_3N_3F$), a fluoroquinolone antibiotic. The relative abundance of the various isotopic signals of an experimental spectrum needs to match the simulated distribution (m/Δm 10,000,000)

Table 10.1 Number of possible molecular formulas for a given mass, as a function of the instrument mass accuracy, and considering isotopic distribution

Without isotope abundance information						2% isotopic abundance accuracy	5% isotopic abundance accuracy
Molecular mass [Da]	10 ppm	5 ppm	3 ppm	1 ppm	0.1 ppm	3 ppm	5 ppm
150	2	1	1	1	1	1	1
200	3	2	2		1	1	1
300	24	11	7	2	1	1	6
400	78	37	23	7	1	2	13
500	266	115	64	21	2	3	33
600	505	257	155	50	5	4	36
700	1046	538	321	108	10	10	97
800	1964	973	599	200	20	13	111
900	3447	1712	1045	345	32	18	196

Only elements C and H used; reproduced with permission – [26]

Table 10.2 Candidate sum formulas for mass 180.1020, the corresponding double bond equivalents (DBE), mass difference in milli-mass units (mmu) and parts per million (ppm) as well as some selected element ratios with the range covered by 99.7% of the chemicals listed in the Wiley mass spectral library (mass from 30 to 1,500 Da)

Sum formula	DBE	Mass [Da]	Difference		H/C[a]	N/C	O/C	F/C	Cl/C	Si/C
			[mmu]	[ppm]	0.2–3.1	0–1.3	0–1.2	0–1.5	0–0.8	0–0.5
$C_{10}H_{14}O_2N$	4.5	180.10191	0.09	**0.526**	**1.3**	**0.1**	**0.2**	–	–	–
$C_6H_{17}O_2N_2P$	0.0	180.10222	−0.22	**−1.198**	**2.7**	**0.3**	**0.3**	–	–	–
$C_5H_{13}O_2N_4F$	1.0	180.10171	0.29	**1.636**	**2.4**	**0.8**	**0.4**	**0.2**	–	–
$C_2H_{14}ON_7Si$	0.5	180.10236	−0.36	−2.005	6.5	3.5	**0.5**	–	–	**0.5**
$C_7H_{17}ON_2Cl$	0.0	180.10239	−0.39	−2.179	**2.3**	**0.3**	**0.1**	–	**0.1**	–
$C_7H_{16}NF_2Si$	0.5	180.10146	0.54	3.004	**2.1**	**0.1**	–	–	–	**0.1**
$C_3H_{14}N_7S$	0.5	180.10259	−0.59	−3.280	4.3	2.3	–	–	–	–
$C_5H_{15}N_5Cl$	0.5	180.10105	0.95	5.276	**2.8**	**1.0**	–	–	**0.2**	–
$H_{12}N_{10}Si$	1.0	180.10102	0.98	5.450	–	–	–	–	–	–
$C_7H_{15}O_3NF$	0.5	180.10305	−1.05	−5.819	**2.0**	**0.1**	**0.4**	**0.1**	–	–

Bold and underlined entries are in agreement with the seven golden rules [24]

[a]Calculated without the hydrogen responsible for the positive charge

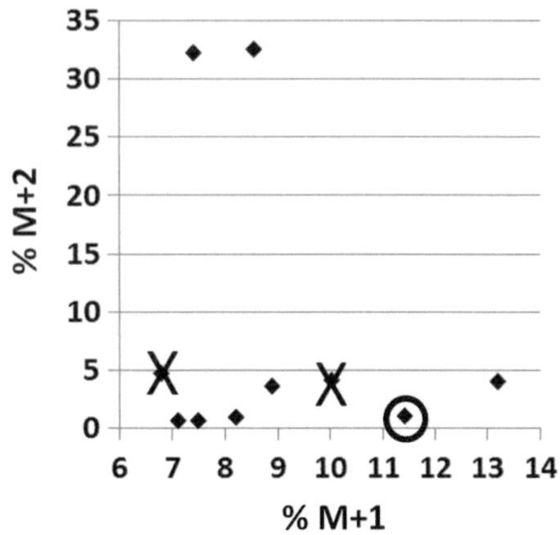

Fig. 10.9 Relative intensities of the M + 1 and M + 2 isotopes of the ten candidate sum formulas. The top candidate is circled. The crossed-out candidates did not comply with Rule 4 and 5. The two candidates containing one chlorine are clearly separated with M + 2 > 30%

10.4 Mass Spectrometry for Characterizing Interactions of Chemicals with Organisms

EDA described above depends on the availability of established bioassays measuring biological responses that can be mapped onto a particular toxicity pathway leading to an apical outcome of interest. However, cellular and physiological changes occurring in response to environmentally relevant chronic exposure to low levels of contaminant mixtures can be rather subtle and not easy to pinpoint with 'traditional' bioassays, as these may suffer from inferior sensitivity. To develop novel bioassays for EDA that cover additional toxicity pathways, in particular those leading to sublethal effects as a result of chronic exposure, a solid understanding of how a chemical or a chemical mixture affects an organism is needed. However, for many chemicals present in the environment the understanding of their effects and mechanisms of toxicity, both individually and in a mixture, is currently incomplete.

Interactions between chemicals and living organisms are typically characterized by two aspects – toxicokinetics and toxicodynamics [27, 28]. Toxicokinetics covers everything that an organism does to and with a chemical, i.e. uptake, biotransformation, and excretion, but also internal distribution to the target sites where the chemical or its metabolite will actually exert its toxic action(s). These are addressed by toxicodynamics, i.e. what a chemical does to an organism. This includes for example, causing DNA adduct formation [29], oxidizing membrane lipids [5], or interacting with a nuclear receptor and by that triggering a molecular cascade resulting in gene expression changes and metabolite alterations [30]. MS-based techniques can be applied to obtain both toxicokinetic and toxicodynamic data.

Internal chemical concentrations of chemicals and their transformation products can be determined for biological specimens with the same approaches used for chemical measurements in environmental compartments [8, 27]. Taking samples at different time points, performing depuration experiments, or carrying out nontargeted or targeted metabolite screening allows constructing time-resolved profiles of chemical uptake, biotransformation, and excretion [31, 32]. Furthermore, MS-imaging techniques such as matrix-assisted laser desorption-ionization

(MALDI) can be used to study the internal distribution of chemicals. This is done by measuring the analyte of interest on histological sections from different body parts and organs. MALDI MS imaging (MALDI MSI) has for instance been successfully applied to study the internal distribution of cocaine in zebrafish larvae. This study showed that, following waterborne exposure, the chemical distributes not only to the expected site of action, the brain, but also to other body sites, notably melanin-containing cells along the body and in the eyes [32]. These data were critical to the interpretation of behavioral changes occurring in this model organism in response to waterborne cocaine, in particular for understanding the differences in responses observed in zebrafish larvae and mammals.

When investigating the toxicodynamics of chemical interaction with the organism, in particular when looking at molecular mechanisms of toxicity, MS can provide data on gene expression and cellular signaling cascades. MS-based approaches are especially useful when looking at proteins (proteomics) and metabolites (metabolomics) [33].

On the protein level, both global and targeted MS-based techniques can be used to characterize the protein abundance as well as post-translational modifications of proteins. For example, Multidimensional Protein Identification Technology (MudPIT) allows the simultaneous characterization of several thousands of protein species as well as their alterations in response to toxicants. Thanks to the advancement of sequencing techniques and the availability of sequencing information, global proteomics analysis can now be carried out not only in the relatively well-characterized species often used in the lab, such as zebrafish (*Danio rerio*) [34] or green alga (*Chlamydomonas reinhardtii*) [5, 35], but also in field-relevant species such as mussels (*Mytilus galloprovincialis*) [4]. Recently, global proteomics was also used to investigate the identity of proteins forming a corona of silver nanoparticles (AgNP) in a gill cell line from rainbow trout (*Oncorhynchus mykiss*) [36]. MS can also be used to study the epigenome and epigenetic alterations, which are often governed by specific interactions between nucleic acids and proteins [37]. One such example are the histone proteins whose post-translational modifications play a crucial role in determining the chromatin state and hence activation or silencing of transcription in particular chromatin regions [38]. Diverse MS-based techniques are being increasingly applied to study these epigenetic marks [39]. Overall, global proteomics is well suited for obtaining initial proteome information needed to investigate diverse ecotoxicological questions in various species.

One important drawback of global proteomics is its general bias towards higher abundance proteins, as this limits its ability to characterize all relevant proteins that could play a role in a particular toxicity mechanism. Meta-analyses performed with mammalian [40] and zebrafish [41] global proteomics data have shown that many proteins commonly reported to be altered in global differential proteomics studies may in fact represent non-specific general stress responses, while more toxicant-specific responses, commonly occurring among the lower-abundance proteins, such as transcription factors or xenobiotic-metabolizing enzymes, cannot be characterized by global proteomics due to insufficient sensitivity.

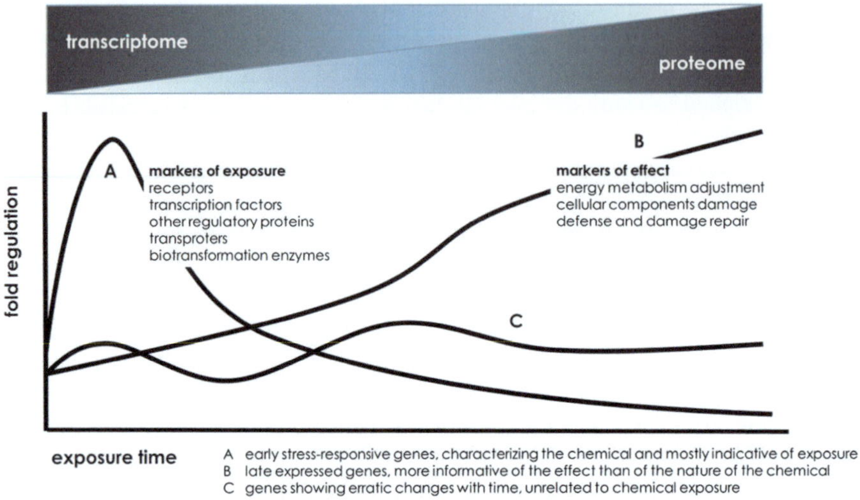

Fig. 10.10 Applicability of transcriptomics and proteomics for studying chemical effects on gene expression (Schematic representation of gene expression trajectories adapted from Van Straalen and Feder [42])

Indeed, many aspects of the general stress response are shared among different species and stressors, since they represent a reaction to any form of macromolecular damage, with common features being manifested once the insult exceeds a certain threshold, regardless of the causative stressor [43]. These features may include gross adjustments in energy metabolism, degradation of cellular components, but also upregulation of defense and damage repair mechanisms. However, in parallel or often even preceding the general stress response developing with an ensuing damage, many other, potentially more stressor-specific responses do occur in the cell. These include rapid gene expression changes, activation or inhibition of specific enzymes or transporters, recruitment of transcription factors, or alterations in the levels of signaling molecules. These multiple toxicity pathways specifically activated in response to different stressors, have a tendency to converge on a few common general stress responses, which will be manifested once the damage becomes too high.

When approached from the gene expression point of view (see Fig. 10.10), the responses that can be grouped under the general stress response umbrella may be considered as markers of effect (e.g. consequences of damaging influences occurring on the cellular level, such as oxidative stress or inhibition of ATP synthesis), while those responses that are stressor-specific may be denoted as markers of exposure. It is this group of responses that may allow gaining detailed insights into specific mechanisms of action for each toxicant, allowing to distinguish between different toxicant groups, but also to predict their toxicity and potency [42]. Meta-analysis of global proteomics studies performed in zebrafish has shown that these datasets tend to over-represent the proteins belonging to the general stress response group, and under-represent those that can be considered to be involved in more specific responses [41]. This bias is a consequence of the overall bias of global proteomics towards better detection of higher abundance proteins. Thus, compared to transcriptomics and sequencing technologies that have a superior sensitivity and can thus deliver information on both markers of exposure and markers of effect, the most commonly applied MS-based proteomics techniques (i.e. global proteomics) appear capable of providing comprehensive information only for the latter group.

However, information on mRNA expression is not a reliable proxy for protein expression [5] due to multiple additional levels of control coming into play after the mRNA synthesis [44]. Therefore, the improvement of sensitivity and versatility of high-throughput proteomics approaches is a crucial prerequisite for improving our ability to study and understand molecular mechanisms of toxicity. Fortunately, sensitivity for protein detection can be significantly improved by using targeted proteomics. With this technique, specific MS/MS transitions of peptides representative of target proteins of interest are monitored, allowing to focus on a few peptides instead of attempting to simultaneously measure all analytes present. Due to significantly shorter MS run durations compared to global approaches, targeted proteomics can also be used to measure panels of select proteins across multiple samples and conditions, thus providing detailed comparative data for proteins of interest. For example, targeted proteomics was used to characterize expression of multiple proteins potentially involved in sex determination and differentiation in zebrafish, at multiple time points during gonad differentiation [45]. Targeted proteomics has also been used to monitor a panel of proteins expected to respond to a particular chemical class, for example glucocorticoids [46], in search of biomarkers of exposure or effects of these chemicals. Furthermore, targeted proteomics methods have been developed for glutathione-S-transferases to study their involvement and significance for phase II biotransformation of xenobiotics in zebrafish [47]. Despite these advances, the sensitivity of MS-based proteomics applied in environmental toxicology currently remains suboptimal and may require further technical advancements, such as the employment of laser microdissection or cellular sorting techniques to allow collecting and analyzing cells of one type, or incorporation of dedicated enrichment workflows to allow increasing the concentration of analytes of interest.

It is commonly accepted that any physiological change in the organism in response to a stressor is accompanied by certain molecular and biochemical alterations, and in fact it can be viewed as an apical adverse outcome of such alterations. Hence, obtaining a better understanding of molecular responses may allow linking to adverse outcomes at higher levels of organization, which may even provide the means for predicting such outcomes. This assumption forms the basis of the adverse outcome pathway (AOP) concept [48] which deals with organizing the knowledge on the progression of toxicity through multiple levels of biological organization, starting with a molecular initiating event and culminating in an adverse outcome of regulatory relevance, occurring on the organism or population level [49]. Data generated by various '-omics' techniques can be used to hypothesize on the mechanisms of action, helping to construct putative AOPs potentially triggered by certain toxicants [50]. However, it is important to realize that a mere demonstration of certain molecular changes is not enough for an AOP, and experimental studies should be designed in a way allowing to understand not only the molecular alterations but ideally also the apical outcomes of relevance perceived to affect the fitness and survival chances of the whole organism [48].

10.5 Conclusions

In conclusion, MS-based techniques have become essential in environmental toxicology and chemistry, (i) for determining chemicals in environmental and biological compartments, (ii) for identifying unknown chemical stressors using effect-directed analysis, scan-dependent and –independent techniques and (iii) for helping to fill the knowledge gaps regarding the molecular mechanisms of toxicity and apical effects caused by a certain chemical. Thus, MS plays an important role in both fundamental and applied research in environmental toxicology, supporting environmental risk assessment and management.

References

1. Bertazzi PA, Bernucci I, Brambilla G, Consonni D, Pesatori AC (1998) The Seveso studies on early and long-term effects of dioxin exposure: a review. Environ Health Perspect 106:625–633
2. Eckermann I (2005) The Bhopal saga – causes and consequences of the world's largest industrial disaster. Universities Press, Hyderabad
3. Giger W (2009) The Rhine red, the fish dead – the 1986 Schweizerhalle disaster, a retrospect and longterm impact assessment. Environ Sci Pollut Res 16:S98–S111
4. Oliveira IB, Groh KJ, Stadnicka-Michalak J, Schonenberger R, Beiras R, Barroso CM, Langford KH, Thomas KV, Suter MJ-F (2016) Tralopyril bioconcentration and effects on the gill proteome of the Mediterranean mussel *Mytilus galloprovincialis*. Aquat Toxicol 177:198–210
5. Pillai S, Behra R, Nestler H, Suter MJ-F, Sigg L, Schirmer K (2014) Linking toxicity and adaptive responses across the transcriptome, proteome, and phenotype of *Chlamydomonas reinhardtii* exposed to silver. Proc Natl Acad Sci 111:3490–3495
6. C.A. Service, CAS assigns the 100 millionth CAS registry number® to a substance designed to treat acute myeloid leukemia, in 2015
7. Schwarzenbach RP, Escher BI, Fenner K, Hofstetter TB, Johnson CA, von Gunten U, Wehrli B (2006) The challenge of micropollutants in aquatic systems. Science 313:1072–1077
8. Ammann AA, Macikova P, Groh KJ, Schirmer K, Suter MJ-F (2014) LC-MS/MS determination of potential endocrine disruptors of cortico signalling in river and waste water. Anal Bioanal Chem 406:7653–7665
9. Kern S, Baumgartner R, Helbling DE, Hollender J, Singer H, Loos MJ, Schwarzenbach RP, Fenner K (2010) A tiered procedure for assessing the formation of biotransformation products of pharmaceuticals and biocides during activated sludge treatment. J Environ Monit 12:2100–2111
10. Ammann AA, Suter MJ-F (2016) Multimode gradient high performance liquid chromatography mass spectrometry method applicable to metabolomics and environmental monitorin. J Chromatogr A 1456:145–151
11. Burkhardt-Holm P, Giger W, Güttinger H, Ochsenbein U, Peter A, Scheurer K, Segner H, Staub E, Suter MJ-F (2005) Where have all the fish gone? The reasons why fish catches in Swiss rivers are declining. Environ Sci Technol 35:441A–447A
12. Friedl C (1999) Fischfangrückgang in schweizerischen Fliessgewässern. Swiss Agency Environ For Landscape (SAFEL) 32
13. Vermeirssen ELM, Korner O, Schoenenberger R, Suter MJ-F, Burkhardt-Holm P (2005) Characterization of environmental estrogens in river water using a three pronged approach: active and passive water sampling and the analysis of accumulated estrogens in the bile of caged fish. Environ Sci Technol 39:8191–8198

14. Brack W (2003) Effect-directed analysis: a promising tool for the identification of organic toxicants in complex mixtures? Anal Bioanal Chem 337:397–407
15. Liedtke A, Schoenenberger R, Eggen RIL, Suter MJ-F (2009) Internal exposure of whitefish (*Coregonus lavaretus*) to estrogens. Aquat Toxicol 93:158–165
16. Thomas KV, Hurst MR, Matthiessen P, Waldock MJ (2001) Characterization of estrogenic compounds in water samples collected from United Kingdom estuaries. Environ Toxicol Chem 20:2165–2170
17. Bandow N, Altenburger R, Streck G, Brack W (2009) Effect-directed analysis of contaminated sediments with partition-based dosing using green algae cell multiplication inhibition. Environ Sci Technol 43:7343–7349
18. Bernet D, Wahli T, Kueng C, Segner H (2004) Frequent and unexplained gonadal abnormalities in whitefish (central alpine coregonus sp.) from an alpine oligotrophic lake in Switzerland. Dis Aquat Org 61:137–148
19. Schymanski EL, Singer HP, Longrée P, Loos M, Ruff M, Stravs MA, Ripollés Vidal C, Hollender J (2014) Strategies to characterize polar organic contamination in wastewater: exploring the capability of high resolution mass spectrometry. Environ Sci Technol 48:1811–1818
20. Singer HP, Wössner A, McArdell CS, Fenner K (2016) Rapid screening for exposure to "non-target" pharmaceuticals from wastewater effluents by combining HRMS-based suspect screening and exposure modeling. Environ Sci Technol 50:6698–6707
21. Liedtke A, Schoenenberger R, Eggen RIL, Suter MJ-F (2009) unpublished results
22. Soto AM, Sonnenschein C, Chung KL, Fernandez MF, Olea N, Serrano FO (1995) The e-screen assay as a tool to identify estrogens – an update on estrogenic environmental-pollutants. Environ Health Perspect 103:113–122
23. Gillet LC, Navarro P, Tate S, Röst H, Selevsek N, Reiter L, Bonner R, Aebersold R (2012) Targeted data extraction of the MS/MS spectra generated by data-independent acquisition: a new concept for consistent and accurate proteome analysis. Mol Cell Proteomics 11: PMID 22261725
24. Kind T, Fiehn O (2007) Seven golden rules for heuristic filtering of molecular formulas obtained by accurate mass spectrometry. BMC Bioinformatics 8:1–20
25. Wiley, Wiley Registry of Mass Spectral Data, 2016
26. Kind T, Fiehn O (2006) Metabolomic database annotations via query of elemental compositions: mass accuracy is insufficient even at less than 1 ppm. BMC Bioinformatics 7:234
27. Hidasi AO, Groh KJ, Suter MJ-F, Schirmer K (2017) Clobetasol propionate causes immunosuppression in zebrafish (*Danio rerio*) at environmentally relevant concentrations, ecotox. Ecotox Environ Safe 138:16–24
28. Ashauer R, Agatz A, Albert C, Ducrot V, Galic N, Hendriks J, Jager T, Kretschmann A, O'Connor I, Rubach MN, Nyman AM, Schmitt W, Stadnicka J, van den Brink PJ, Preuss TJ (2011) Toxicokinetic-toxicodynamic modeling of quantal and graded sublethal endpoints: a brief discussion of concepts. Environ Toxicol Chem 30:2519–2524
29. Madureira DJ, Weiss FT, Van Midwoud P, Helbling DE, Sturla SJ, Schirmer K (2014) Systems toxicology approach to understand the kinetics of benzo(a)pyrene uptake, biotransformation, and DNA adduct formation in a liver cell model. Chem Res Toxicol 27:443–453
30. Mottaz H, Schoenenberger R, Fischer S, Eggen RIL, Schirmer K, Groh KJ (2017) Dose-dependent effects of morphine on lipopolysaccharide (LPS)-induced inflammation, and involvement of multixenobiotic resistance (MXR) transporters in LPS efflux in teleost fish. Environ Pollut 221:105–115
31. Di Paolo C, Groh KJ, Zenneg M, Vermeirssen E, Murk A, Eggen RIL, Hollert H, Werner I, Schirmer K (2015) Early life exposure to PCB126 results in delayed mortality and growth impairment in the zebrafish larvae. Aquat Toxicol 169:168–178
32. Kirla KT, Groh KJ, Steuer AE, Poetzsch M, Banote RK, Stadnicka-Michalak J, Eggen RI, Schirmer K, Kraemer T (2016) Zebrafish larvae are insensitive to stimulation by cocaine: importance of exposure route and toxicokinetics. Toxicol Sci 154:183–193

33. Groh KJ, Suter MJ-F (2014) Mass spectrometry in environmental toxicology. Chimia (Aarau) 68:140–145
34. Groh KJ, Nesatyy VJ, Segner H, Eggen RI, Suter MJ-F (2011) Global proteomics analysis of testis and ovary in adult zebrafish (*Danio rerio*). Fish Physiol Biochem 37:619–647
35. Nestler H, Groh KJ, Schoenenberger R, Eggen RI, Suter MJ-F (2012) Linking proteome responses with physiological and biochemical effects in herbicide-exposed *Chlamydomonas reinhardtii*. J Proteome 75:5370–5385
36. Yue Y, Behra R, Sigg L, Suter MJ-F, Pillai S, Schirmer K (2016) Silver nanoparticle-protein interactions in intact rainbow trout gill cells. Environ Sci Nano 3:1174–1185
37. Vandegehuchte MB, Janssen CR (2011) Epigenetics and its implications for ecotoxicolog. Ecotoxicology 20:607–624
38. Bannister AJ, T. Kouzarides T (2011) Regulation of chromatin by histone modifications. Cell Res 21: 381-395
39. Brunner AM, Tweedie-Cullen RY, Mansuy IM (2012) Epigenetic modifications of the neuroproteome. Proteomics 12:2404–2420
40. Petrak J, Ivanek R, Toman O, Cmejla R, Cmejlova J, Vyoral D, Zivny J, Vulpe CD (2008) Deja vu in proteomics: a hit parade of repeatedly identified differentially expressed proteins. Proteomics 8:1744–1749
41. Groh KJ, Suter MJ-F (2015) Stressor-induced proteome alterations in zebrafish: a meta-analysis of response patterns. Aquat Toxicol 159:1–12
42. van Straalen NM, Feder ME (2012) Ecological and evolutionary functional genomics–how can it contribute to the risk assessment of chemicals? Environ Sci Technol 46: 3-9
43. Kultz D (2005) Molecular and evolutionary basis of the cellular stress response. Annu Rev Physiol 67:225–257
44. Schwanhaeusser B, Busse D, Li N, Dittmar G, Schuchhardt J, Wolf J, Chen W, Selbach M (2012) Global quantification of mammalian gene expression control. Nature 473:337–342
45. Groh KJ, Schoenenberger R, Eggen RI, Segner H, Suter MJ-F (2013) Analysis of protein expression in zebrafish during gonad differentiation by targeted proteomics. Gen Comp Endocrinol 193:210–220
46. Hidasi AO (2016) Personal communication, Eawag
47. Tierbach A (2016) Personal communication, Eawag
48. Groh KJ, Carvalho RN, Chipman JK, Denslow ND, Halder M, Murphy CA, Roelofs D, Rolaki A, Schirmer K, Watanabe KH (2015) Development and application of the adverse outcome pathway framework for understanding and predicting chronic toxicity: I. Challenges and research needs in ecotoxicology. Chemosphere 120:764–777
49. Ankley GT, Bennett RS, Erickson RJ, Hoff DJ, Hornung MW, Johnson RD, Mount DR, Nichols JW, Russom CL, Schmieder PK, Serrano PK, Tietge JE, Villeneuve DL (2010) Adverse outcome pathways: a conceptual framework to support ecotoxicology research and risk assessment. Environ Toxicol Chem 29:730–741
50. Sturla SJ, Boobis AR, FitzGerald RE, Hoeng J, Kavlock RJ, Schirmer K, Whelan M, Wilks MF, Peitsch MC (2014) Systems toxicology: from basic research to risk assessment. Chem Res Toxicol 27:314–329

Chapter 11
Enhancing the Analysis of Complex Lipid Samples Through Developments in Chromatography and Chemical Derivatization

Samuel W.J. Shields, Carlos R. Canez, Karl V. Wasslen, Hyunmin Lee, Danisz Stalinski, Lennart Trouborst, Shira Joudan, Sarah Whitton, Hillary P. Weinert, Jeffrey M. Manthorpe, and Jeffrey C. Smith

Abstract Lipids are one of the main building blocks of cellular life and represent the most abundant class of biomolecule by weight. Some of their most obvious roles include providing a source of energy storage within cells and creating barriers to compartmentalize cells by forming boundaries between cells and organelles. More recently, the roles that lipids and lipid metabolites play in cellular signaling have been increasingly revealed by the scientific community, fueling interest in this field. Chromatography coupled to mass spectrometry has emerged as an eminent analytical tool with which to study lipids, however not without challenges along the way. This chapter first summarizes some of the recent developments in the use of chromatography to separate complex lipid samples in an efficacious manner including the presentation of new techniques to increase precision and avoid sample carry-over. The latter part of the chapter details recent strategies to chemically derivatize lipids to enhance their analytical characteristics for separation and analysis by mass spectrometry with the aim to gain greater insight into the roles that these biomolecules play in living systems.

Although this chapter largely summarizes previously published work from a variety of sources, some of the information presented here represents novel research that has been previously unpublished; credit for this research falls to the authors listed above.

S.W.J. Shields • C.R. Canez • K.V. Wasslen • H. Lee • D. Stalinski • L. Trouborst • S. Joudan
S. Whitton • H.P. Weinert • J.M. Manthorpe • J.C. Smith (✉)
Department of Chemistry, Carleton University, Ottawa, ON K1S 5B6, Canada
e-mail: JeffCSmith@cunet.carleton.ca

J.H. Banoub, R.M. Caprioli (eds.), *Molecular Technologies for Detection of Chemical and Biological Agents*, NATO Science for Peace and Security Series A: Chemistry and Biology, DOI 10.1007/978-94-024-1113-3_11

11.1 Lipidomics, Lipid Structure and Function

Lipidomics is an ever-expanding field that studies lipids and their interactions within biological systems [1]. A large portion of the field is devoted to the structural characterization and quantitation of lipids to gain insight into their function. The goal of lipidomics research is to obtain the full characterization of lipid species and their specific biological roles within the context of all other other cellular mechanisms (e.g. gene transcription, protein translation and other metabolomic processes). This goal can be realized by several different methods but few as sensitive and powerful as mass spectrometry (MS). MS has been used in lipidomics research to find and identify potential lipid biomarkers, compare two or more lipid profiles, discover new lipids and lipid metabolites and provide quantitative information on these analytes in the context of complex biological systems [1].

Abnormal lipid behavior in humans has been implicated in a variety of pathologies including diabetes, cancer, and some neurodegenerative diseases. Apart from being energy storage molecules, lipids are involved in many biochemical functions related to disease states such as: signal transduction pathways, proliferation, apoptosis and membrane trafficking in the cell [2–5]. Accordingly, the field of lipidomics is of great interest to many factions of the health sciences.

Lipids are generally non-polar or amphiphilic molecules and usually possess large hydrocarbon (non-polar) chains and in some cases a polar head group (e.g. carboxyl, phosphate, amino, or carbohydrates). Figure 11.1 depicts some of the most common lipid classes. Although lipids are composed of very simple building blocks they have the potential for ~100,000 unique molecular species [6]. With each of the different lipid structures and classes having important biological roles to play, lipidomics has become one of the leading fields in health research with the potential for many significant clinical applications.

Each structural variant of lipids can have many different functions within a biological system. For example, fatty acyl lipids can be used in fatty acid oxidation for the generation of energy for cells but also are the main building block for other higher order biomolecules [7]. *In vivo*, fatty acyl chains can be converted to mono-, di-, or tri- acylglycerols which are the main long term energy storage biomolecules for mammals (Scheme 11.1). Along the same pathway glycerophospholipids can also be produced from the 1,2-diacyl glycerols. Glycerphospholipids are comprised of a polar phosphate head group and two fatty acyl chains labeled as *sn*-1 and *sn*-2 (Fig. 11.1) and are subdivided into sub-classes depending on the phosphate head group and the nature of the fatty acyl chains. Phospholipids make up ~60% of a cells lipidome with phosphatidylethanolamine (PE) and phosphatidylcholine (PC) classes composing 75 mol% of eukaryotic membranes [8].

Two other glycerophospholipids that play important roles in nervous system and cardiovascular system cellular membranes are sphingolipids and plasmalogens. Sphingomyelins (SM) are characterized by having a polar phosphate group, a fatty N-acyl group at the *sn*-2 position, and a fatty allylic alcohol moiety at *sn*-1. These

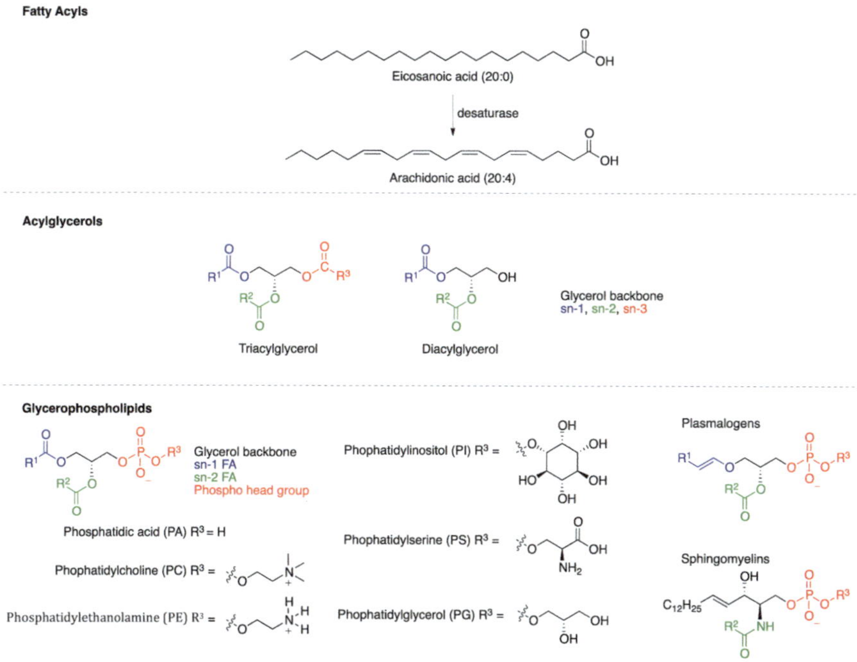

Fig. 11.1 Some common lipid classes

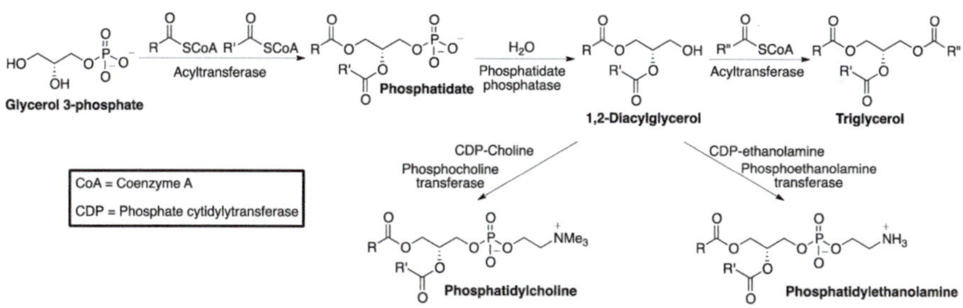

Scheme 11.1 Biosynthesis of the most abundant lipids in humans [7]

lipids make up a large percentage of the myelin sheath of neurons that insulates the axon allowing for signal transduction [9]. Abnormalities in sphingolipid composition of neuronal cells can lead to neurodegenerative disorders like multiple sclerosis [9]. Plasmalogens are characterized by having a polar phosphate group, a vinylic fatty ether component at *sn*-1 position, and a fatty acyl substituent at the *sn*-2 position. Aberrant plasmalogen behavior can be linked to Alzheimer's disease [10]. It is hypothesized that many other lipids play roles in disease states in the human body and therefore new technologies and analytical techniques are required to accurately determine lipid structure and ascertain the composition of complex lipidomics in order to progress towards a full understanding of the diseases that plague humankind.

11.2 Lipid Extraction, Separation and Analysis

To analyze lipids and gain insight into cellular function, they must first be extracted from a biological sample. Biological samples contain many interfering compounds such as carbohydrates, nucleic acids, proteins, and salts that must be first separated from the lipids. The most common approach to extract lipids is to use some variation of a micro-scale liquid-liquid extraction. Briefly, the biological sample is homogenized and suspended in aqueous and alcoholic media with a non-polar solvent (e.g. chloroform, ethanol, benzene) to form a monophasic system. The mixture is then agitated vigorously and centrifuged at high speeds to obtain a biphasic system. The non-polar layer can then be removed and the remaining aqueous layer re-extracted with more non-polar solvent [11, 12]. The most common solvent extraction technique that follows these steps is the Bligh and Dyer extraction and the resulting extract is often complex and contains a wide variety of different lipids. Variations on the Bligh and Dyer extraction include modifications designed to remove some lipid classes over others to obtain a more simplified sample, while others control pH using buffers or alter the ratio of alcohol to water to modulate the disruption of lipid-protein complexes [13].

11.2.1 MS Analysis of Lipid Samples

MS is a powerful tool that may be used to analyze complex lipid samples containing hundreds to thousands of analytes. Electrospray ionization (ESI) and atmospheric pressure chemical ionization (APCI) have demonstrated their utility to ionize non-volatile lipid analytes and are most commonly cited in the lipidomics literature [13, 14]. In the case of glycerophospholipids, triple quadrupole or hybrid quadrupole instruments are used to analyze the composition of the lipidome by single MS scans and tandem MS analysis (MS/MS or MS^2). MS^2 analysis helps to provide information on the fatty acyl components (product ion scan in negative ion mode), the polar head group (product ion scan in positive ion mode), and analyze the composition of a particular lipid class (precursor and neutral loss scans). There are many challenges in analyzing lipids via MS such as their amphipathic nature, low proton affinity, vast and often subtle structural diversity, wide range of abundances and low relative abundance of some lipid biomarkers.

In the case of a triple quadrupole instrument, MS^2 data is collected by selecting a small bandwidth (~1 Da) of ions in the first quadrupole (Q1), colliding these ions in the second quadrupole (q2) with an inert gas to induce fragmentation (collision induced dissociation (CID)) and scanning the *m/z* values of the resulting fragments in the third quadrupole (Q3). There are four different MS^2 scanning modes that this instrument geometry affords that are most commonly employed in lipidomics analyses, product ion scanning, precursor ion scanning, neutral loss scanning and multiple reaction monitoring (MRM). Production ion scanning uses Q1 to select a single ion of interest and fragment it in q2 via CID where Q3 scans and reports the *m/z* values of the resulting fragments. The mass spectrum of a product ion scan displays the ion fragments that originate from a single ion in the sample, which provides structural information about the molecule. Precursor ion scanning is the inverse of product ion scanning in that Q1 scans a range of *m/z* ratios that in turn are fragmented in q2 whereas Q3 is set to select only one of the fragment ions. The final mass spectrum displays the parent ions that fragmented in Q2 to yield a specific *m/z* fragment that passed through Q3. Neutral loss scanning employs both Q1 and Q3 scanning a range of *m/z* values offset by a constant amount; ions that fragment in q2 and lose a specific neutral mass will successfully transmit through Q3. The final mass spectrum will display the *m/z* of ions that passed through Q1 and fragmented to lose a specific neutral mass. MRM does not employ *m/z* scanning and fixes Q1 and Q3 on specific *m/z* values, ions at a specific *m/z* value pass through Q1, fragment in q2 and a specific fragment passes through Q3. MRM is a highly sensitive scanning mode as the instrument duty cycle is focused on one or more specific ion-ion fragment transitions rather than scanning for *m/z* values that may not exist in the sample. MRM simply displays the intensity of the ion signal for each transition and therefore is most useful when coupled with chromatography. Lipids extracts are most commonly analyzed by product ion scans to help probe structural features (head group and fatty acyl chains), and by precursor ion scans to simplify complex biological samples to just specific lipid classes.

Figure 11.2 illustrates the general workflow for a lipidomics experiment initiated with lipid extraction leading to analysis via MS. The use of chromatography for further separation of lipid extracts prior to MS analysis is debated in the literature [14]. Lipid extracts are complex and warrant chromatographic separation; however, complications with lipid chromatography are non-trivial.

In the case of glycerophospholipids, analyte molecules are amphipathic which means they possess properties that allow adsorption to both polar and non-polar surfaces. As a result, some literature suggests that direct infusion of complex lipid extracts into a mass spectrometer via electrospray ionization (ESI) is the best course of action [14]. Direct infusion minimizes the exposure that a lipid sample has to capillary tubing leading up to ESI, decreasing sample loss and non-specific adsorption to sample handling surfaces. Direct infusion necessitates the need for high resolution MS to distinguish lipids of similar mass and literature has emerged that has demonstrated the analysis of thousands of lipids from a single extract via direction infusion high resolution MS [14]. Some of the limitations of direct infusion however include the fact that high resolution MS is not always available and that many

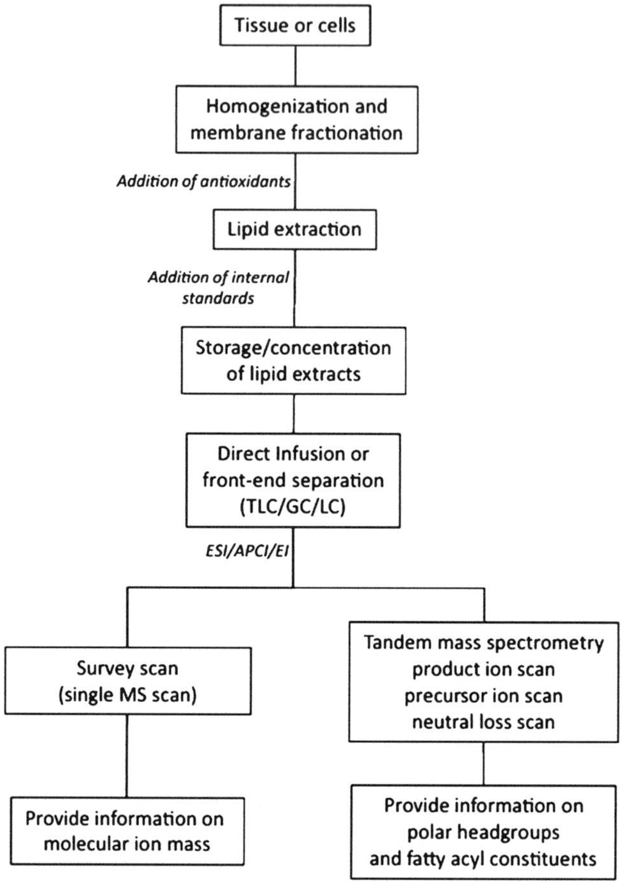

Fig. 11.2 A general workflow for lipid analysis [13]

isobaric lipid species exist that may be chromatographically separable but are indistinguishable based on mass alone. Furthermore, chromatography offers the benefit of desalting a sample and reducing the number of analytes being simultaneously ionized via ESI. Salts such as sodium are prevalent in biological samples and persist in lipid preparations; the presence of salt will lead to signal splitting in the mass spectrum as lipid analytes may be protonated or cationized by a salt to achieve ionization. ESI operates on the principle that analytes are ionized based on having a higher proton affinity in comparison to the solvent within which they are contained; when samples are increasingly complex, ion suppression occurs as analytes of higher proton affinity are protonated while analytes of lower proton affinity are not and remain neutral [14]. Chromatography remediates both of these problems. By binding lipid analytes to a stationary phase, salts are removed and analytes are eluted in order of increasing hydrophobicity reducing sample complexity during the ESI process and remediating ion suppression. Furthermore, chromatographic separation concentrates analytes into a narrow elution window and therefore increases the sensitivity of analyses; quantitation is easily achieved by integrating the area under the peak produced from MS signal as the peak elutes from the column. These benefits are the main driving force behind the use of chromatography in most HPLC-ESI-MS applications. However, the amphipathic nature of lipids causes considerable sample carryover that interferes with the precision of analyses and for this reason many research groups in this field have discontinued the use of chromatography. In light of this, we collected data to document the extent to which precision is hampered in HPLC-ESI-MS analyses of complex lipid samples and investigated novel chromatographic strategies to remediate these issues so that the fundamental benefits of using chromatography may be realized in the field of lipidomics.

11.2.2 Optimization of Lipid Chromatography

The vast majority of chromatographic separations are conducted using an octadecyl or C_{18} functionality covalently bonded to the stationary phase and a mixture of aqueous and less polar organic mobile phases to create a reversed phase separation. Within this chromatography scheme, analytes partition to the less polar stationary phase and are eluted either over time isocratically or when the eluent strength of the mobile phase is altered in a gradient to collapse the sample and elute it as a narrow band of eluate. The columns and hydraulic path of an HPLC are generally a mixture of stainless steel, polyetheretherketone and/or fused silica tubing. Therefore, lipid analytes are exposed to both polar surfaces within the hydraulic path as well as non-polar surfaces on the stationary phase. In the case of glycerophospholipids, the separation capabilities of liquid chromatography are complicated by the amphipathic nature of lipid molecules as adsorption to any component of the hydraulic path via their polar head groups or hydrophobic tails is possible. To demonstrate this principle and the extent to which sample carryover is a result of polar interactions with the hydraulic path, we conducted several experiments where the column was removed in a reversed phase HPLC separation and sample carryover was observed, Fig. 11.3.

The four panels in Fig. 11.3 demonstrate that glycerophospholipids (PC and SM) have a propensity to adsorb to the hydraulic path of an HPLC. Although the elution time for all species is within the first 15 min of the solvent gradient (Fig. 11.3a), the lipids had persisted despite several volumes of column washing following injection onto the empty column. The lipids shown in Fig. 11.3b are an average of the entire chromatographic run and demonstrate that dozens of phosphocholine-containing species elute from the column. The intensity of these peaks are roughly two orders of magnitude higher when examining an individual time point, their intensities have been diminished by averaging the signal at every *m/z* over 60 min. Following the acquisition of this data, a 20-min wash of 100% acetonitrile was flushed through the system. Figures 11.3c and 11.3d repeat this experiment, injecting distilled water instead of lipid extract and demonstrate the extent of sample carryover despite

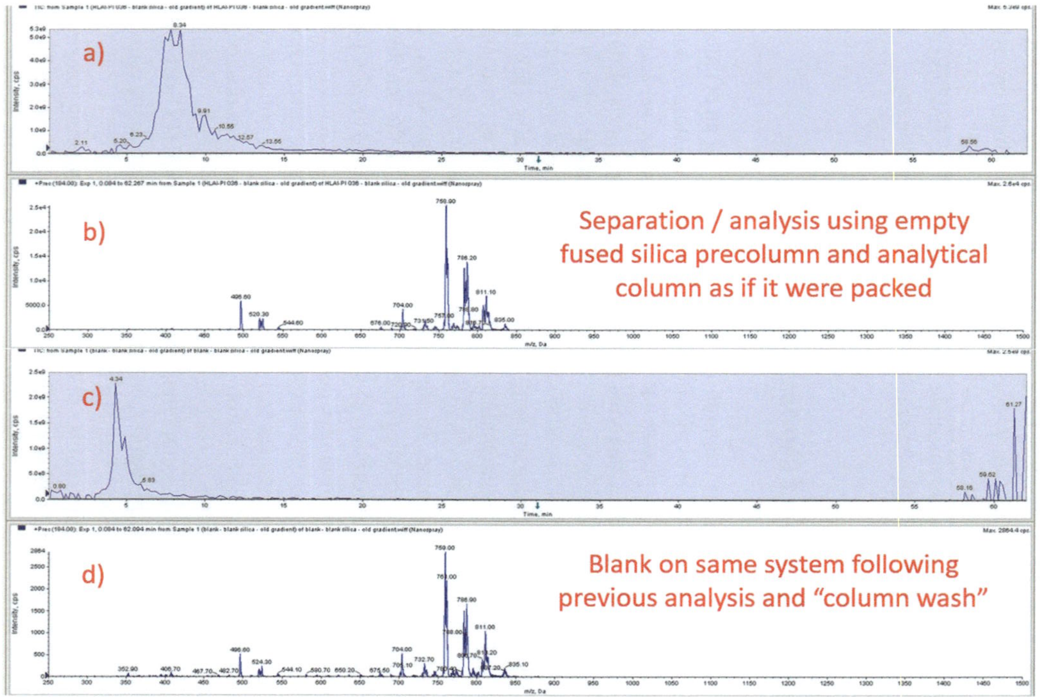

Fig. 11.3 Observation of lipid carryover within an HPLC system that has no column attached. Polar lipids were extracted from PC12 cells using the Bligh and Dyer method and 20 μL were injected through an HPLC where the column had been replaced by a 75 μm × 10 cm length of empty fused silica. The sample was eluted from the tubing using a reversed phase gradient (100% H_2O to 100% acetonitrile, both with 0.1% formic acid) and analyzed via ESI MS using a precursor ion scan for *m/z* 184 in positive ion mode, highlighting all PC and SM species present. (**a**) the ion current over the 60 min from the initial injection of lipid extract, (**b**) the average precursor ion mass spectrum over the course of the entire run, (**c**) the ion current over a 60-min gradient after subsequently injecting 20 μL of water through the system, (**d**) the average precursor ion mass spectrum over the course of the entire blank run

injecting a blank and conducting a "column wash" (in quotations since there was actually no column present). These results demonstrate the extent to which polar lipids may adsorb to fused silica and stainless steel and highlight the challenge of separating amphipathic analytes. Interestingly, amphipathic cationic surfactants such as didodecyldimethylammonium are intentionally used to bind to the surface of fused silica in capillary electrophoresis (CE) and create a bilayer that reverses the charge on the wall of the capillary and therefore the flow of the electrolyte through the column [15]. The parallels between this experimental practice in CE and the injection of polar lipids in a chromatographic system are striking; it is of little surprise that polar lipids such as PC and SM adsorb strongly to fused silica. The challenge in this result is that the methods that were used to obtain the data in Fig. 11.3 mirrored the most commonly used mobile phase system that appeared in the literature for lipid separation at the time. Due to these reasons the field as a whole had begun to move towards direct injection for MS analysis to avoid carryover issues. Sample carryover and chromatographic performance was made even worse when a reversed phase column was set within the hydraulic path, Figs. 11.4 and 11.5.

Figure 11.4 demonstrates the performance of both C_{18} and C_4 reversed phase chromatography with lipid analytes by highlighting the elution profile for *m/z* 524 via a precursor ion scan. The panels a) and b) of Fig. 11.4 reveal a broad tailing band for this peak with the C_4 resin allowing the lipid to elute earlier than the C_{18} resin due to relatively weaker hydrophobic interactions between the analyte and stationary phase. Even on a new column, the chromatographic performance is less than desirable with bandwidths in excess of 10 min. The performance is compromised further with multiple injections however, as revealed in panel c) of Fig. 11.4. Despite washing the column with 100% acetonitrile and 0.1% formic acid for 20 min between analyses, carryover is so significant that chromatographic performance is essentially lost. The precision of replicate injections using this quality of separation is low to the point that extrapolation of biological meaning from a dataset is highly unlikely. To demonstrate the amount of lipid extract that remained on the column between runs, a blank run was conducted after three injections on a C_{18} column, Fig. 11.5.

As revealed in Fig. 11.5, adding a stationary phase to the chromatographic separation complicated the carryover issues further compared to Fig. 11.3. The peak at *m/z* 807 was chosen randomly as an example of a lipid analyte that demonstrated persistent ion current throughout the entire separation, other lipids showed the same separation behavior with XICs having significant intensities throughout the entire run (data not shown). The poor chromatography and carryover hampered the precision of this analytical method and ultimately rendered datasets impossible to interpret. To exemplify this, lipids from PC12 cells were extracted four times and each extract was analyzed in triplicate, Fig. 11.6.

The complications of carryover and poor chromatography are illustrated in Fig. 11.6 as precision is hampered to the point of obscuring any biological trend and

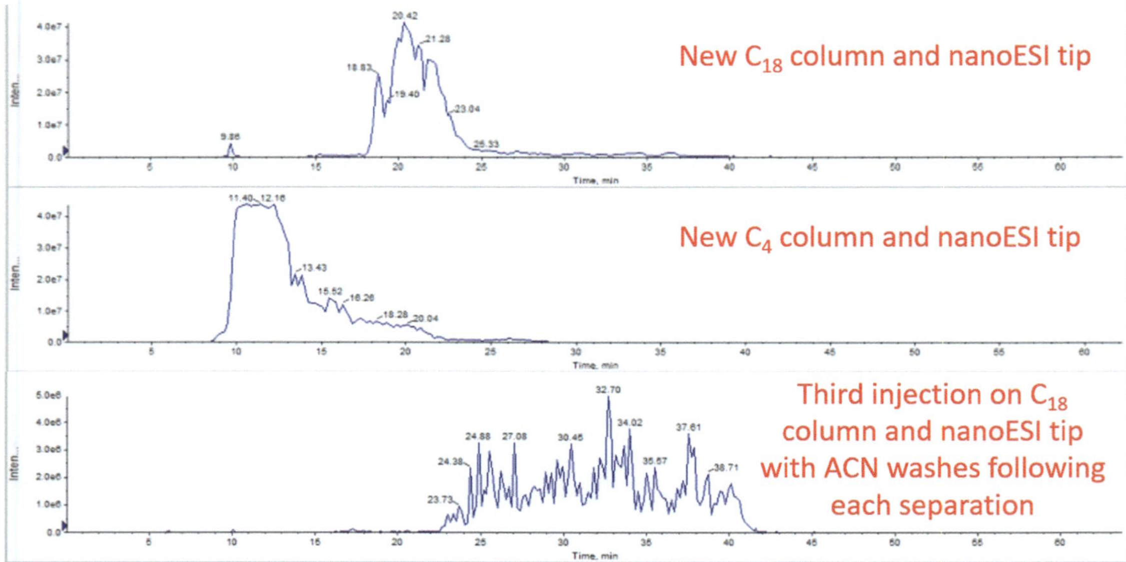

Fig. 11.4 Observation of lipid chromatography within an HPLC system equipped with 5 cm × 200 μm reversed phase columns. Lipids were extracted from PC12 cells and separated as described in Fig. 11.3. A precursor ion scan of *m/z* 184 revealed many different phosphocholine-containing lipid species and *m/z* 524 was selected to demonstrate chromatographic performance. (**a**) the extracted ion current (XIC) for *m/z* 524 over a 60 min separation using a new C_{18} column reveals an approximately 8 min elution window and some tailing, (**b**) the experiment was repeated using a new C_4 column and *m/z* 524 is observed to elute over a longer time period but emerge from the column at an earlier timepoint, (**c**) XIC for *m/z* 524 on a used C_{18} column after two previous separations of lipid extract. The chromatographic performance observed in all panels is poor and significant deterioration of chromatographic performance is observed following multiple injections of sample compared to the first injection on a new column

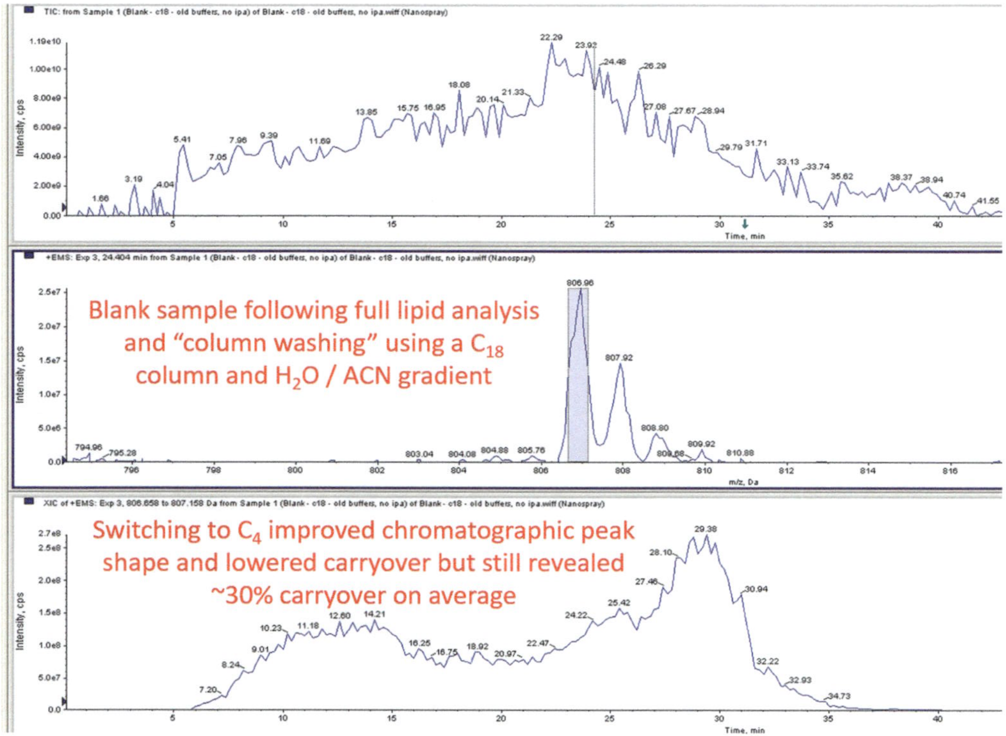

Fig. 11.5 Observation of lipid carryover within an HPLC system equipped with a C_{18} column. Water was injected and separated as described in Fig. 11.3. (**a**) the total ion current over the 60 min separation, (**b**) enhanced mass scan (EMS) at 24.4 min revealing a prominent lipid peak at *m/z* 807, (**c**) XIC of *m/z* 807 over the course of the experiment. Signal is observed for the majority of the experiment indicating extremely low quality chromatographic performance and significant carryover. The chromatography was slightly improved by switching to a C_4 stationary phase yet still demonstrated poor chromatographic performance and an average carryover of approximately 30% (data not shown)

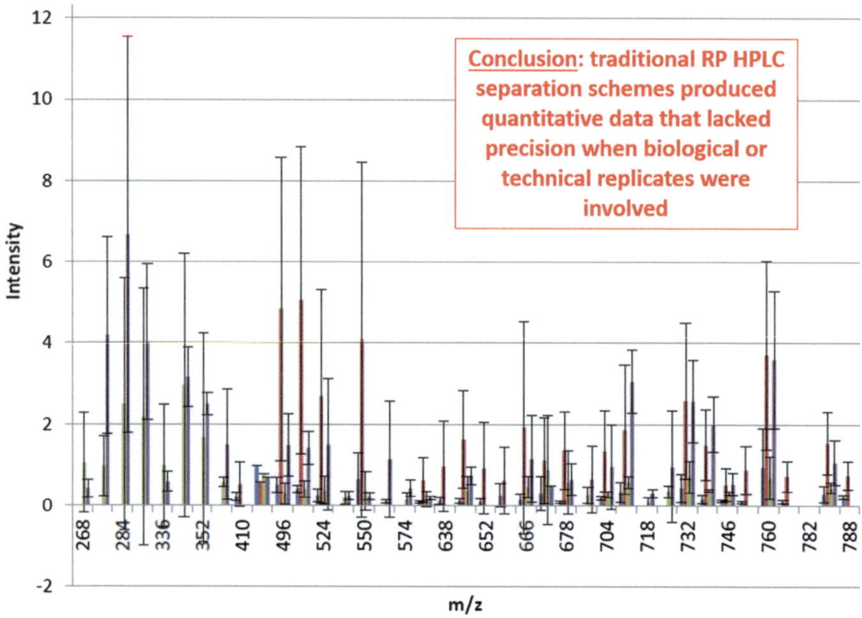

Fig. 11.6 The lipids from PC12 cells were extracted four times and each extract was analyzed in triplicate as described in Fig. 11.3. MS analysis included a precursor ion scan for *m/z* 184 to detect phosphocholine-containing lipids. The figure indicates the *m/z* values of PC and SM lipids from the extract and indicates their intensities normalized to an internal standard that was spiked in at constant concentration. Lipids from different extractions are indicated by the *colour* of the bar in the histogram, standard deviations on the measured relative intensities are indicated by the error bars for each lipid. Each extraction is identical and the intensities of each lipid should be consistent; the differences in intensities with significant error bars highlights the deficiencies that were prevalent in the lipid chromatography that was used by the field at that time

defeating the purpose of the experiment. In the field, some lipidomics laboratories had reverted to the costly practice of changing columns for every analysis to avoid stationary phase carryover or randomizing samples to condition chromatographic pathways with consistent carryover with the aim that precision would be negatively affected, but not completely compromised (personal communications). Most in the field however, switched to direct infusion of lipid extracts for MS analysis and avoided chromatography altogether.

To our mind however, the potential benefits of using chromatography were too great to overlook and our team conducted a series of experiments where different mobile phases were tested in our microflow reversed phase chromatographic system to remediate the poor chromatographic separation of lipid extracts that had been observed until this point in time. The optimization data is beyond the scope of this chapter; however, optimized conditions will be described and presented below.

In order to eliminate carryover of an amphipathic analyte, the mobile phase that is used must also possess amphipathic character in that it must have a polar character as well as a non-polar character. To complicate this further, the mobile phase must be compatible with ESI meaning any salt must be volatile and not contribute to or interfere with the mass spectrum being generated. Moreover, any salt must not adduct to the lipid analyte and split the signal between different cationization (e.g. Na^+, K^+, Li^+) states (chromatography is used to remove salts). The polar character of the mobile phase must be tuned to create ion pairs with the analytes and disrupt any hydrophilic activity, specifically between the lipids and the chromatographic hydraulic path. The non-polar character of the mobile phase must be tuned to disrupt hydrophobic adsorption of the lipids to the stationary phase and eliminate carryover by cleaning the column effectively between separations.

Many mobile phase compositions were tested for their effect on chromatographic performance, all of them being a non-polar solvent containing a polar modifier. Ammonium-based salts were investigated for their effect on the system and it was determined that 10 mM ammonium acetate was the most effective at disrupting hydrophilic interactions within the system. Higher concentrations of this salt did not yield an increase in performance and as the concentration was increased the likelihood of flow stoppages increased as well, particularly in the ESI tip. It was also determined that 10 mM ammonium acetate was beneficial to add to all mobile phases to continually disrupt polar interactions throughout the separation. Several modifications were made to the non-polar solvents that were used for analyte elution as well as system cleaning following analyses. The first modification was made following a series of experiments looking at the solubility of lipid extracts in aqueous buffers operating on the hypothesis that lipid analytes dissolved in ethanol injected into a column running aqueous buffers may render the lipids insoluble through the formation of micelles or aggregates. Although the nature of the insolubility was not determined, it was concluded that chromatographic performance diminished when the mobile phase solvents were composed of anything greater than 70% water. Furthermore, it was determined that no sample was lost via immediate elution at 70% water (not being retained by the column). Ultimately, it was determined that the best mobile phase starting condition (buffer A) was 30% methanol in water with 10 mM ammonium acetate. A variety of solvents were tested to elute lipid analytes including 100% acetonitrile or methanol but the carryover was significant demonstrating insufficient eluent strength for these solvents. Solvents with no polarity whatsoever were evaluated for lipid elution as well (e.g. hexanes) and it was observed that chromatographic elution was effective however ESI performance was deleteriously affected. Furthermore, the immiscibility of hexanes with aqueous buffers limited their use in a binary solvent system. Ultimately what was determined to elute the lipids most effectively was a ternary solvent system that provided excellent chromatographic performance and low carryover while maintaining ESI stability and remained miscible at all times. Buffer A was 30% methanol with 10 mM ammonium acetate as described above, buffer B was isopropanol with 10 mM ammonium acetate and buffer C was 100% hexanes. Lipids would effectively elute as the mobile phase composition was ramped to 100% buffer B and continue to elute as buffer B was gradually replaced with buffer C. The mobile phase composition was then ramped to 100% buffer C and flow rates were increased to increase column backpressure for effective column washing to minimize carryover. Surprisingly, our data indicated that long periods of high pressure hexane washes were imperative to reduce carryover to acceptably low levels; optimization experiments in this regard ultimately settled on 60 min of washing time at 100% buffer C. The gain in chromatographic performance may be observed in Fig. 11.7.

As observed in panel (c) of Fig. 11.7, the ternary solvent system with 10 mM ammonium acetate more effectively elutes the lipid analytes and eliminates tailing arising from hydrophilic interactions between the analytes and hydraulic path. These chromatographic improvements translated to a more robust separation scheme that had a positive impact on the reproducibility of replicate analyses and minimized carryover to acceptably low levels, Fig. 11.8.

Figure 11.8 demonstrates that run to run reproducibility is feasible when chromatographically separating complex lipid samples. The ternary solvent system effectively eluted analytes at an efficiency where carryover was minimized to acceptable levels. The net effect of such reproducibility is a gain in the precision of replicate measurements and therefore confidence that the trends observed in the data relate to a biological variable(s) rather than random variation in the experiment, Fig. 11.9.

Figure 11.9 demonstrates the ultimate success of the optimized ternary gradient in that precision is significantly improved permitting the observation of biological variation in a given experimental system. Although a discussion of the differences between the normoxic and hypoxic samples are beyond the scope of this document, the precision of the measurement of each lipid is sufficient to draw out statistically significant differences and gain biological insight. Overall, we believe that the challenges associated with lipid chromatography are surmountable and that the benefits of chromatographic separation prior to ESI MS analysis may be realized using an optimized ternary solvent system.

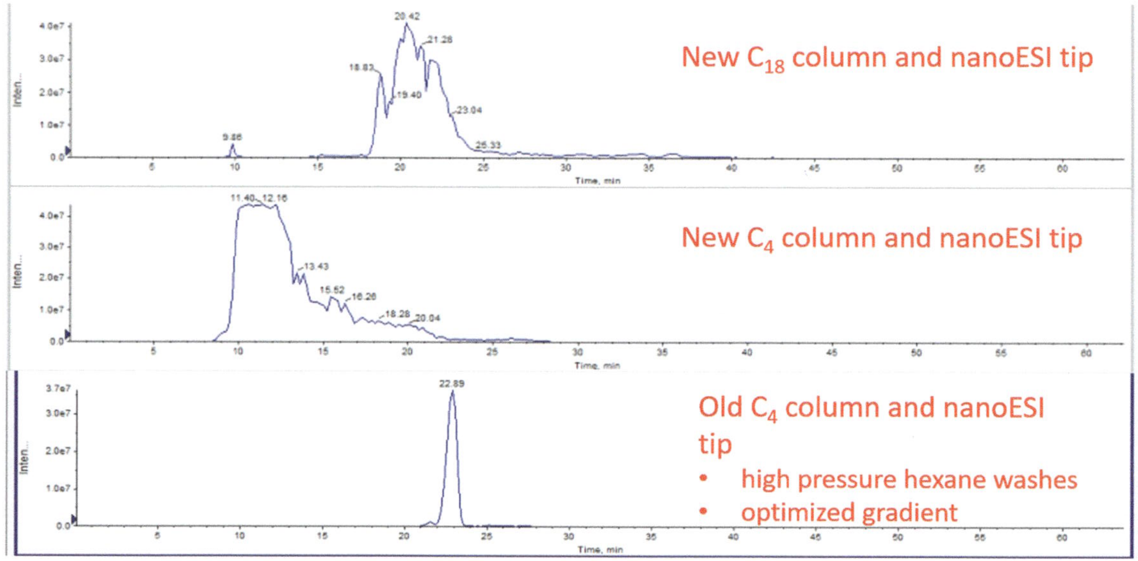

Fig. 11.7 A repeat of the experiment described in Fig. 11.4. Lipids were extracted from PC12 cells and separated as described in Fig. 11.3. A precursor ion scan of *m/z* 184 revealed many different phosphocholine-containing lipid species and *m/z* 524 was selected to demonstrate chromatographic performance. (**a**) the extracted ion current (XIC) for *m/z* 524 over a 60-min separation using a new C_{18} column reveals an approximately 8-min elution window and some tailing, (**b**) the experiment was repeated using a new C_4 column and *m/z* 524 is observed to elute over a longer time period but emerge from the column at an earlier time point, (**c**) the experiment was repeated using a C_4 column that had been used for numerous previous analyses and using the optimized ternary solvent system. The column was also cleaned with high pressure hexane washes between analyses. The improvement in chromatographic performance is evidenced by an approximately 1.5 min, Gaussian-shaped elution curve with minimal to no observable tailing

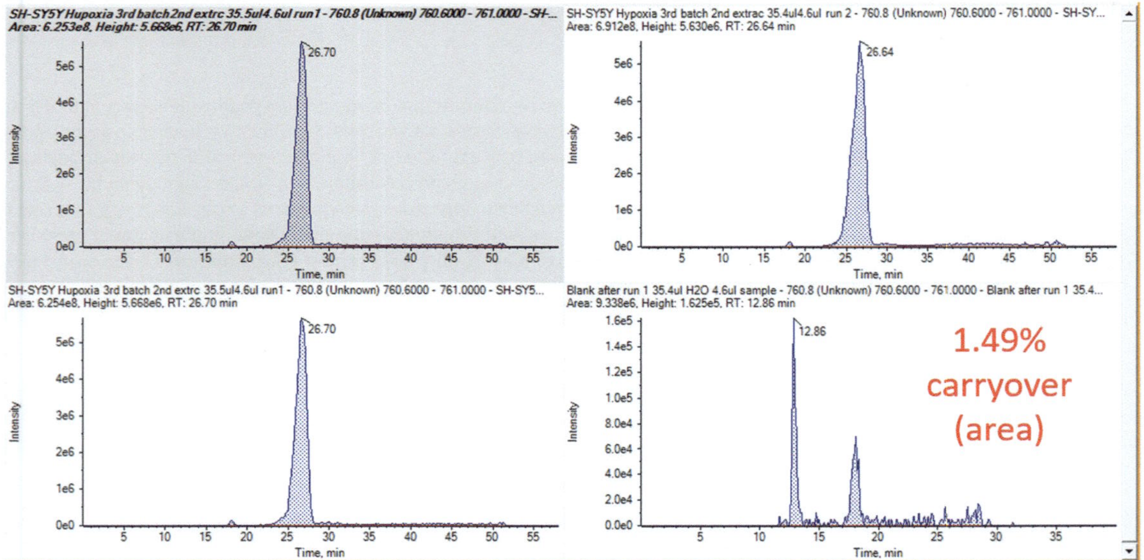

Fig. 11.8 Demonstration of the reproducibility of replicate analyses using new ternary gradient and hexane washes. SH-S5Y5 cells were cultured and harvested and their lipids were extracted using the Bligh and Dyer technique. Three aliquots of the lipid extract were analyzed via HPLC ESI MS using the ternary gradient described above and a precursor ion scan for *m/z* 184 on a triple quadrupole mass spectrometer. Within these three analyses, the prevalence of one of the most abundant PC lipids, 1-palmitoyl-2-oleoyl-sn-glycero-3-phosphocholine (POPC), was evaluated by conducting an XIC for *m/z* 760.8. Panels (**a**), (**b**) and (**c**) illustrate the XICs for POPC from the technical replicates and reveal consistent chromatographic performance in terms of retention time and peak areas/intensities. Panel (**d**) illustrates the XIC for POPC from a blank injection following the third technical replicate to demonstrate a run to run carryover of less than 1.5%

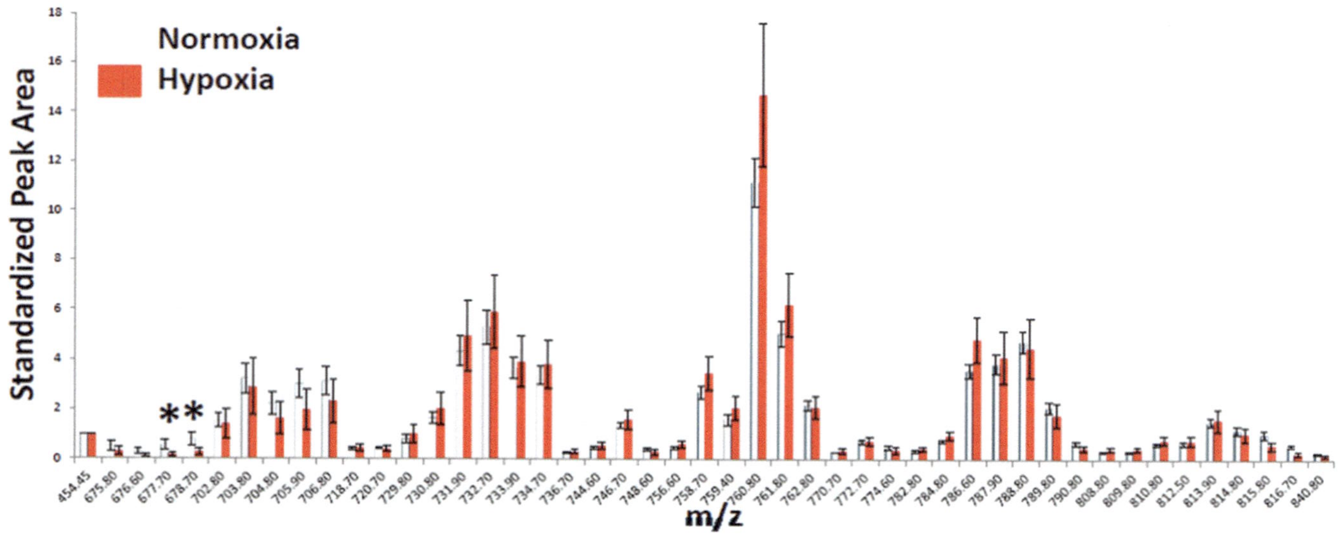

Fig. 11.9 Demonstration of the precision attained using the new ternary solvent gradient with hexane washes. SH-S5Y5 cells were cultured normally (normoxia, *white* bars) or at 5% oxygen (hypoxia, *red* bars) and harvested and their lipids were extracted using the Bligh and Dyer technique. Three aliquots of each lipid extract were analyzed via HPLC ESI MS using the ternary gradient described above and a precursor ion scan for *m/z* 184 on a triple quadrupole mass spectrometer. The histogram plots the relative peak areas of different phosphocholine-containing lipids normalized to an internal standard based on their mass to charge ratio. The precision of the technical replicates is evident in the size of the standard deviation error bars

11.3 Chemical Derivatization to Improve Sensitivity in Lipidomics Analyses

Chemical derivatization of lipids is one approach to increase sensitivity, separate isobaric masses, probe double bond location, and allow for methods of absolute quantitation of certain lipid classes. Here we review of some of the chemical derivatization techniques that have been developed to overcome challenges in MS analysis of lipids.

11.3.1 Fatty Acyl Derivatization

One of the earliest methods of chemical derivatization of fatty acids is methylation followed by GC-MS analysis. Methylation of fatty acid lipids (Scheme 11.2), improves gas chromatographic properties such as volatility and polarity [16]. The three main methylation derivatization procedures are acid catalyzed Fischer esterification, alkylation via diazomethane and base catalyzed transesterification. Briefly, acid catalyzed esterification and base catalyzed transesterification relies on a large excess of methanol to push the equilibrium to completion, while alkylation via diazomethane proceeds irreversibly to the corresponding fatty acid methyl ester (FAME).

A modern example of preparing fatty acid methyl esters (FAMEs) to improve their chromatographic separation is shown below (Fig. 11.10). This experiment uses methanolic potassium hydroxide to transesterify the lipids extracted from eggs [17]. Figure 11.10 (left) shows the gas chromatograph of several fatty acid methyl esters,

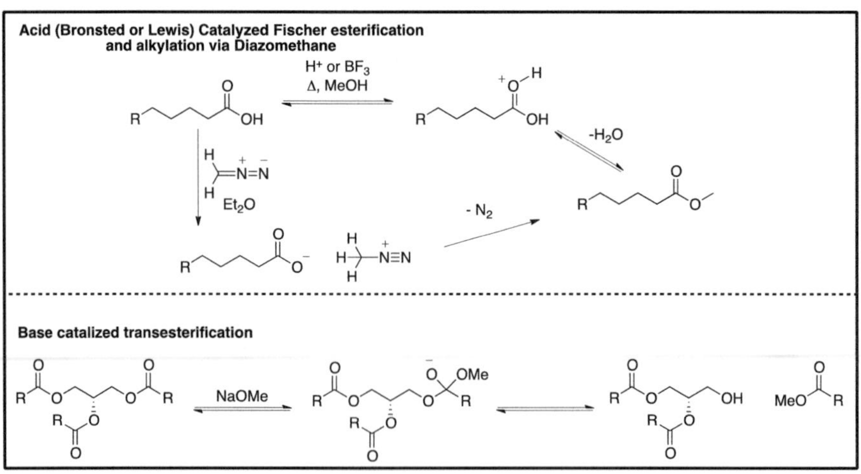

Scheme 11.2 Synthesis of fatty acid methyl esters

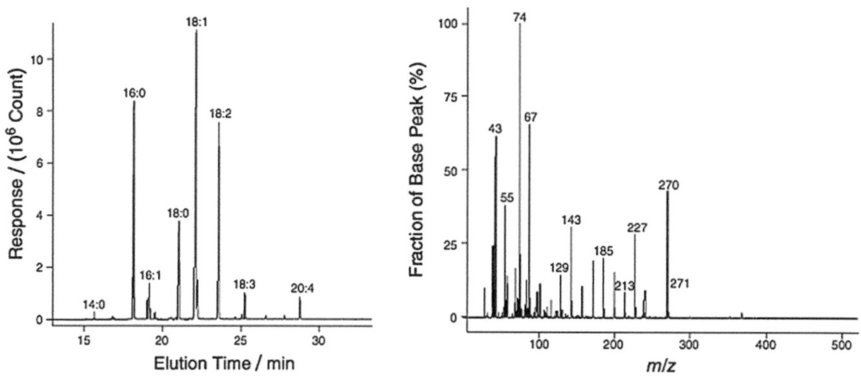

Fig. 11.10 Gas chromatograph of FAMEs derived from eggs (*left*) and EI-MS of methyl palmitate 16:0 (*right*) [17]

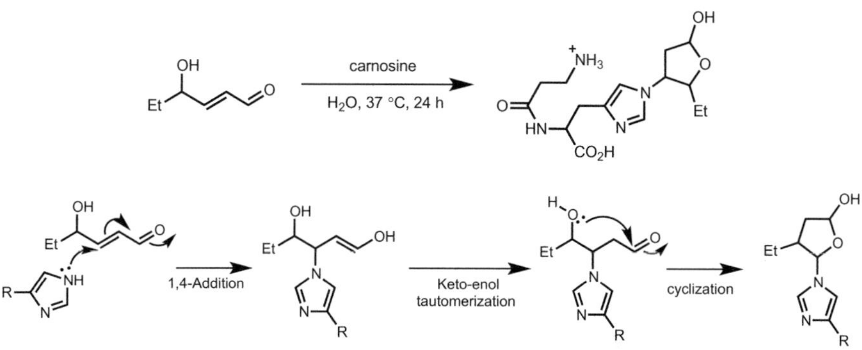

Scheme 11.3 General reaction scheme and mechanism of adduct formation of 4-hydroxyalkenals and carnosine

which are denoted by number of carbons in the chain and number of unsaturations in the chain. Figure 11.10 (right) shows an electron impact MS of methyl palmitate (16:0) that was successfully derivatized, extracted and separated via GC. This is a standard chemical derivatization technique to improve the gas chromatographic properties of fatty acyl components.

Han et al. have also employed a novel chemical derivatization technique for fatty acyls. The first example is the derivatization of 4-hydroxyalkenals to improve their quantitation from complex biological samples [18]. 4-hydroxyalkenals are known lipid peroxidation metabolites that can be used as lipid biomarkers for estimating the level of oxidative stress within a biological sample. However the analysis of these compounds by MS is relatively difficult since they readily form adducts with proteins and peptides from cell extracts. The general reaction scheme and mechanism is shown below (Scheme 11.3), the 4-hydroxylalkeanl is incubated at body temperature in water for 24 h before a microscale liquid-liquid extraction. These

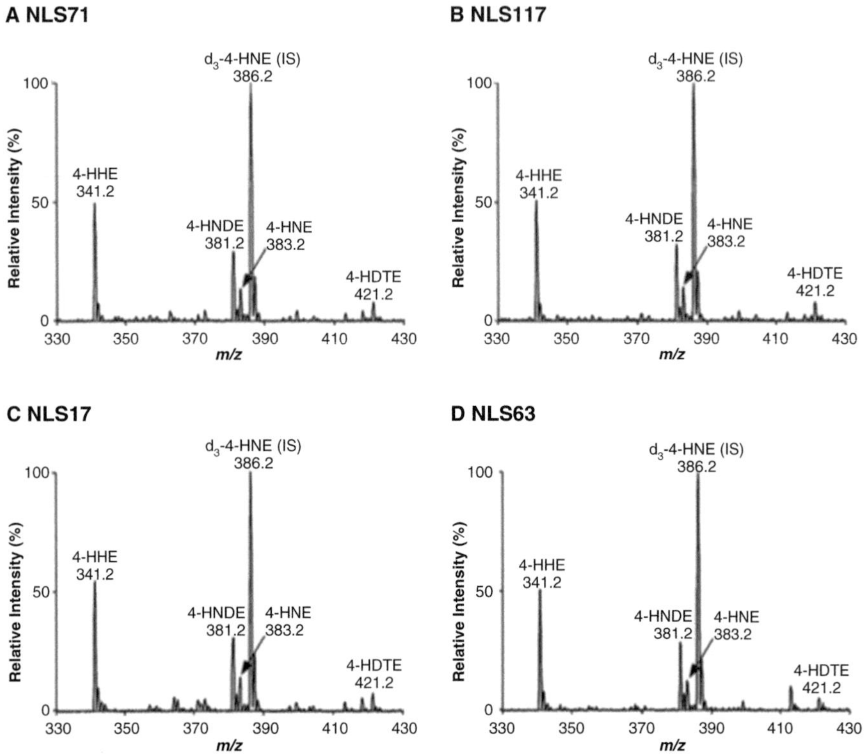

Fig. 11.11 Quantitation of a variety of 4-hydroxyalkenals using neutral loss scans [18]

adducted products could then be analyzed using a triple quadrupole mass spectrometer. Each of the samples also contained 4-hydroxynoneneal-d_3 as an internal standard where the deuterium is added to give a mass shift of 3 Da to avoid complicating the spectra.

After analysis of 4-hydroxyalkenal standards using product ion scanning, it was observed that the carnosine moiety had extremely reliable fragmentation. The loss of 17, 63, 71, and 117 Da corresponded to the neutral loss of ammonia, carbon dioxide plus ammonia, carbonyl chain (fragmentation at amide bond), and carbonyl chain plus carbon dioxide respectively. It was noted by the experimenters that the same fragments occurred with the deuterated standard as with the non-deuterated compound and thus it could be used as a relative standard for quantitation purposes. Figure 11.11 illustrates the results of neutral loss scans from derivatized 4-hydroxyalkenals from rat liver. By forming carnosine adducts, four different 4-hydroxyalkenals could be observed and precisely quantified by the requisite neutral loss scans.

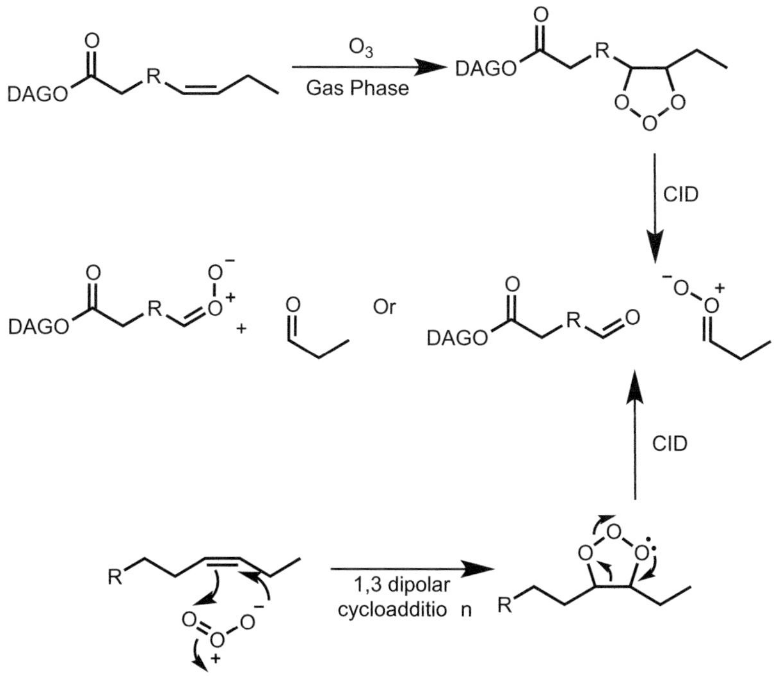

Scheme 11.4 General reaction and mechanism of the ozonolysis of triacylglycerols

11.3.2 Di-, and Triacylglycerol Derivatization

The analysis of di- and triacylglycerols is important as these lipids are some of the most ubiquitous classes found in mammalian biology. The primary function of these lipids is energy storage and they often serve as precursors to more complex lipid classes (e.g. glycerophospholipids); however, they can play roles in signal transduction pathways as well [13]. The elucidation of absolute lipid structure (location of the unsaturations and position of fatty acyl chains with respect to *sn*-1 and *sn*-2) is an obstacle that cannot be entirely surmounted by MS alone highlighting the importance of breakthroughs in chemical derivatization in this area.

In 2010 Blanksby et al. devised a chemical derivatization technique to help elucidate the absolute structures of triacylglycerols via gas phase ozonation (Scheme 11.4) [19]. Using a modified linear ion trap, methods were developed to use ozone-induced dissociation (OzID) to yield characteristic fragments that indicated the position of the double bonds within fatty acyl chains. Ozone was generated externally from O_2 (~11% by volume) and was flowed into the collision cell along with nitrogen gas at pressures of ~10^{-3} Torr. Ions that were generated from a lipid solution in methanolic sodium acetate were trapped in the collision cell for 10–25 ms before being analyzed by the final quadrupole to generate product ion mass spectra.

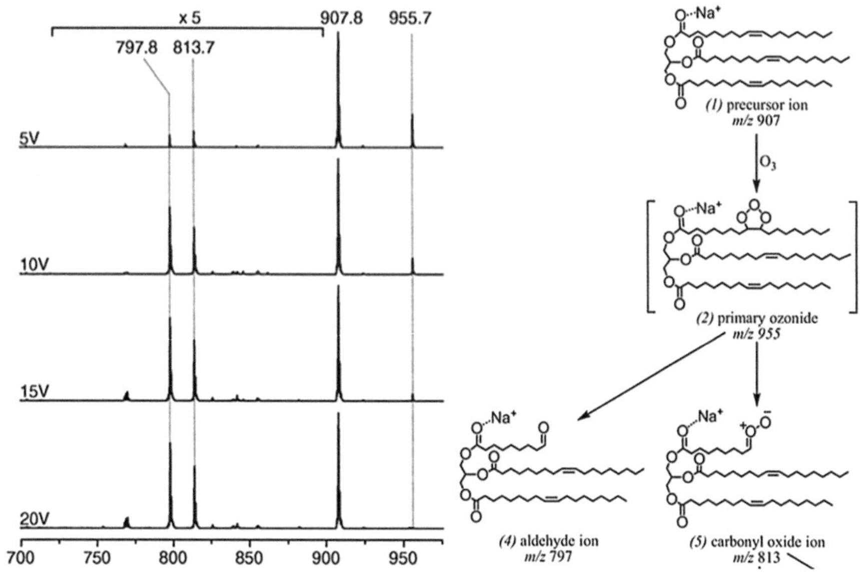

Fig. 11.12 Optimization of collision energy for OzID (*left*), possible products of OzID (*right*) [19]

A triacylglycerol standard was used to optimize collision energies to fragment the primary ozonide into two main products, an aldehyde and carbonyl oxide fragment. Both fragments can be used to determine double bond location since ozone will selectively react with an alkene in the lipid and predictable mass losses will ensue. The collision energy was adjusted until all of the primary ozonide was fragmented (Fig. 11.12). From the mass losses of 158 ($C_9H_{18}O_2$) and 142 ($C_9H_{18}O$) Da from the primary ozonide, the location of the unsaturation was determined to be at C9 (ω-9 unsaturated fatty acyl). Using this method of chemical derivatization, predictable fragmentation of OzID is a useful tool in the structural elucidation of triacylglycerols.

Although diacylglycerols (DAG) play very important roles in biological mechanism, their quantitation has been challenging due to low abundance and poor proton affinity [20, 21]. An example of chemical derivatization of acylglycerols is presented by Han et al. in 2014, where the use of a common peptide coupling reaction (esterification/amidation) is used to allow for enhanced ESI-MS ionization efficacy [22]. The general reaction conditions and mechanism are shown in Scheme 11.5, whereby adding a tertiary amine to the DAG, a large increase in proton affinity (ionization efficiency) was observed. The MS analysis was performed on a triple quadrupole instrument to provide product, precursor and neutral loss scans.

Using an equimolar amount of 16 different DAG standards at fmol/μL concentrations, this group showed extremely favorable fragmentation of dimethylglycine (DMG) modified DAG that gave a reproducible ion at *m/z* 110 corresponding to lithiated DMG. This predictable fragmentation could then be used in precursor and

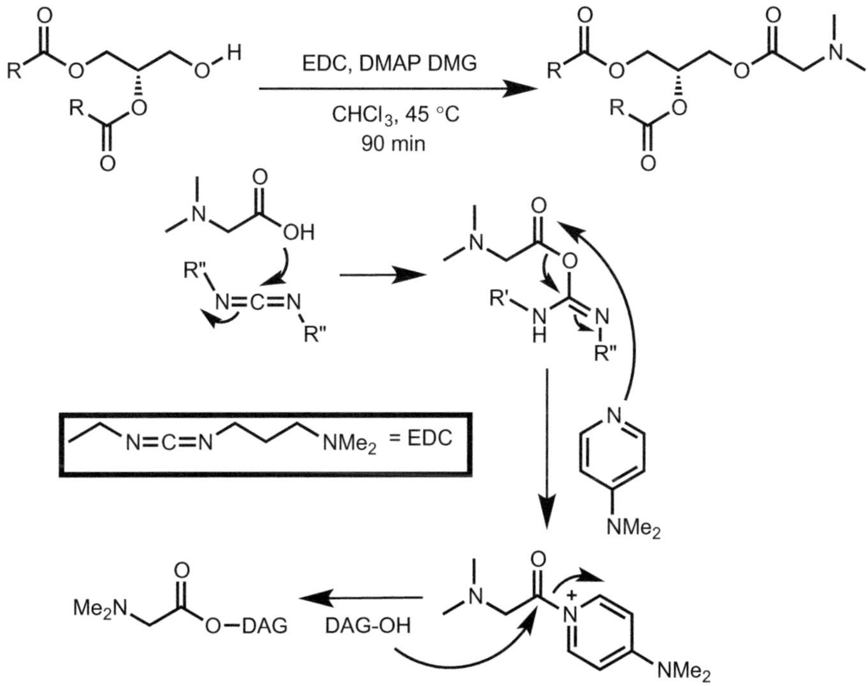

Scheme 11.5 Esterification of diacylglycerols

neutral loss scans to simplify complex biological samples for relative quantitation of many DAG species. Figure 11.13 shows the results of both protonated and lithiated neutral loss scans. The spectrum on the right illustrates all peak intensities are approximately equal which is advantageous for the quantitation of each of the lipid species; the spectrum on the left however (using LiOH to initiate the ionization), showed a variety of peak intensities. The reason for using a lithium additive is that lithium facilitates fatty acyl cleavage upon CID and thus some structural information is obtainable. In summary, this chemical derivatization method greatly increases the ionization efficiency of DAGs, permitting quantitation when formic acid or structural elucidation when Li^+ are added as additives, respectively.

11.3.3 PE and PC Derivatization

As described above, PE and PC constitute the majority of the total lipids in a cell and therefore require rigorous analytical methods. Challenges encountered when analyzing PE and PC include sensitivity loss via ion suppression in ESI and isobaric interference. For example, a PC with fatty acyl chains of 16:0 and 22:6 would have the exact same *m/z* value as a PC with 18:1 and 20:5. The absolute structural

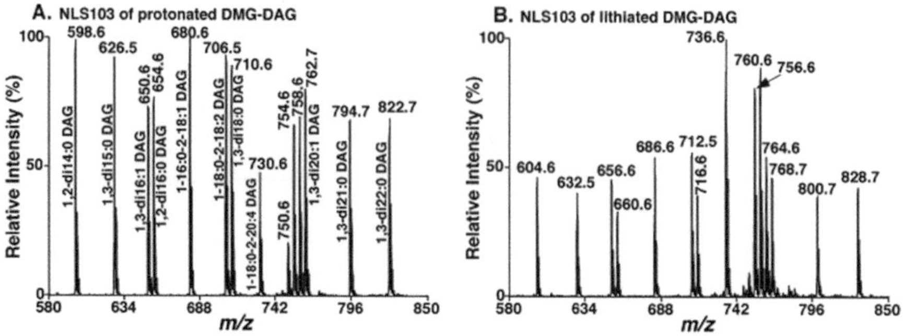

Fig. 11.13 Neutral loss of 103 of an equimolar mixture of DAG standards [22]

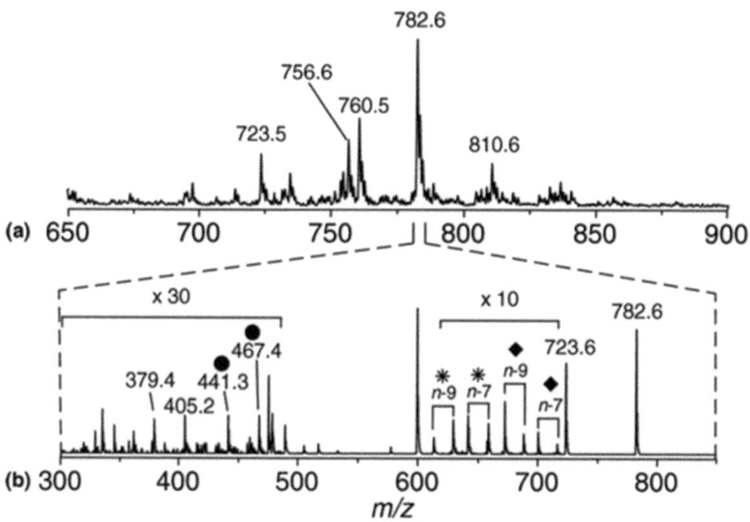

Fig. 11.14 OzID of triacylglycerols isolated from cow brain [19]

elucidation of PE and PC is crucial to fully understand lipid biochemistry and the roles they play in cellular life highlighting the need and importance of chemical derivatization techniques in this area.

The Blanksby group used their OzID methodology to analyze PCs from lipids extracted from cow brain [19]. Figure 11.14b illustrates the resulting spectrum when OzID is used on a lipid observed at 782.6 Da. Two fragmentation products are observed for each structural isomer, where the higher mass fragment corresponds to a sodiated carbonyl oxide and the lower mass fragment corresponds to a sodiated aldehyde fragment. The peaks denoted with a diamond are directly from ω-7 and ω-9 oxidative cleavage where peaks marked with asterisks are from the loss of 59 Da corresponding to neutral loss of NMe_3 from the choline head group. In this way, the location of the double bond within fatty acyl substituents can be obtained as well as the differentiation of isobaric PC species.

PE lipids have a free amine (at higher pH) that can act as a nucleophile for activated electrophiles such as acid chlorides and *N*-hydroxy succinimide (NHS) esters. In 2012, the Reid group developed a novel chemical derivatization procedure for PEs based on a common nucleophilic substitution of an activated NHS ester, Scheme 11.6 [23]. Interestingly, d_6-dimethylthibutanoylhydroxysuccinimide ester (d_6-DMBNHS) bears a fixed positive charge in the form of a sulfonium ion. Therefore when d_6-DMBNHS is reacted with PE species to form a new amide bond, the lipid will also bear a fixed positive charge improving its ionization efficiency greatly. This will greatly increase the sensitivity of the lower abundance PE species as well as separating the isobaric lipid species due to the mass shift of the d_6-DMB moiety (136 Da). Again, some of the challenges with quantitation and isobaric lipid species can be overcome by a facile one step chemical derivatization procedure.

This group were interested in obtaining the global lipid profile of metastatic colon adenocarcinoma cell lines by using their novel technique to identify and quantify some low abundant PE species as well as the more abundant ones. Figure 11.15 illustrates the results of a non-derivatized lipid extract and d_6-DMB modified lipid extract from SW480 cancer cell lines. The emergence of higher mass peaks as well as the decrease in intensity of many of the residual peaks from the unmodified lipid extract indicates that a significant number of isobaric lipid species were separated. Many of the modified lipids were identified and accurately quantified using high-resolution/accurate-mass analysis, permitting a global lipid comparison of the two cancer cell lines. This information can be used to monitor the differences between non-cancerous and metastatic cell and perhaps be used to help identify colon cancer at its early stages.

Other modifications of PE and PC head group can be accomplished by methylation via an activated methyl electrophile. Some examples of methyl electrophiles used to alkylate the amine moiety of ethanolamine phospho-lipids includes, methyl iodide [24] and reductive amidation with formaldehyde/sodium cyanoborohydride [25]. Our group has developed the used of acid initiated diazomethane (DZM) alkylation [26] as a simple, fast, and cost effective method of PC and PE alkylation entitled trimethylation enhancement using diazomethane (TrEnDi, Scheme 11.7). This method is observed to methylate primary amines as well as the phosphate oxygen giving a truly permanent positive charge residing on derivatized lipid. The scheme below shows the general reaction of diazomethane with PE. This reaction requires the use of an acid to mediate the eventual protonation of diazomethane leading to the loss of N_2 Gas.

The signal of unmodified PE/PC lipids can be split between protonated and sodiated species that inevitably decreases sensitivity as well as further complicating spectra. The methylation of PE and PC via TrEnDi requires the use of an acid with a non-coordinating counter ion because the counterion can react with the DZM in an unsatisfactory way. This type of modification is very efficient in that, the reactions go to 99% completion as well as only producing N_2 and the sole byproduct. Thus a post extraction protocol is not required.

The reaction of diazomethane with PE and PC lipid standards yielded a significant increase in sensitivity (Table 11.1). This increase was measured by comparing

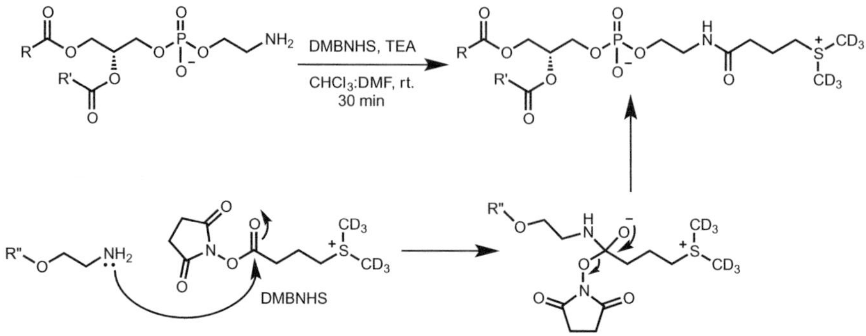

Scheme 11.6 Chemical derivatization of PEs with d_6-DMBNHS general reaction and mechanism

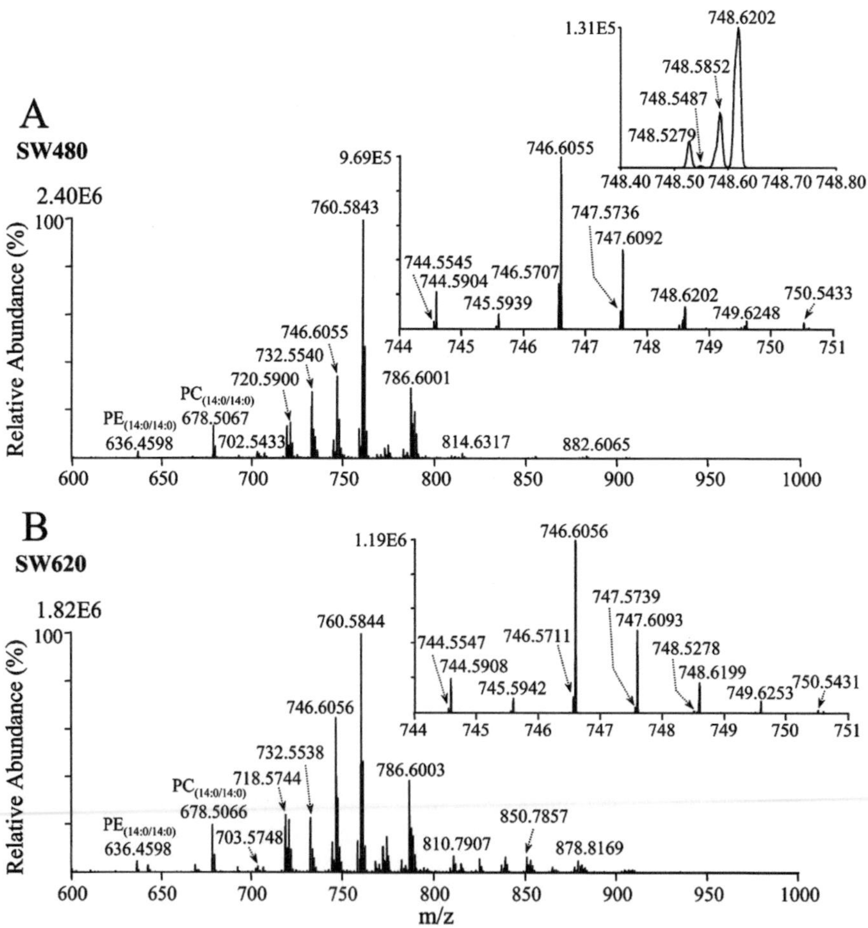

Fig. 11.15 Unmodified lipid extract (*top*) and d_6-DMB modified lipid extract (*bottom*) [23]

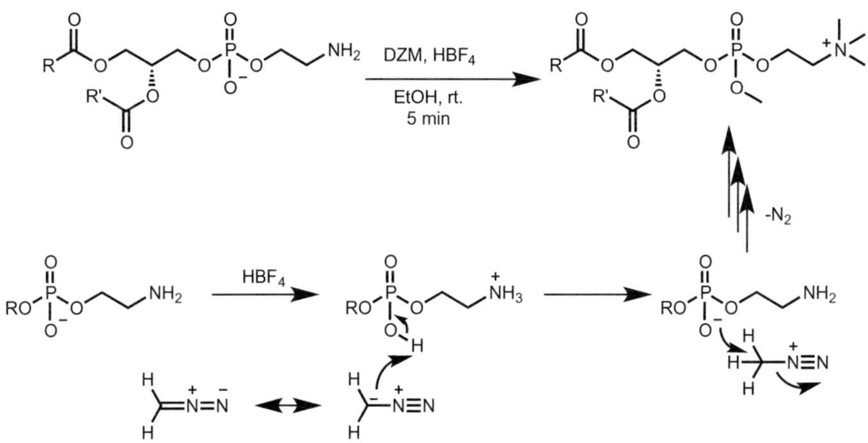

Scheme 11.7 Reaction of Diazomethane (DZM) with PE general reaction and mechanism

Table 11.1 Sensitivity comparison of methylated PE/PC against non-modified lipids [26]

	Unmodified			TrEnDi-modified		
Lipid	Species	*m/z* (Th)	Relative intensity (%)	$[M^{Tr}]^+$ (Th)	Relative intensity (%)	Sensitivity increase[a]
PE (16:0/18:1(9Z))	$[M + H]^+$	718.52	0.16	774.59	100	625×
	$[M + Na]^+$	740.52	45.0			2.22×
	sum		45.2			2.21×
PS (18:0/18:2(9Z,12Z))	$[M + H]^+$	788.59	0.09	858.67	100	1110×
	$[M + Na]^+$	810.59	1.16			86.2×
	$[M - H+ 2Na]^+$	832.57	1.84			54.3×
	sum		3.09			32.4×
PC (18:1(9Z)/14:0	$[M + H]^+$	732.56	0.69	746.57	100	145×
	$[M + Na]^+$	754.54	96.2			1.04×
	sum		96.9			1.03×
SM (d18:1/16:0)	$[M + H]^+$	703.60	3.79	717.60	100	26.4×
	$[M + Na]^+$	725.58	36.1			2.77×
	sum		39.9			2.51×

[a]Fold increase calculated by dividing the relative intensity of the TrEnDi-modified lipid species to the relative intensity of the unmodified lipid species

the relative intensities of modified lipid with non-modified in the same sample solution (i.e. the modified and non-modified samples were mixed in a equimolar ratio before analysis). It was observed that protonated PE shows a 625-fold sensitivity increase, and sodiated PE shows a 2.2-fold increase. For PC the results were similar where protonated and sodiated PC gave 145- and 1.04-fold increases. These results were the first of their kind in the chemical derivatization of glycerophospholipids.

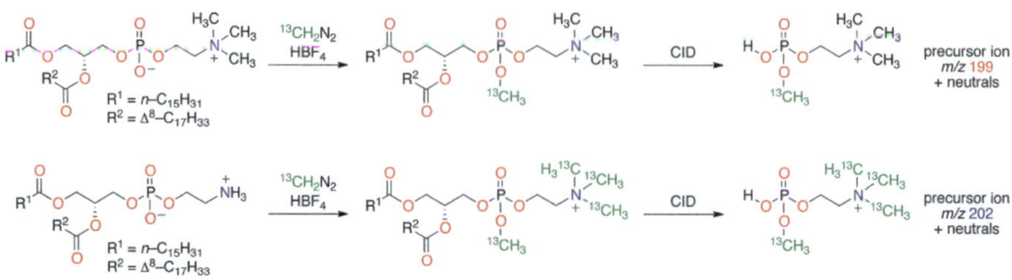

Scheme 11.8 ^{13}C–TrEnDi derivatization of PC(16:0/18:1(9Z)) and PE(16:0/18:1(9Z)) with $^{13}CH_2N_2$ [27]

More recently, we have developed a strategy using ^{13}C–labelled diazomethane to isotopically label glycerophospholipids via TrEnDi, termed ^{13}C–TrEnDi [27]. This modification to the derivatization remediates the isobaric overlap of PE and PC lipids with identical acyl chains, Scheme 11.8.

The utility of this derivatization was demonstrated on a lipid extract from HeLa cells where the number of PE lipids observed above the limit of detection almost doubled and several phosphatidylserine lipids were observed that were previously invisible to MS analysis

11.4 Conclusion

Through carful optimization, the benefits of chromatography may be applied to the separation of complex lipid samples. Using a ternary gradient, the precision of lipid separations is maintained with good chromatographic peak shape and minimal carryover. Deeper insight into lipid biochemistry is empowered via chemical derivatization techniques that help overcome challenges in analyzing lipids by MS. Derivatization techniques increase sensitivity, separate isobaric masses, and elucidate absolute structure of a variety of different lipid classes. Successful derivatization techniques are typically simple organic reactions performed on a very small scale and are effective on standard solutions as well as complex lipid samples. Early indicators in the literature reveal that additional developments in the chemical derivatization of lipids will continue to emerge in this field as the importance of lipids in cellular biochemistry becomes more fully realized.

References

1. Wenk MR (2005) The emerging field of lipodomics. Nat Rev Drug Discov 4:594–610
2. Shimizu T (2009) Lipid mediators in health and disease: enzymes and receptors as therapeutic targets for the regulation of immunity and inflammation. Annu Rev Pharmacol Toxicol 49:123–150

3. Meyer HD, Jakobs KH (2007) Lysophospholipid receptors: signalling, pharmacology and regulation by lysophospholipid metabolism. Biochim Biophys Acta 1768:923–940
4. Kirschnek S, Paris F, Weller M, Grassme H, Ferlinz K, Riehle A, Fuks Z, Kolesnick R, Gulbins E (2000) CD95-mediated apoptosis in vivo involves acid Sphingomyelinase. J Biol Chem 275:27316–27323
5. Lemmon MA (2008) Membrane recognition by phospholipid-binding domains. Nat Rev Mol Cell Biol 9:99–111
6. Meer G (2005) Cellular lipidomics. EMBO J 24:3159–3165
7. Horton HR, Moran LA, Scrimgeour KG, Perry MD, Rawn JD (2006) Principles of biochemistry, 4th edn. Pearson, Upper Saddle River, pp 479–509
8. Pulfer M, Murphy RC (2003) Electrospray mass spectrometry of phospholipids. Mass Spectrom Rev 22:332–364
9. Bartke N, Hannun Y (2009) Bioactive sphingolipids: metabolism and function. J Lipid Res 50:S91–S96
10. Goldfine H (2010) The appearance, disappearance and reappearance of plasmalogens in evolution. Prog Lipid Res 49:493–498
11. Bligh EG, Dyer WJ (1959) A rapid method of total lipid extraction and purification. Can J Biochem Physiol 37:911–917
12. Folch J, Lees M, Sloane-Stanley GH (1957) A simple method for the isolation and purification of total lipides from animal tissues. J Biol Chem 226:497–509
13. Kalil MB, Hou W, Zhou H, Elisma F, Swayne LA, Blanchard AP, Yao Z, Bennett SA, Figeys D (2010) Lipidomics era: accomplishments and challenges. Mass Spectrom Rev 29:877–929
14. Yang K, Han X (2016) Lipidomics: techniques, applications, and outcomes related to biomedical sciences. Trends Biochem Sci 41:954–969
15. Harris DC (2016) Quantitative chemical analysis, 9th edn. W. H. Freeman and Co., New York
16. Christie WW (1993) Preparation of ester derivatives of fatty acids for chromatographic analysis. Adv Lipid Methodol – Two 2:69–111
17. Alty LT (2009) Analysis of fatty acid methyl esters in egg yolk using GC-MS. J Chem Educ 86:962–965
18. Wang M, Fang H, Han X (2012) Shotgun lipidomics analysis of 4-hydroxyalkenal species directly from lipid extracts after one-step in situ derivatization. Anal Chem 84:4580–4586
19. Poad BL, Pham HT, Thomas MC, Nealson JR, Campbell JL, Mitchell TW, Blanksby SJ (2010) Ozone-induced dissociation on a modified tandem linear ion-trap: Observations of different reactivity for isomeric lipids. J Am Soc Mass Spectrom 21:1989–1999
20. Preiss J, Loomis CR, Bishop WR, Stein R, Niedel JE, Bell RM (1986) Quantitative measurement of sn-1,2-diacylglycerols present in platelets, hepatocytes, and ras- and sis-transformed normal rat kidney cells. J Biol Chem 261:8597–8600
21. Billah MM, Eckel S, Mullmann TJ, Egan RW, Siegel MI (1989) Phosphatidylcholine hydrolysis by phosphinolipase D determines phosphatidate and diglyceride levels in chemotactic peptide-stimulated human neutrophils. J Biol Chem 264:17069–17077
22. Wang M, Hayakawa J, Yang K, Han X (2014) Characterization and quantification of diacylglycerol species in biological extracts after onestep derivatization: a shotgun lipidomics approach. Anal Chem 86:2146–2155
23. Fhaner CJ, Liu S, Ji H, Simpson RJ, Reid G (2012) Comprehensive lipidome profiling of isogenic primary and metastatic colon adenocarcinoma cell lines. Anal Chem 84:8917–8926
24. Wang M, Kim GH, Wei F, Chen H, Altarejos J, Han X (2015) Improved method for quantitative analysis of methylated phosphatidylethanolamine species and its application for analysis of diabetic-mouse liver samples. Anal Bioanal Chem 407:5021–5032
25. Wang X, Wei F, Xu J, Lv X, Dong X, Han X, Quek S, Huang F, Chen H (2016) Profiling and relative quantification of phosphatidylethanolamine based on acetone stable isotope derivatization. Anal Chim Acta 902:142–153
26. Wasslen KV, Canez CR, Lee H, Manthorpe JM, Smith JC (2014) Trimethylation enhancement using diazomethane (TrEnDi) II: rapid insolution concomitant quaternization of

glycerophospholipid amino groups and methylation of phosphate groups via reaction with diazomethane significantly enhances sensitivity in mass spectrometry analyses via a fixed, permanent positive charge. Anal Chem 86:9523–9532

27. Canez CR, Shields SWJ, Bugno M, Wasslen KV, Weinert HP, Willmore WG, Manthorpe JM, Smith JC (2016) Trimethylation enhancement using (13)C-diazomethane ((13)C- TrEnDi): increased sensitivity and selectivity of phosphatidylethanolamine, phosphatidylcholine, and phosphatidylserine lipids derived from complex biological samples. Anal Chem 88:6996–7004

Chapter 12
Advanced Mass Spectrometric Methodologies in the Evaluation of Health Risk Assessment Associated to Exposure to Drugs and Fraudulenty Modified Foods

Giovanni Sindona

Abstract Environmental prevention strategies include the development of suitable analytical protocols to assess health risk associated to exposure to drugs and fraudulently modified foods. Mass spectrometry (MS) is the analytical methodology of choice since it is based on the identification and assay of organic and inorganic components present, and eventually fraudulently added, to foods. An important achievement of the methodology is represented by the so-called *ambient ionization* approach that allows the direct evaluation of species present in fluids and/or solids without any previous chemical manipulation.

12.1 The Absolute Method of Analysis

The role of Mass Spectrometry (MS) in the identification and assay of ionizable species present in complex mixture is well established in the world of chemistry scientists. It is still problematic, however, to promote a successfull interplay of MS scientific results with real life tradition and demands. An emblematic case could be represented by the results recently published by german scientists proving that non-stick parchment papers transfer polydimethylsiloxane (PDMS) species to baked cookies and pizzas, as shown by the presence of ions of general structure $[(C_2H_6SiO)_n + NH_4]^+$ in the m/z 800–1900 range [1]. The MS spectra were obtained by applying a relatively new ambient ionization system (DART) [2] coupled with electrospray ionization (ESI).

G. Sindona (✉)
Department of Chemistry and Chemical Technologies, University of Calabria, Arcavacata di Rende (CS), Italy
e-mail: giovanni.sindona@unical.it

J.H. Banoub, R.M. Caprioli (eds.), *Molecular Technologies for Detection of Chemical and Biological Agents*, NATO Science for Peace and Security Series A: Chemistry and Biology, DOI 10.1007/978-94-024-1113-3_12

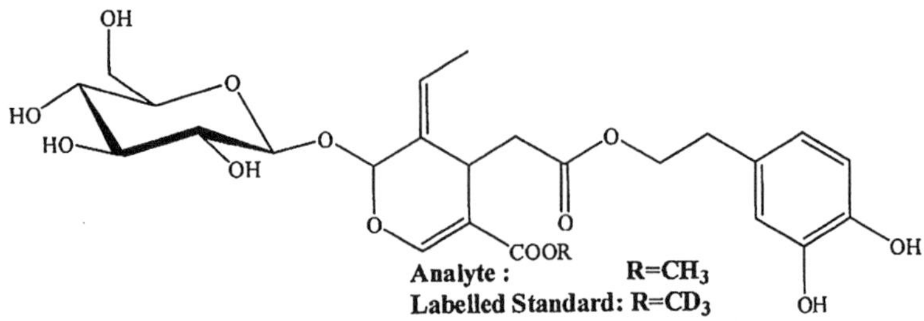

Fig. 12.1 The deuterated analog is added to olive oil samples to determine the real amount of the antioxidant by MS/MS absolute method

Unfortunately, no notice has appeared in the news about private or public enterprises interested in the evaluation of the risk that PDMS contaminated foods can cause to consumers.

Either the identification, but more important the assay of a specific molecule in any environment can confidently be performed by MS using stable, labelled internal standards which have the same structure of the analyte to be checked.

The labelling should not affect the chemistry of the reference compound with respect to the analyte in the experimental conditions typical of the methodology here examined.

Figure 12.1 reports the structure of the secoiridoid Oleuropein whose presence and relative abundance in foodstuffs derived from olives has been determined easily by the methodology here presented [3–5].

A very important field of application is represented by the so-called newborn screening (NBS) [6] which allows the identification and quantitation by MS/MS, of tens of Inborn Errors of Metabolism (IEM) in babies, in runs of 2 min or less, from dried blood spots collected on a filter paper in the first 3 days after the birth.

Concerns about the use of sophisticated analytical instruments are often arisen by *well-informed* people about the cost of analysis requiring the use of isotopically labelled compounds as reference standards. Although this is not true (the cost of a full NBS analysis is in the order or tens of dollars), it should be emphasized that the synthesis of labelled compounds is a routine application of synthetic organic chemistry and that suppliers in the research and fine chemicals products are present in the worldwide market.

12.2 Quality Safety and Tracebility of Foods [7]

12.2.1 Babies Injured by Melamine Fraudulently Added in Milk

The milk scandal that occurred in China in 2008 was a food safety incident which caused 300,000 victims, most of them infants and babies. Milk was deliberately adulterated with melamine (2,4,6-triamino-1,3,5-triazine) a low molecular weight compounds (126,12 uma) with a relatively high content of nitrogen atoms. This was not only a chinese problem, since nowadays the determination of the proteins present in dairy products is still based (*worldwide!*) on the determination of the total nitrogen content, by applying the Kjeldahl analysis developed in **1883** [8]. It does not provide a measure of true protein content, since non-protein nitrogen deriving from the degradation of natural or fraudulently added compounds, such as melamine, contribute as well to the total nitrogen counts. As a matter of fact, Chinese scientists clearly have demonstrated, soon after the scandal appears on the international press, that melamine could be easily detected in milk by the absolute analytical method whereby tandem mass spectrometry, in the multiple Reaction Monitoring (MRM) mode, was applied to milk samples containing ^{13}C and ^{15}N labelled standard.

The spectra reported in Fig. 12.2 clearly show how molecular and specific product ion species can be easily identified and evaluated.

It is important to underline that the limits of quantification were 8 ng/g for liquid milk and 15 ng/g for dry milk powder and the run time of the measurement was 5.5 min [9].

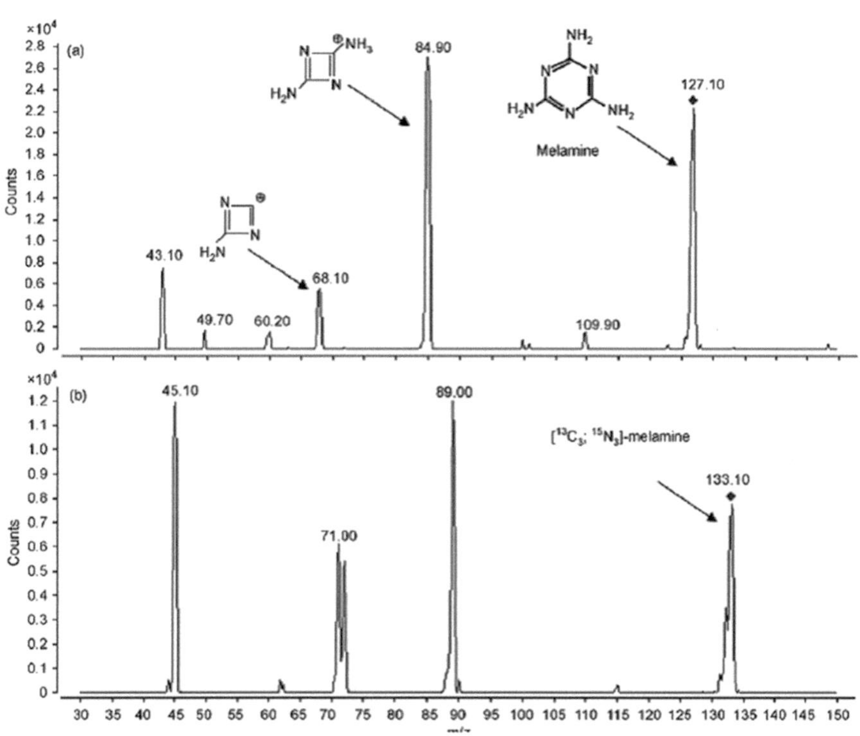

Fig. 12.2 MRM spectra of (**a**) melamine and (**b**) isotopically labelled analogue

Table 12.1 Evaluation of sudan dyes in 2004 in chili pepper and in typical Calabrian foods (*fresa and nduja*)

Reproducibility of the measurements: recovery, LOQ, and LOD from foodstuff matrixes					
Typical product	Mean value (ppm)	RSD%	LOQ (ppm)	LOD (ppm)	Recovery
"fresa bread" I hot chili pepper "nduja"	0.777 ± 0.025	3.24	0.008	0.007	84%
	909.267 ± 37.434	4.12	0.008	0.007	85%
	93.263 ± 3.154	3.38	0.010	0 009	70%

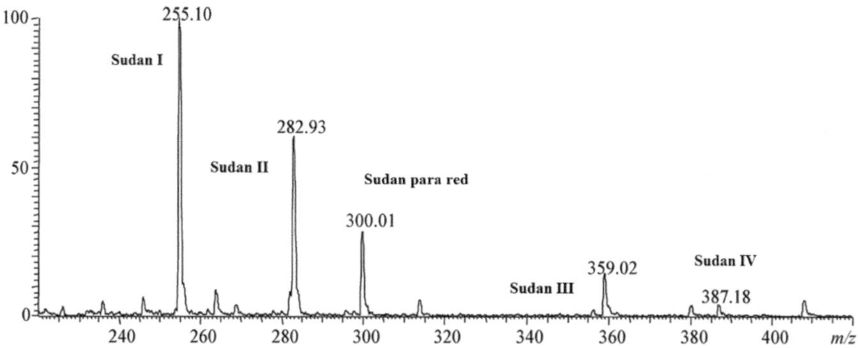

Fig. 12.3 Mass spectrum of the family of labelled (d_6) Sudan molecules used as internal standard obtained by PS-MS

12.2.2 Cancerogenic Sudan Dyes Molecules Added to Sausages

A family of azo dyes named Sudan was illegaly added to chili powder and chili products to extend the lifespan of the typical reddish colour produced by the natural chili added, for centuries, to sausages and some meat foods prepared in Italy. Sudan dyes, have been classified as category 3 carcinogens by the International Agency for Research on Cancer (IARC), and are unsafe for human consumption.

The Department of Chemistry of our university was required by the regional Federation of Italian Employers (Confindustria) to provide an absolute method for the detection and assay of the dangerous sudan dyes poisons in Italian food. Table 12.1 reports on the application of the MS/MS method here discussed to commercially available powdered chili pepper and to some typical Calabrian foods. The results [10–12] were shocking because unaware consumers were deliberately exposed to very dangerous poisons.

A modern fast method of determination such as Paper Spray (PS) MS was recently applied to promote the application of this unique procedure, on site (Fig. 12.3) [13].

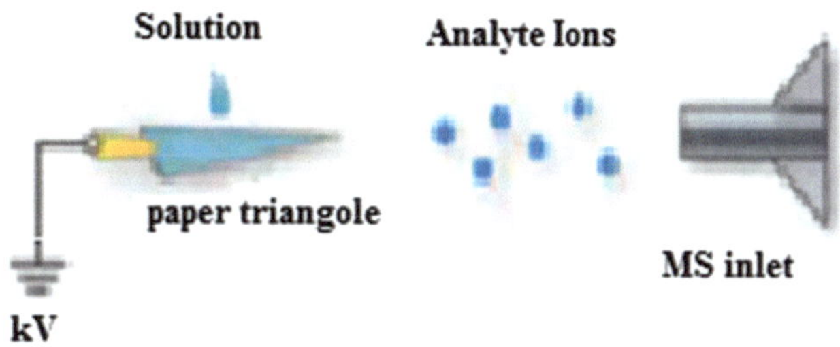

Fig. 12.4 Schematics of PS/MS operating procedure

PS-MS, developed by Cooks (Fig. 12.4) [14] allows the generation of ions by applying a high voltage to a paper triangle wetted with a small volume of solution. In particular, the technique appears applicable to a wide range of molecules, such as small organic compounds, peptides or proteins. In this study, we report a paper spray analysis applied to the assay of Sudan azo dyes, a class of colorants widely used in household commodities and in the textile [13].

12.2.3 PS/MS Discrimination Among Citrus Essential Oils

A novel approach for a rapid molecular profiling of citrus fruits essential oils, bypassing chromatography, was introduced using PS-MS. Oxygen heterocyclic compounds such as psoralens, coumarins and some flavonoids were easily determined (Fig. 12.5), whereby the coupling of the analytical methodology with chemometric analysis provided a reliable and molecularly based comprehensive discrimination of essential oils from different citrus fruits [15].

The employment of PS-MS profiling as marker for the classification of bergamot essential oils was evaluated by using multivariate statistical analysis. Two pattern recognition techniques (LDA and SIMCA), each with different characteristics, were applied to MS data, and prediction ability of these chemometric techniques was tested using cross-validation procedure. Both approaches allowed us to obtain excellent results because the constructed models have correctly

12.2.4 Analysis of Sexual Assault Evidence

Francese and coworkers have introduced the potentiality of MALDI-MS in sexual assault forensic analysis [16], with a specific reference, in particular, to the identification of condom-doped latent fingerprints, an evidence when condoms are used in a crime. The maintenance of the sample for further examination can be achieved by reducing or eliminating the sample preparation, and, in that perspective, the use of ambient ionization techniques is particularly well suited.

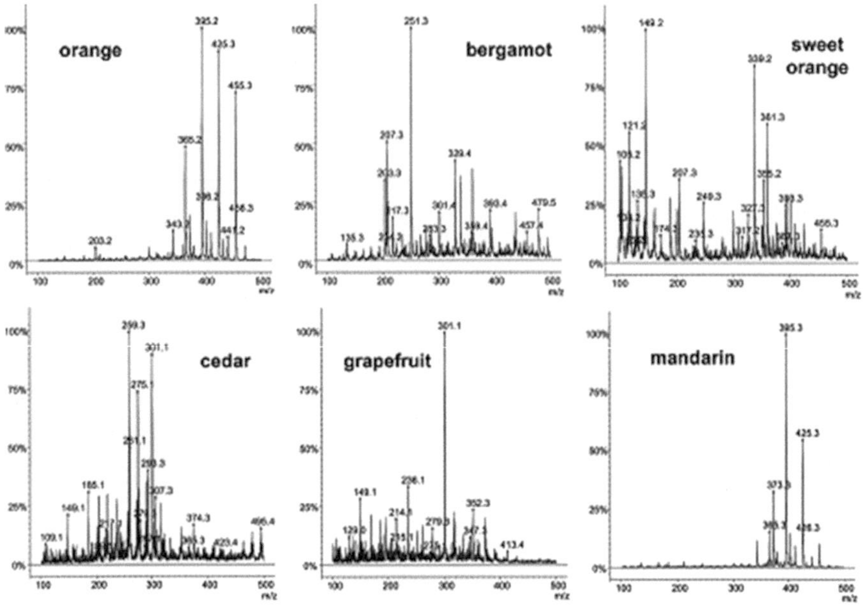

Fig. 12.5 PS-MS molecular profiling within 100–500 mass range of different essential oil samples from different citrus fruits

Even if analysis of the same sample with complementary techniques can be useful for discrimination, it is important to investigate the possibility to achieve an effective discernment of different condom samples by using mass spectrometric techniques only. Here, we present a statistical approach to differentiate and identify condoms analyzed by DESI-MS and EASI-MS, an analytical approach showing the feasibility and the potential of *in situ* analysis [17]. Each condom model had shown a peculiar spectrum allowing the differentiation one from the others, it was important, however, to find a systematic and reproducible way to differentiate them based on their chemical composition and this goal was achieved by using multivariate statistical analysis.

12.2.5 Identification of Minor Components in Food by LTP Mass Spectrometry

A relatively new ambient plasma-based ionization technique has been introduced by Cooks group termed low temperature plasma (LTP), which permits the direct analysis of trace compounds on solid surfaces and in complex matrices, without the need for solvents or reagents besides a low flow of carrier/discharge gas [18]. Olive oil has been used as model to show the identification, and in some case quantitation, of different species at different concentration scales from a particular foodstuff

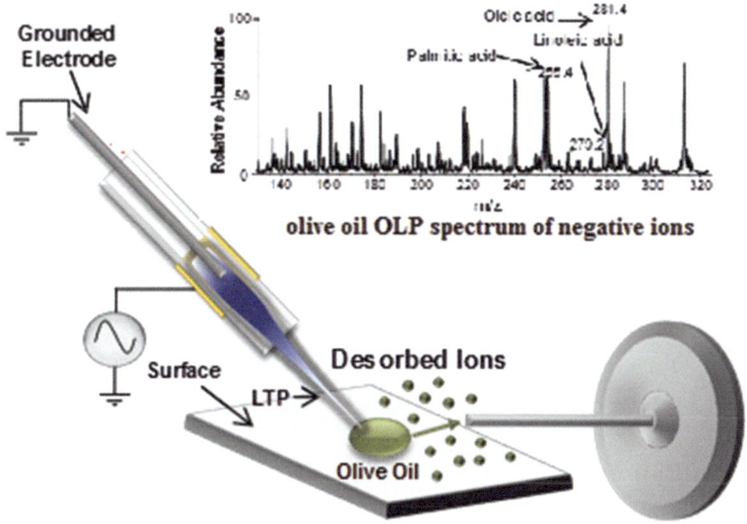

Fig. 12.6 LTP-MS full scan spectrum of crude virgin olive oil in the negative ionization mode

containing more than 98% of fats while the rest is characterized by volatile and relatively polar antioxidants species. This very rapid screening allows, however, a rough estimation of the concentration range of particular analytes in terms of antioxidant activity (total phenolic content), fatty acids and volatiles (Fig. 12.6) [19, 20].

12.2.6 Clinical Applications of Ambient Ionization Procedures

Caffeine is a natural alkaloid present in many drinks and drugs classified by FDA as *generally recognized as safe* (GRAS), the toxic doses for an adult being over 10 g per day. A very structurally similar species is represented by Theophylline usually conjugated with Amphetamine in the poisoning drug Fenethylline a drug of abuse known as Capatagon, or more often recently indicated as the jihadist drug, easili found in the middle east countries.

A reliable, fast, and green method for the assay of caffeine present in commercial beverages and drugs has been developed to provide alternative and faster strategies to the classic approaches for quantitative analysis commonly used and reported in literature, such as LC-UV methods (Table 12.2) [21].

PS-MS/MS absolute method was employed using the $^{13}C_3$ Trimethyl derivative of caffeine as internal standard for the quantitative assays of the analyte. The data reported in table two matched with the results obtained by LC-UV data, enforced the exceptionality of the PS-MS technique. Since ambient MS is quite adaptable to rapid and reproducible molecular screenings and quantitative analyses, the method could be extended to the assay of other similar species such as Fenethylline.

Table 12.2 Amount of caffeine in real samples

Samples	Caffeine found (PS-MS) μg/mL	RSD %	HPLC analysis μg/mL
Cola	115.9 ±8.2	4.14	123
Coffee espresso 1	6887.8 ± 203.9	3.38	6935
Coffee espresso 2	4921 ± 154.1	3.13	4650
Coffee espresso 3	4074 ± 201.5	4.94	4190
Coffee espresso 4	7864 ± 305.6	3.9	8177
Coffee espresso 5	6742 ± 287.2	4.2	6600
Energy drink	381.5 ± 12.0	3.15	378
Tea	119.7 ± 15.3	12.7	121
Drug 1	126.8 ± 8.9	7.03	122
Drug 2	164.3 ± 9.5	9.48	165

12.3 Conclusion

Most of the MS/MS analytical approach previously discussed deal with the characterization of a specimen, in terms of identification and assay, based on the use of labeled internal standard. This approach belongs to the so-called *absolute methods of analysis*. Quality safety and traceability of foods can be determined not only for environmental health but also to assess unambiguously the presence of fraudulently added molecules, such as the Sudan or melamine, extremely dangerous to people. The use of paper spray (PS) for the determination of Caffein in coffee drinks can be used for the evaluation of the presence of drugs like Captagon in the environment. The QUASIORA laboratory active at the University of Calabria is providing facilities to safeguarding people from environmental risks.

References

1. Jakob A, Crawford EA, Gross JH (2016) Detection of Polydimethylsiloxanes transferred from silicone-coated parchment paper to baked goods using direct analysis in real time mass spectrometry. J Mass Spectrom 51:298–304
2. Cody RB, Laramée JA, Durst HD (2005) Versatile new ion source for the analysis of materials in open air under ambient conditions. Anal Chem 77:2297–2302
3. De Nino A, Di Donna L, Mazzotti F, Perri E, Raffaelli A, Sindona G, Urso E (2001) Quantitative determination of Oleuropein in organic olive oil by ESI-MS/MS and Isotope Dilution. In: Pfannhauser W, Fenwick GR, Khokar SR (eds) Biologically-active phytochemicals in food. Analysis, metabolism, bioavailability and function. The Royal Society of Chemistry, Cambridge, p 131
4. De Nino A, Di Donna L, Mazzotti F, Muzzalupo E, Perri E, Sindona G, Tagarelli A (2005) Absolute method for the assay of Oleuropein in olive oils by atmospheric pressure chemical ionization tandem mass spectrometry. Anal Chem 77(18):5961
5. Sindona G (2010) Chapter 11: A marker of quality of olive oils: the expression of Oleuropein. In: Preedy V, Watson R (eds) Olives and olive oil in health and disease prevention. Academic Press, London, pp 95–100

6. La Marca G (2014) Mass spectrometry in clinical chemistry: the case of newborn screening. J Pharm Biomed Anal 101:174–182
7. Mcgorrin RJ (2009) One hundred years of progress in food analysis. J Agric Food Chem 57:8076–8088
8. Lynch JM, Barbano DM (1999) Kjeldahl nitrogen analysis as a reference method for protein determination in dairy products. J AOAC Int 82:1389–1401
9. Wu Q, Fan KX, Sha W, Ruan HQ, Zeng R, Shieh CH (2009) Highly sensitive detection of melamine based on reversed phase liquid chromatography mass spectrometry. Chin Sci Bull 54:732–737
10. Di Donna L, Maiuolo L, Mazzotti F, De Luca D, Sindona G (2004) ***Assay of Sudan I*** contamination of foodstuff by atmospheric pressure chemical ionization tandem mass spectrometry and Isotope Dilution. Anal Chem 76:5104–5108
11. Di Donna L, De Nino A, Maiuolo L, Mazzotti F, Napoli A, Salerno R, Sindona G (2007) High-throughput mass spectrometry: the mechanism of sudan azo dye fragmentation by ESI tandem mass spectrometry and extensive deuterium labeling experiments. J Mass Spectrom 42:1057–1061
12. Mazzotti F, Di Donna L, Maiuolo L, Napoli A, Salerno R, Sajjad A, Sindona G (2008) Assay of the set of all Sudan Azodyes (I, II, III, IV, and Para-red) contaminating agents by liquid chromatography-tandem mass spectrometry and Isotope Dilution methodology. J Agric Food Chem 56:63–67
13. Taverna D, Di Donna L, Mazzotti F, Policicchio B, Sindona G (2013) High-throughput determination of Sudan Azo-dyes within powdered chili pepper by paper spray mass spectrometry. J Mass Spectrom 48:544–547
14. Liu J, Wnag H, Nanicke NE, Lin J-M, Cooks RG, Ouyang Z (2010) Development, characterization, and application of paper spray ionization. Anal Chem 82:2463–2471
15. Taverna D, Di Donna L, Mazzotti F, Tagarelli A, Napoli A, Furia E, Sindona G (2016) Rapid discrimination of bergamot essential oil by paper spray mass spectrometry and Chemometric analysis. J Mass Spectrom 51:761–767
16. Bradshaw R, Wolstenholme R, Blackledge RD, Clench MR, Ferguson LS, Francese S (2011) A novel matrix-assisted laser Desorption/ionization mass spectrometry imaging based methodology for the identification of sexual assault suspects. Rapid Commun Mass Spectrom 25:415–422
17. Mirabelli MF, Ifa DR, Sindona G, Tagarelli A (2015) Analysis of sexual assault evidence: statistical classification of condoms by ambient mass spectrometry. J Mass Spectrom 50:749–755
18. Mulligan CC, Talaty N, Cooks RG (2006) Desorption electrospray ionization with a portable mass spectrometer: in situ analysis of ambient surfaces. Chem Commun 16:1709–1711
19. Harper JD, Charipar NA, Mulligan CC, Zhang X, Cooks RG, Ouyang Z (2008) Low-temperature plasma probe for ambient desorption ionization. Anal Chem 80:9097–9104
20. Garcıa-Reyes JF, Mazzotti F, Harper JD, Charipar NA, Oradu S, Ouyang Z, Sindona G, Cooks RG (2009) Direct olive oil analysis by Low-Temperature Plasma (LTP) ambient ionization mass spectrometry. Rapid Commun Mass Spectrom 23:3057–3062
21. Taverna D, Di Donna L, Bartella L, Napoli A, Sindona G, Mazzotti F (2016) Fast analysis of caffeine in beverages and drugs by paper spray tandem mass spectrometry. Anal Bioanal Chem 408:3783–3787

Chapter 13
A Tandem Mass Spectrometry Strategy for Validating the Synthesis of Glycoconjugate Vaccines

Wael L. Demian, Farid Jahouh, and Joseph H. Banoub

Abstract In principle, various types of vaccines that are commercially used as "antibacterial or anticancer vaccines" can be produced using various types of antigenic carbohydrate haptens containing relevant epitopes.

This review highlights the strategies used for the characterization of such glycoconjugate vaccines as well as a complete mass spectrometry-based strategy for validating their synthesis. The location of the covalently linked antigenic saccharide haptens on the protein carrier is usually termed as "glycation sites" and these are determined using tandem mass spectrometry. The ratio of carbohydrate hapten: BSA is determined by availing of the matrix-assisted laser desorption/ionization-TOF-MS analyses of the glycoconjugates. The outcome of our different studies was that all the glycated residues were located mainly near the outer surface of the carrier protein.

Keywords Protein carrier • TACA • MALDI-TOF-MS • LC-ESI-QqTOF-MS/MS • Glycoconjugate vaccines • Glycation sites

W.L. Demian
Department of Biochemistry, Memorial University of Newfoundland, St. John's, NF A1B 3X9, Canada

F. Jahouh
Department of Chemistry, Memorial University of Newfoundland, St. John's, NF, 232 Elizabeth Avenue, A1B 3X7, Canada

J.H. Banoub (✉)
Fisheries and Oceans Canada, Science Branch, Special Projects, Chemistry Department, Memorial University, St John's, NL, Canada
e-mail: banoubjo@dfo-mpo.gc.ca

J.H. Banoub, R.M. Caprioli (eds.), *Molecular Technologies for Detection of Chemical and Biological Agents*, NATO Science for Peace and Security Series A: Chemistry and Biology, DOI 10.1007/978-94-024-1113-3_13

13.1 Introduction

A vaccine is a substance used to confer protection by stimulating the immune system to secrete antibodies against diseases [1, 2]. In 1796, Edward Jenner discovered that the human inoculation with cowpox was able to protect against smallpox infection [1, 2]. Since then, different vaccines were developed to prevent infectious diseases [1, 2].

There are different forms of vaccines: attenuated vaccines, inactivated vaccines, toxoid vaccines, subunit vaccines, DNA vaccines, recombinant vector vaccines and conjugate vaccines [3]. This review is more focused on the science behind the formation of synthetic neoglycoconjugate vaccines.

Lipopolysaccharides or LPSs are one of the major constituents of the cell walls which coat the Gram negative bacteria [4, 5]. LPSs consist of three main parts: an external part called the *O*-antigen, composed of repeats of identical sugar oligosaccharide units, and an internal core which is covalently linked to the third part, the "lipid A" [4, 5]. The lipid A represents the toxic part of the LPS, and it consists of a β-D-(1 → 6) glucosamine (GlcN) disaccharide in which O-3, O-3′, O-4′, C2-N and C2-N'are acylated with different C:12 and C:14 fatty acids [4, 5]. The carbohydrate part of the LPS, such as the core oligosaccharide or the *O*-antigen part, can be conjugated with biomolecules like proteins or dendrimers, to form elegant branched molecules to produce the glycoconjugate vaccine [4–7].

A carbohydrate antigen neoglycoconjugate vaccine is composed of the antigenic carbohydrate part called the "hapten", covalently linked by a spacer to a carrier biomolecule (usually a protein, Fig. 13.1).

The synthesis of efficient glycoconjugate vaccines depends on many factors, and their efficacy relies on the saccharide size, the average number of saccharide chains per conjugate molecule, the nature of the carrier and the distance between the saccharide and the protein in the formed glycoconjugate [8–11].

It is worthy to mention that antigenic carbohydrate-protein vaccines are not limited to fighting bacteria, as they can also be developed as anticancer drugs by conjugation of a tumor associated carbohydrate antigen (TACA) to a carrier protein [12]. The first carbohydrate-protein vaccine was reported by Landsteiner's group [13, 14], and later, it was discovered that they can induce strong antibody reaction [15].

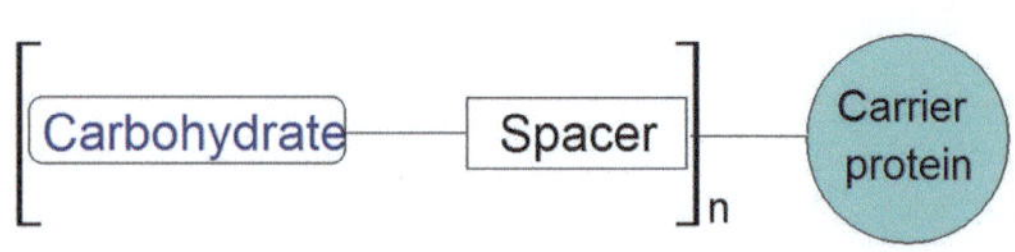

Fig. 13.1 Composition of the glycoconjugate vaccine

13.2 Methods for Neoglycoconjugates Synthesis

Different methods have been used for the synthesis of carbohydrate antigen-protein neoglycoconjugates. One of these methods consisted of the use of the squaric acid chemistry to perform a single-point attachment of carbohydrates to proteins [15–18]. Tietze et al. used squaric acid diethyl esters for conjugation [16]. It has also been reported that squaric acid dimethyl esters [19, 20], as well as didecyl squarate, can be used for the single point attachment of carbohydrates to proteins [19–22]. In addition, Kamath *et al.* used squaric acid amide ethyl esters for the conjugation of oligosaccharides to protein and monitored the conjugation using Matrix Assisted Laser Desorption Ionization-Time Of Flight-Mass Spectrometry (MALDI-TOF-MS) [23].

More recently, the group of Kováč utilized the squaric acid chemistry to conjugate different carbohydrate antigens to protein carriers [24–26]. They used this strategy to conjugate the synthetic tetrasaccharide side chain of the *Bacillus anthracis* exosporium to the bovine serum albumin (BSA) protein [25].

Furthermore, the development of glycoconjugate vaccines can also be achieved by the Michael addition of carbohydrate antigens to the carrier protein. This has been used for the synthesis of anticancer vaccines by targeting the sulfhydryl group of the cysteine residues, and the ε- amino groups of the lysine residues [27–30]. Basically, the Michael reaction or Michael addition is the nucleophilic addition of a carbanion or nucleophile to an α,β-unsaturated carbonyl compound.

13.3 Role of the Biomolecule Protein Carrier

In essence, the carrier protein used in the glycoconjugate vaccine synthesis is very important, since it is responsible for the production of memory cells that confers the immune protection [33]. Recently, Roy's group provided evidences of the role of the carrier protein [34]. The carrier protein had been replaced with repetitively branched molecules called dendrimers. Conversely, dendrimers do not elicit the memory B cells, although they provoke plasma B cells to secrete enough antibodies [34].

In case of infection, the immune system produces antibodies to kill the infected cells. However, the memory pool is depleted and the memory B cells are not secreted [33]. Previous studies showed the molecular mechanisms of the generation of immune responses against protein-polysaccharide conjugate vaccines [33]. The protein carrier is known to stimulate helper T-cells that elicit the memory B cells production. Two possible theories are reflecting the vital role of the protein carrier for stimulating the helper T-cells. The first theory suggests that the processed peptide epitope derived from the carrier protein is presented by MHC II and recognized by a peptide-specific T cell (Fig. 13.2a).

As the hydrophilic polysaccharide cannot enter the MHC II cavity, the polysaccharide itself cannot be presented to T cells by MHC II on the surface of the B cell in order to induce T cell help. The other mechanism suggests that the peptide

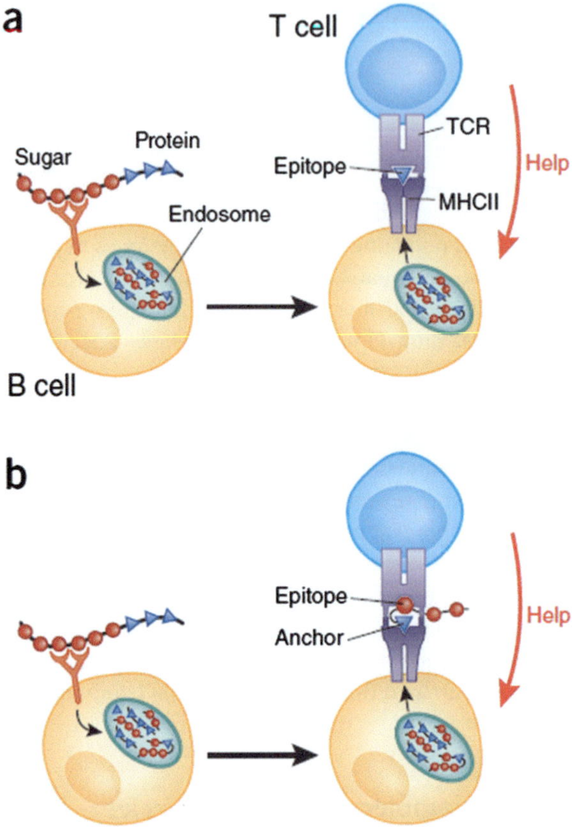

Fig. 13.2 Two possible mechanisms for the generation of immune responses against protein-polysaccharide conjugate vaccines [33]. (**a**) The processed peptide epitope derived from the carrier protein is presented by MHCII and recognized by a peptide-specific T cell. (**b**) The peptide anchors the sugar epitope to the MHC and allows presentation of the sugar epitope to a polysaccharide-specific T cell

anchors the sugar epitope to the MHC II and allows the presentation of the sugar epitope to a polysaccharide-specific T cell (Fig. 13.2b) [33].

To summarize, polysaccharide vaccines (without protein carriers conjugation) stimulate primarily an IgM response. However, conjugation with protein carriers stimulates both IgM and IgG immune responses. Different protein carriers such as BSA, keyhole limpet hemocyanine (KLH), diphtheria toxoid (DT), cross-reactive material 197 (CRM197) and tetanus toxoid (TT) have been used for the synthesis of various anticancer carbohydrate-based vaccines [35–37].

BSA is a plasma protein in cattle that is one of the most stable and soluble albumins available. It has a MW of 67 × 103 Da, and includes 59 lysines. There are ~30–35 lysine residues accessible for use in linker conjugation, and so BSA is a popular carrier protein for weakly antigenic compounds [38]. In this review, we discuss two examples of the conjugation of carbohydrates to BSA.

13.4 Role of the Antigenic Carbohydrate Hapten in the Glycoconjugate Vaccine

Haptens are small molecules "antigens" that cannot elicit an immune response unless they are attached to a large carrier such as a protein. Although haptens are able to bind to the secreted antibodies, they usually do not initiate any immune response. Karl Landsteiner first pioneered the use of different synthetic haptens to study immunochemical phenomena [13, 14]. The first example of hapten used was an diazonium salt obtained from either o-, m- and p-aminobenzoic acid, that readily formed a covalent bond with nucleophilic amino acid residues on carrier proteins [13, 14]. The usage of the hapten-carrier protein conjugates is still commonly practiced today, in order to obtain hapten-specific antibodies. Bacterial polysaccharides have been used successfully as hapten for glycoconjugate vaccines production. For instance, direct conjugation of polyribosyl ribitoyl phosphate (PRP) oligomer to the protein carrier CRM197, a nontoxic variant of diphtheria toxin, was used to generate specific antibodies against *Haemophilus influenza* type b (Hib) [39]. Epitopes of capsular polysaccharides offer promising possibilities for the development of vaccines for the prevention of infectious diseases such as *Haemophilus influenzae* type b, human immunodeficiency virus (HIV), *Plasmodium falciparum*, *Vibrio cholerae*, *Shigella dysenteriae*, *Cryptococcus neoformans*, *Streptococcus pneumoniae*, *Neisseria meningitides*, *Bacillus anthracis* and *Candida albicans* [17, 39–44].

The potency of neoglycoconjugate vaccines are determined by changing different lengths of the oligosaccharide chain. The effectiveness of synthetic polysaccharide type-3 related di- tri- and tetrasaccharide-CRM197 conjugates were investigated to provide protection against *S. pneumoniae* type 3 infection [44]. These synthetic oligosaccharides were coupled via the squarate method. Mice were injected with the di- tri- and tetrasaccharide-CRM197 conjugates [45]. All mice immunized with the tri- or tetrasaccharide conjugates had IgG antibodies that bound the capsular polysaccharide [41]. Another study showed that a hexasaccharide derived from polysaccharide type 3 coupled to KLH was capable to induce the immune response against type 3 *pneumococci* in mice [45].

An interesting example of glycoconjugate vaccine against *H. influenzae* type B showed the potency of conjugates of tri- and tetrameric 3-β-D-ribose-(1 → 1)-ribitol-5-phosphate and tetanus or diphtheria toxoid. These conjugates were capable to initiate an immune response against capsular-polysaccharide responses in adult mice and monkeys with an increasing in IgG/IgM ratio [41].

Lately reliable vaccines against *B. anthracis* has been developed using different synthetic protocols, as a precautionary measure for prevention of Anthrax infection [25, 47, 48]. These different protocols were based on the preparation of immunogenic conjugates of the capsular polypeptide (polyglutamic acid) of *B. anthracis*, and anthrax capsular vaccine. Kováč *et al.* synthesized the tetrasaccharide isolated from *B. anthracis* exosporium and linked it to the BSA using the squaric acid chemistry [25]. For instance, *B. anthracis* neoglycoconjugate vaccines were investigated by MALDI-TOF/TOF-CID-MS/MS and LC-ESI-QqTOF-tandem mass spectrometry [48–51].

13.5 Role of Antigenic Haptens Conjugates Against Cancer

The identification of distinct glycoproteins on the cell surface of malignant cells has spurred intense research into exploiting these tumor-associated carbohydrate antigens (TACA) for the development of anticancer vaccines. These antigens shared by several cancer types, or expressed by the normal tissue from which the tumor grows [12, 52, 53]. Some of these antigens are carbohydrates and several trials are being carried out to explore their possibilities to be used for the development of vaccines.

The cytotoxic T cells (CD8+ T cells) play an important role in fighting cancer cells through recognizing the TACAs in a conventional class I MHC-restricted fashion [12, 54–56]. Accordingly, in some forms, the resulting increase in the numbers of glycosylation sites are reflected by the presence of the TACAs such as: Tn (GalNAcα1 → Ser/Thr), sialyl-Tn (Neu5Acα2 → 6GalNAc) and Thomsen-Friedenreich (TF: Galβ1 → 3GalNAcα1 → Ser/Thr) antigens [12, 55–58].

13.6 Role of Mass Spectrometry in Validation of Glycoconjugate Vaccines Synthesis

The analysis and the localization of the glycation sites in carbohydrate-protein neoglycoconjugates are very important features for the quality control of conjugate vaccines. Different strategies have been developed for the analysis of glycoproteins, mainly based on the mass spectrometry analysis of enzymatic digests. Herein, we did explore strategies for the characterization of glycoconjugate vaccines using MS, and for the determination of the glycation sites [48–51].

13.6.1 Strategies for the Characterization of Glycoconjugate Vaccines Using MS

Initially, the use of MALDI-TOF-MS for the analysis of different hapten-BSA glycoconjugate vaccines allowed the determination of the molecular weights of the glycoconjugate and concurrently permitted the determination of the hapten-to-BSA ratios [50–53]. This is followed by the enzymatic digestion of the glycoconjugate vaccine and the MALDI-TOF/TOF-MS/MS and LC-ESI-QqTOF-MS/MS analysis of the digests for the determination of the glycation sites using *de novo* sequencing. The enzymatic digestion is usually done with two different enzymes. The first one, trypsin, does not digest or react with the glycated lysine residues of the protein, but cleaves the lysine and asparagine residues. The second enzyme that is commonly used is the GluC V8 endoproteinase which is known to digest proteins at C-terminus of aspartic acid and glutamic acid residues [42, 50].

13.6.2 Molecular Mass and Carbohydrate-to-Protein Ratio Determination

The purpose of determining the molecular mass and the carbohydrate-to-protein ratio of a carbohydrate-protein neoglycoconjugate is to define the number of carbohydrates that are incorporated into the protein carrier, as a result of conjugation. Two main methods are currently used for the molecular mass determination of carbohydrate-protein glycoconjugates: MALDI-TOF-MS [61] and Surface Enhanced Laser Desorption Ionization Time-Of-Flight Mass Spectrometry (SELDI-TOF-MS) [23–25, 60–62].

Both of these methods allow the carbohydrate-to-protein ratio determination of the neoglycoconjugates by comparing the molecular mass of the protein before and after the conjugation.

13.6.3 Glycation Sites Determination

One of the first method for the glycation sites determination of carbohydrate-protein neoglycoconjugates ever carried out in the literature, involved the digestion the neoglycoconjugate using a protease, such as trypsin or GluC V8 endoproteinase, followed by MALDI-TOF-MS/MS or LC-MS/MS [48–51].

It is primordial to note that during the tandem mass spectrometry analysis of the glycoconjugate digests, the identification of the glycated peptides are usually initiated by distinguishing the presence of diagnostic carbohydrate product ions in the CID-spectrum. In addition, the tandem mass spectrometry analysis allows also the unraveling of sequence of the peptide through diagnostic product ions of the peptide moiety of the glycated peptide. The combined information allows the unambiguous characterization of the carbohydrate-peptide and glycation site identification.

During tandem mass spectrometry analyses of the glycated peptides, the product ions corresponding to the fragmentation of the peptide portion are identified using the nomenclature established by Roepstorff et al. and lately is modified by Johnson and coworkers [63, 64], whereas, the product ions resulting from the fragmentation of the carbohydrate moiety are assigned using the nomenclature introduced by Domon and Costello, as: A, B, C, X, Y and Z [65].

13.6.3.1 Neoglycoconjugate Vaccine of the Biological Agent *Bacillus anthracis*, the Etiological Agent of Anthrax

Bacillus anthracis is a Gram-positive bacterium that causes anthrax to both humans and animals [66]. The formation of endospores at the maturity stage of the bacterium allows a protection against severe conditions, such as extreme temperatures, radiations, physical damages and chemicals [67, 68]. *Bacillus anthracis* is the etiologic agent of anthrax that can be used as a biological weapon [69, 70]. Indeed, *Bacillus anthracis* a pathogen that is lethal in most cases for both humans and animals. There are different *Bacillus anthracis* strains, among those, 89 of them were identified, such as the *Sterne* strain [76], the *Vollum* strain [77], the *Ames* strain [75, 78] and the H9401 strain [74]. The last strain has also been studied for the development of the anthrax vaccine [74]. In addition, *Bacillus anthracis* has been extensively studied in the aim of understanding its pathogenesis, identifying new biomarkers and vaccines design [75]. The capsular polypeptide (polyglutamic acid) of the *Bacillus anthracis* has been targeted for the development of synthetic vaccines [46].

Daubenspeck et al. reported the structure of the tetrasaccharide side chain of the collagen like region of the major glycoprotein of the *B. anthracis* exosporium [47]. Their findings was that the upstream terminal of the tetrasaccharide corresponds to the sugar anthrose [4,6-dideoxy-4-(3-hydroxy-3-methylbutyramido)-2-O-methyl-D-glucopyranose].

The group of Kováč prepared a vaccine composed of a synthetic tetrasaccharide side chain of the collagen like region of the major glycoprotein of the *B. anthracis* exosporium (MW = 950.43 Da) attached to the BSA using the squaric acid chemistry [25]. The conjugation leaded to the formation of vaccines with different carbohydrate: BSA ratios. The carbohydrate: BSA ratio and glycation sites were well validated and determined in our previous work [51].

- **Carbohydrate:BSA ratio of the anthrax vaccine**

MALDI-TOF-MS analysis of the synthetic vaccine allowed to observe the following protonated molecular ions: $[M + H]^+$ at *m/z* 71,448.36 and $[M + 2H]^{2+}$ at *m/z* 35,730.08. The molecular weight of this vaccine was thus found to be 71,447.36 Da. Similarly, the MALDI-TOF-MS analysis of the BSA revealed the following protonated molecular ions: $[M + H]^+$ at *m/z* 66,318.88 and $[M + 2H]^{2+}$ at *m/z* 33,161.81, which allowed to determine the molecular weight of BSA, being 66317.88 Da. It was observed in our previous that the squaric acid chemistry allows the attachment of the synthetic carbohydrate antigens to the ε-amino groups of the lysine residues in BSA with loss of a molecule of ethanol [48–50]. Thus, the molecular weight of the synthetic tetrasaccharide side chain of the collagen like region being 950.43 Da, the carbohydrate: protein ratio of the synthetic vaccine was determined to be 5.4:1.

- **Glycation sites determination of the anthrax neoglycoconjugate vaccine**

The determination of the glycation sites of the hapten-BSA vaccine neoglycoconjugate was carried out by its enzymatic digestion with the trypsin or GluC V8 proteases, followed by the MALDI-MS/MS and LC-MS/MS analysis of the digests.

The enzymatic digestion of the hapten-BSA glycoconjugate and MALDI-MS/MS as well as LC-MS/MS analyses were carried out as previously described [49].

- **MALDI-CID-MS/MS of the anthrax neoglycoconjugate vaccine**

The spectra obtained during the MALDI-CID-MS/MS analysis of the tryptic and GluC V8 digests were submitted to the MASCOT library to identify by PMF the peptides matching to the BSA (Fig. 13.3). Two Serum Albumin protein isoforms from the *Bos taurus* species were identified for the tryptic and GluC V8 digests: the serum albumin precursor (gi|1351907) and the serum albumin (gi|74267962). The detected peptides that were not identified in the database were analyzed by tandem mass spectrometry. The MALDI-MS/MS analysis of the tryptic digests allowed to identify three glycated peptides (Fig. 13.3) corresponding to BSA peptides with an increment of 950 Da to their original molecular mass, namely: ALK*AWSVAR at *m/z* 1951.0130, VTK*CCTESLVNR at *m/z* 2416.1372, and QNCDQFEK*LGE YGFQNALIVR at *m/z* 3478.6429 (glycation represented with an asterisk on the lysine residue).

The high-energy CID-MS/MS analysis of the glycopeptide ALK*AWSVAR (Fig. 13.4) at *m/z* 1951.0130 afforded a series of product ions corresponding to the entire precursor ion that losses different carbohydrate portions, the monosaccharide B_1 (−259 Da), disaccharide B_2 (−405 Da), trisaccharide B_3 (−551 Da) and/or the tetrasaccharide B_4 (−697 Da), leading respectively to the formation of the following product ions: Y_3^+ at *m/z* 1691.8829, Y_2^+ at *m/z* 1545.8376, Y_1^+ at *m/z* 1399.7768 and Y_0^+ at *m/z* 1253.7010 (Fig. 13.4). The mass difference between Y_3^+ and Y_2^+, Y_2^+ and Y_1^+, Y_1^+ and Y_0^+ was found to correspond to one α-L-rhamnopyranosyl unit (146 Da). Moreover, product ions resulting from the fragmentation of the same tetrasaccharide were observed: $C_{17}H_{26}N_3O_4^+$ at *m/z* 336.1242, B_1^+ at *m/z* 260.0739, $[B_1 - H_2O]^+$ at *m/z* 242.0709, $^{2,5}A_1^+$ at *m/z* 230.0719, $C_{11}H_{16}N_3O_2^+$ at *m/z* 222.0670 and $[C_1 - C_5H_{11}NO_2]^+$ at *m/z* 159.0486 (Fig. 13.4).

Accordingly, the fragmentation of the carbohydrate portion allowed to confirm its structure and contributed to establish a diagnostic fragmentation signature of the synthetic tetrasaccharide. In addition, the fragmentation of the peptide portion leaded to the formation of b- and y- product ions that allowed the determination of the sequence of the glycated peptide. However, some peptide product ions had the particularity to have lost the carbohydrate portion but were still attached to the spacer-squaric acid chain: $[b_3 - B_4]^+$ at *m/z* 565.2661, $[b_4 - B_4]^+$ at *m/z* 636.2715 and $[y_7 - B_2]^+$ at *m/z* 1361.7265. Thus, the identified glycated peptides ALK*AWSVAR at *m/z* 1951.0130, VTK*CCTESLVNR at *m/z* 2416.1372, and QNCDQFEK*LGEYG FQNALIVR at *m/z* 3478.6429 allowed determining the glycation site on the following lysine residues: Lys 235, Lys 498 and Lys 420, respectively.

Similarly, the MALDI-MS/MS analysis of the GluC V8 digests afforded different glycated peptides: LCK*VASLRE at *m/z* 2025.0186, YAVSVLLRLAK*E at *m/z* 2311.2476 and YAVSVLLRLAK*EYE at *m/z* 2603.3484, allowing the identification of the following glycation sites on lysine residues: Lys 100 and Lys 374. To sum it up, only 5 glycation sites were identified during the MALDI-MS and MS/MS analyses of the tryptic and GluC V8 digests of the hapten-BSA glycoconjugate: Lys 100, Lys 235, Lys 374, Lys 420 and Lys 498.

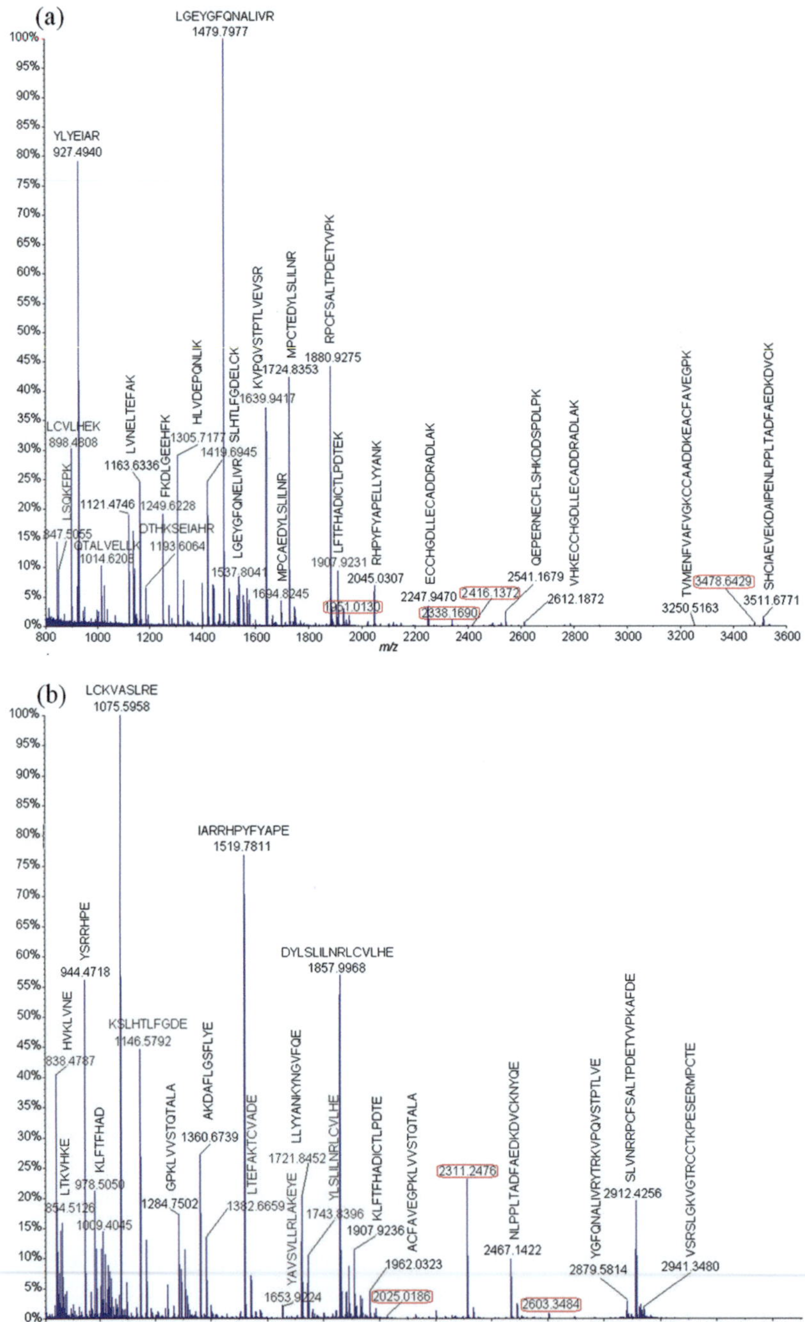

Fig. 13.3 MALDI-MS analysis of the glycoconjugate trypsin digests (**a**) and GluC V8 endoproteinase digests (**b**) [51]

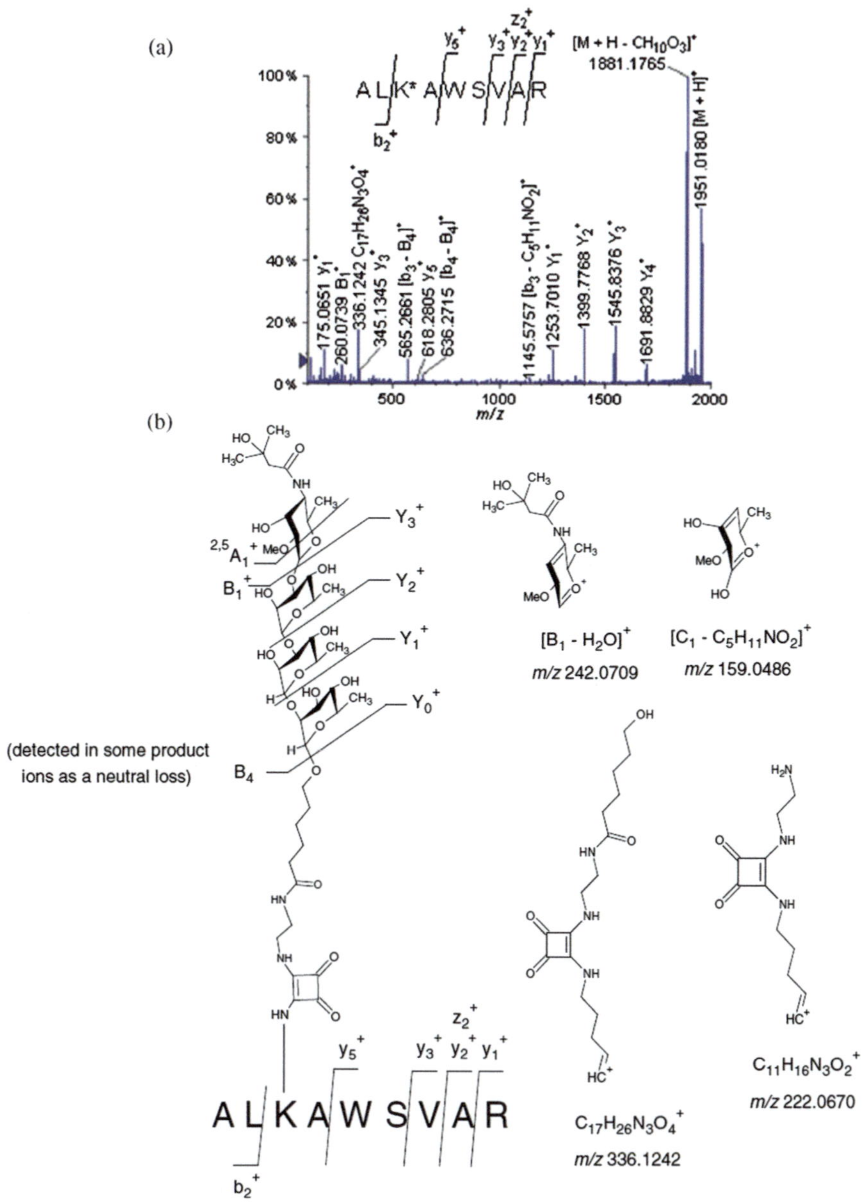

Fig. 13.4 (**a**) MALDI-TOF/TOF-MS/MS spectra of the glycated peptide ALK*AWSVAR (Lys 235) at *m/z* 1951.0130. (**b**) Different product ions involving the fragmentation of the carbohydrate hapten observed during the MALDI-TOF/TOF-MS/MS analysis of the glycated peptide ALK*AWSVAR (Lys 235) at *m/z* 1951.0130 [51]

- **LC-ESI-CID-MS/MS of the anthrax neoglycoconjugate vaccine**

The second approach for the determination of the glycation sites of the vaccine neoglycoconjugate was the LC-MS/MS analysis of the tryptic and GluC V8 digests. It has to be noted that the LC-MS/MS analysis of peptides has the advantage of minimizing the ionization suppression effect comparing to the MALDI-MS/MS analysis [76, 77].

The LC-MS/MS data of the analysis of the tryptic and GluC V8 digests were submitted to the MASCOT library and also matched two serum albumin protein isoforms: the serum albumin precursor (gi|1351907) and serum albumin (gi|74267962) from *Bos taurus*.

For the tryptic digests, the BSA sequence coverage was found to be 57% for the serum albumin precursor from *Bos taurus* (gi|1351907) and 58% for serum albumin protein from *Bos taurus* (gi|74267962). The LC-MS/MS analysis of the hapten-BSA tryptic digests allowed the identification of 18 glycated peptides, reported in Table 13.1.

The low-energy CID-MS/MS analysis of the extracted precursor ions of the glycated peptides allowed to localize the glycation sites on the following 18 lysine residues: Lys 140, Lys 155, Lys 156, Lys 204, Lys 211, Lys 228, Lys 235, Lys 304, Lys 374, Lys 401, Lys 420, Lys 437, Lys 455, Lys 463, Lys 495, Lys 498, Lys 547 and Lys 559.

The LC-MS/MS analysis of the hapten-BSA vaccine GluC V8 digest allowed the identification of the serum albumin from *Bos taurus* (gi|74267962) with a sequence coverage of 42% and the precursor serum albumin from *Bos taurus* (gi|1351907) with a sequence coverage of 45%, in the MASCOT database. Table 13.2 displays the identified glycated peptides during the LC-MS/MS analysis of the GluC V8 digests. Thus, the CID-MS/MS analysis of theses glycated peptides allowed to discover 17 glycation sites, localized on the following lysine residues: Lys 65, Lys 75, Lys 88, Lys 100, Lys 117, Lys 151, Lys 183, Lys 197, Lys 256, Lys 266, Lys 304, Lys 309, Lys 336, Lys 374, Lys 420, Lys 455 and Lys 495 (Table 13.2).

In summary, the LC-MS/MS analysis of both tryptic and GluC V8 digests allowed the identification of a total of 30 glycation sites on the lysine residues (Fig. 13.5). Mapping these glycation sites on the 3D representation of the BSA (Fig. 13.5), lysines highlighted in red) permitted to observed that they correspond to lysine residues located at the outer surface of the BSA. In addition, the number of the identified glycation sites (30 lysine residues) being higher than the determined hapten: BSA ratio 5.4:1) of the tetrasaccharide-BSA neoglycoconjugate, it was concluded that the vaccine is composed of a mixture of glycoforms.

Table 13.1 Tryptic glycopeptides identified of the bovine serum albumin protein by LC-ESI-QqTOF-MS/MS analysis of the hapten-BSA glycoconjugate [51]

Precursor ion *m/z* (charge)	Mr(expt)	Mr(calc)	Deviation (Da)	Missed cleavage	Peptide sequence
729.8826 (+2)	1457.7506	1457.7389	0.0117	1	KHK*P (Lys 559)
733.3170 (+2)	1464.6194	1464.7698	−0.1504	1	QIK*K (Lys 547)
770.3757 (+2)	1538.7369	1538.7338	0.0031	1	ADEK*K (Lys 155)
807.9138 (+2)	1613.8129	1613.7964	0.0165	0	K*FWGK (Lys 156)
828.0811 (+3)	2481.2213	2481.2005	0.0208	2	LKECCDK*PLLEK (Lys 304)
832.7770 (+3)	2495.3092	2495.3146	−0.0054	1	LK*HLVDEPQNLIK (Lys 401)
863.7965 (+3)	2588.3677	2588.3572	0.0105	0	K*VPQVSTPTLVEVSR (Lys 437)
883.9655 (+2)	1765.9164	1765.9085	0.0079	1	SLGK*VGTR (Lys 455)
922.0941 (+3)	2763.2605	2763.2493	0.0112	1	LAK*EYEATLEECCAK (Lys 374)
969.5086 (+2)	1937.0026	1936.9868	0.0158	1	TPVSEK*VTK (Lys 495)
970.4900 (+2)	1938.9738	1938.9772	−0.0034	1	EK*VLTSSAR (Lys 211)
976.0173 (+2)	1950.0201	1950.0085	0.0116	1	ALK*AWSVAR (Lys 235)
990.4710 (+3)	2968.3911	2968.3886	0.0025	2	LK*PDPNTLCDEFKADEK (Lys 140)
1073.0144 (+2)	2144.0143	2144.0082	0.0061	1	CASIQK*FGER (Lys 228)
1058.4652 (+2)	2114.9159	2114.9123	0.0036	1	CCTK*PESER (Lys 463)
1208.5857 (+2)	2415.1569	2415.1284	0.0285	1	VTK*CCTESLVNR (Lys 498)
1160.2172 (+3)	3477.6297	3477.6385	−0.0088	1	QNCDQFEK*LGEYGFQNALIVR (Lys 420)
1169.5978 (+2)	2337.181	2337.1583	0.0227	1	GACLLPK*IETMR (Lys 204)

Table 13.2 Glycopeptides identified in the bovine serum albumin protein by LC-ESI-QqTOF-MS/MS analysis of the hapten-BSA glycoconjugate digested with the endoproteinase GluC V8 [51]

Precursor ion *m/z* (charge)	Mr(expt)	Mr(calc)	Deviation Da	Missed cleavage	Peptide
698.3598 (+2)	1394.7051	1394.6803	0.0248	0	K*LGE (Lys 420)
790.3774 (+2)	1578.7403	1578.7288	0.0115	1	K*QEPE (Lys 117)
844.3721 (+2)	1686.7297	1686.7499	−0.0202	1	EFK*ADE (Lys 151)
894.4637 (+2)	1786.9128	1786.8976	0.0152	0	HVK*LVNE (Lys 65)
862.9539 (+2)	1723.8933	1723.8754	0.01179	1	VTK*LVTD (Lys 256)
897.4200 (+2)	1792.8255	1792.8176	0.0079	0	K*SHCIAE (Lys 309)
902.4801 (+2)	1802.9456	1802.9289	0.0167	0	LTKVHK*E (Lys 266)
962.1552 (+2)	2883.4439	2883.4093	0.0346	0	VSRSLGK*VGTRCCTKPE (Lys 455)
987.9658 (+2)	1973.9171	1973.8949	0.0222	0	K*VTKCCTE (Lys 495)
989.9697 (+2)	1977.9248	1977.9154	0.0094	2	K*QEPERNE (Lys 117)
992.4568 (+2)	1982.8989	1982.884	0.0149	1	CCDK*PLLE (Lys 304)
995.4702 (+2)	1988.9258	1988.8912	0.0346	1	FAK*TCVADE (Lys 75)
1013.0142 (+2)	2024.0138	2024.0123	0.0015	0	LCK*VASLRE (Lys 100)
1048.5225 (+2)	2095.0305	2094.9984	0.0321	1	K*SLHTLFGDE (Lys 88)
1097.0595 (+2)	2192.1045	2192.0909	0.0136	0	DK*GACLLPKIE (Lys 197)
1124.5018 (+2)	2246.9890	2246.9876	0.0014	1	DK*DVCKNYQE(Lys 336)
1336.1446 (+2)	2670.2746	2670.2728	0.0018	0	LLYYANK*YNGVFQE (Lys 183)
1156.132 (+2)	2310.2495	2310.2345	0.0150	0	YAVSVLLRLAK*E (Lys 374)

13.6.3.2 Anti-tumor Thomsen-Friedenreich Neoglycoconjugate Vaccine Prepared by Michael Addition

Thomsen-Friedenreich (TF) is a TACA recognized by major histocompatibility protein class (MHC) I [86–90]. TF is a disaccharide (β-D-Galactose and

(a)

```
  1 MKWVTFISLL LLFSSAYSRG VFRRDTHKSE IAHRFKDLGE EHFKGLVLIA
 51 FSQYLQQCPF DEHVK*LVNEL TEFAK*TCVAD ESHAGCEK*SL HTLFGDELCK*
101 VASLRETYGD MADCCAK*QEP ERNECFLSHK DDSPDLPKLK* PDPNTLCDEF
151 K*ADEK*K*FWGK YLYEIARRHP YFYAPELLYY ANK*YNGVFQE CCQAEDK*GAC
201 LLPK*IETMRE K*VLTSSARQR LRCASIQK*FG ERALK*AWSVA RLSQKFPKAE
251 FVEVTK*LVTD LTKVHK*ECCH GDLLECADDR ADLAKYICDN QDTISSKLKE
301 CCDK*PLLEK*S HCIAEVEKDA IPENLPPLTA DFAEDK*DVCK NYQEAKDAFL
351 GSFLYEYSRR HPEYAVSVLL RLAK*EYEATL EECCAKDDPH ACYSTVFDKL
401 K*HLVDEPQNL IKQNCDQFEK* LGEYGFQNEL IVRYTRK*VPQ VSTPTLVEVS
451 RSLGK*VGTRC CTK*PESERMP CAEDYLSLIL NRLCVLHEKT PVSEK*VTK*CC
501 TESLVNRRPC FSALTPDETY VPKAFDEKLF TFHADICTLP DTEKQIK*KQT
551 ALVELLKHK*P KATEEQLKTV MENFVAFVGK CCAADDKEAC FAVEGPKLVV
601 STQTALA
```

(b)

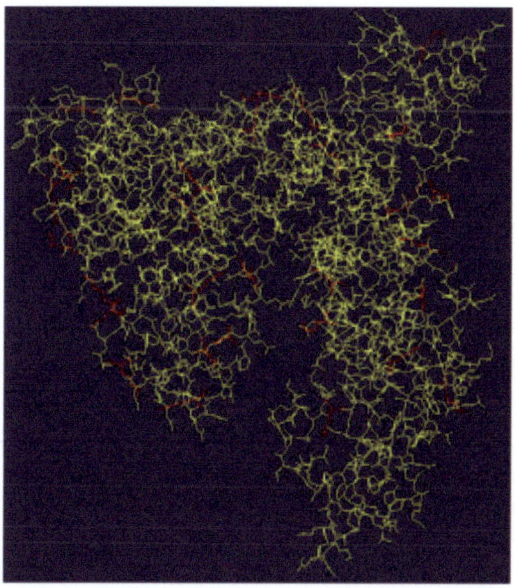

Fig. 13.5 (**a**) BSA sequence where the glycation sites are indicated by an asterisk (*red* = identified on tryptic digests, *blue* = identified on GluC V8 digests and *red* and *underlined* = identified on both tryptic and GluC V8 digests) and (**b**) 3D structure of the BSA. The glycated lysine residues are highlighted in red (Swiss-Pdb Viewer software) [51]

α-D-Galactoseamine) bound to the hydroxyl group of threonine and serine residues on the core of mucin protein [86–90]. This antigen was found on different cancers such as breast, colon, prostate or bladder to metastasize, and it was also responsible for causing aberrant *O*- glycosylation of the mucin protein. In case of cancer, 50% of mucin *O*-sites gain *O*-glycosylation by TF and T_N (N-acetylglucoseamine) antigens. Additionally, sialic acid glycoproteins are overexpressed by cancer cells. Consequently, in case of cancer, sialyl TF and silayl T_N are formed, causing aberrant glycosylation [57–60, 78].

Several studies were reported for a drug design against the "TF" antigen. Monoclonal antibodies had been produced from mouse, such as the JAA-F11 (IgG3) and the C5 (IgM). More recently, Hoffmann-Röder et al. prepared several synthetic antitumor vaccines composed of BSA conjugates of mucin 1 (MUC1) glycopeptides with the TF (TF -MUC1-BSA conjugate) and a fluorine-substituted analogue (difluoro- TF -MUC1-BSA conjugate) [57–60, 78].

In the following example, the TF:BSA protein conjugate based vaccine was synthesized by Michael reaction. To the best of our knowledge, the systematic use of the L-lysine ε-amino groups of the protein carrier has scarcely been investigated as nucleophilic partners for the chemical ligation by a 1,4-conjugate addition [79–81]. Furthermore, the detailed structural identity of the glycoforms which result from the 1,4-conjugate addition (Michael reaction) has never been used for the conjugation of the anticancer TF to BSA as a protein carrier.

- **Synthesis of the anti-tumor Thomsen-Friedenreich neoglycoconjugate vaccine prepared by the Michael addition.**

The Michael addition reaction was used to target the free sulfhydryl group of cysteine 34 (Cys 34) of the BSA protein carrier during the synthesis of this neoglycoconjugate vaccine. BSA is rich in potential glycation sites, because it contains 59 lysine residues and 35 cysteine residues, 34 of which form 17 disulfide groups, leaving one free sulfhydryl group on the Cys 34 [31].

The TF-antigen was prepared according to a published procedure [81]. Briefly, the photocatalyzed reaction of the allyl glycoside of Thomsen–Friedenreich (TF) disaccharide [allyl 3-*O*-(β-D-galactopyranosyl)-2-acetamido-2-deoxy-α-D-glucopyranoside] (**1**) with cysteamine gave the TF glycoside derivative **2** with a terminal primary amino group in 83% yield (Scheme 13.1, Fig. 13.6).

Subsequently, compound **2** was treated with acryloyl chloride in methanol to yield the α,β-unsaturated amide derivative **3** via the nucleophilc displacement of the chloride with the amino function. This reaction was carried out in the presence a strong base anion resin to remove the formed HCl. Finally, the acrylamide-ending TF-antigen **3** was anchored to the BSA carrier protein by treating the α,β-unsaturated amide with BSA at room temperature and at 40 °C, respectively in a 0.2 M carbonate buffer for 3 days to afford the two desired TF-BSA anti-tumor vaccine conjugates having a low and a high TF-content. A colorimetric phenol-sulfuric acid test was performed onto the BSA-TF conjugates using compound **1** a standard [81]. The analysis showed the conjugates to contain 2 ± 1 and 8 ± 2 TF-antigens, respectively.

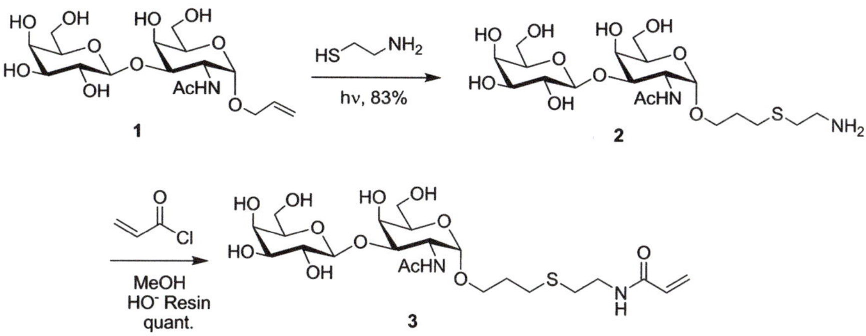

Scheme 13.1 Structure of the TF antigen and its chemical transformation into a suitable Michael acceptor 3 [83]

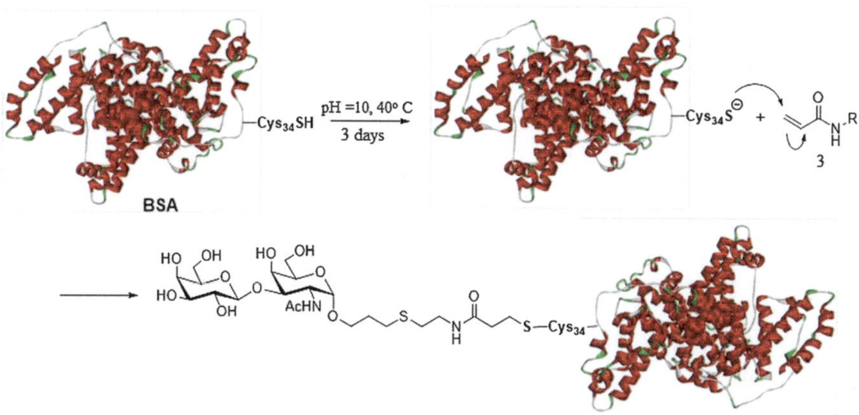

Fig. 13.6 Synthesis of Thomsen–Friedenreich antigen–BSA glycoconjugate vaccines using a Michael addition [83]

- **MALDI-TOF-MS analysis of two TF-BSA glycoconjugates**

The group of Rene Roy prepared two TF-BSA glycoconjugate vaccines, differing from each other on the targeted hapten-to-BSA ratio as described in the above section [83]. For the determination of the average number of carbohydrate-spacer moieties linked to BSA (TF-antigen: BSA), the two glycoconjugates were analyzed by MALDI-TOF-MS [83]. The MALDI-TOF mass spectra were characterized by the formation of a protonated molecular ions $[M + H]^+$ at *m/z* 67599.23 and m/z 70904.76. Knowing that TF-antigen is composed of the disccharide Galβ1 → 3GalNAcα1 → O-linked to spacer $[-(CH_2)_3S(CH_2)_2NH\text{-}CO\text{-}CH = CH_2]$ with a molecular mass of 555.22 Da, the TF-BSA ratios were calculated to be 2:1 and 8:1, respectively. Please, note that the molecular mass of the synthesized neoglycoconjugates were measured four times and are given with a standard error of the mean measurements (SEM). Henceforth, the neoglycoconjugate prepared with a hapten: BSA ratio of 2:1 produced the protonated molecule $[M + H]^+$ at *m/z* 67599.23, 67599.22, 67599.22 and 67599.22. Accordingly, its actual molecular weight is 67598.22 ± 0.0014 Da. On the other hand, the molecular weight of the neoglycoconjugate with a hapten: BSA ratio of 8:1 gave protonated molecule $[M + H]^+$ at *m/z* 70904.76, 70904.76, 70904.76, 70904.76, 70904.76.

The molecular mass calculations of these glycoconjugates (Hapten:BSA ratios of 2:1 and 8:1) were based on the presumption that one molecule of the TF-antigen will be covalently linked to the free sulfhydryl (Cys-34) group forming a C-*S* glycoside bond, by a Michael addition. The remaining stoichiometric amounts of the TF-antigen, that is one remaining molecule for the glycoconjugate with a Hapten:BSA of ratio of 2:1 and seven molecules of the glycoconjugate with a Hapten:BSA ratio of 8:1), respectively, will conjugate by direct targeting of the ε-amino groups of the various lysine residues of BSA, forming a *C-N* bond which occur by loss of ethanol molecules.

- **LC-ESI-CID-QqTOF-MS/MS analysis of tryptic digests of the glycoconjugate vaccines with TF: BSA ratios of 2:1 and 8:1**

The use of high-energy collision induced dissociation MALDI-MS/MS was circumvented in exploring the tryptic digests MS/MS data. This was completed in order to improve the glycopeptides chromatographic separation and to hinder the ionization suppression effect observed during MALDI-TOF-MS analysis [78, 79]. For these reasons, peptide identification was achieved by nano-LC-ESI-MS/MS analysis. This was completed in order to improve the glycopeptides chromatographic separation and to hinder the ionization suppression effect observed during MALDI-TOF-MS analysis [76, 77]. Hereafter, nano-LC-ESI-QqTOF-MS/MS analysis was carried out on tryptic digests of glycoconjugates with TF: protein ratios of 2:1 and 8:1, and this was followed by *de novo* peptide sequencing using low-energy CID-MS/MS. The obtained low-energy CID-MS/MS were submitted to the Mascot library to indentify the proper sequences of the peptides and glycopeptides released from BSA.

The Mascot reports of LC-MS/MS data of the tryptic digests of both glycoconjugates demonstrated the identification of two serum albumin isoforms from *Bos taurus* species. The serum albumin precursor (gi|1,351,907) was identified and had a sequence coverage of 46% and 21% for the glycoconjugate vaccines formed with TF-antigen: BSA ratios of 2:1 and 8:1, respectively. In addition, the serum albumin protein isoform (gi|229,552) was also identified for the tryptic digests of both glycoconjugates and this allowed us to observe a sequence coverage of 42% and 19% for the glycoconjugates with hapten: BSA ratios of 2:1 8:1, respectively.

It is fundamental to remark that the tryptic glycopeptide *m/z* values of the precursor ions that were not matched to that of the predicted BSA peptides were calculated by the addition of the exact mass of the carbohydrate TF-antigen (Molecular Weight, 555.22 Da) to that of the predicted BSA peptide.

As mentioned previously, de *novo* sequencing of the glycopeptides was achieved by measuring the product ion spectra of the potential glycopeptides, identified as described previously [48–51]. Hence, we were able to identify 3 glycated peptides for the glycoconjugate prepared with an antigen-TF:BSA ratio of 2:1 (Table 13.3a):

Table 13.3a Identified glycated peptides during the LC-MS/MS analysis of the trypsin digests of the TF-BSA vaccine with TF: BSA ratio 2:1 [83]

Glycated peptide sequence	Missed Cleavage	TF: BSA ratio 8:1 Observed *m/z* (charge)	Deviation (Da)
FK*DLGEEHFK (Lys 12)	1	902.1155(+2)	−0.0214
GLVLIAFSQYLQQC*PFDEHVK (Cyst 34)	0	1015.4871 (+3)	0.0052
HKPK*ATEEQLK (Lys 535)	1	620.5411(+3)	−0.0044

Table 13.3b Identified glycated peptides during the LC-MS/MS analysis of the trypsin digests of the TF-BSA vaccine with TF: BSA ratio 8:1 [83]

Glycated peptide sequence	Missed Cleavage	TF: BSA ratio 8:1 Observed *m/z* (charge)	Deviation (Da)
FK*DLGEEHFK (Lys 12)	1	902.1251(+2)	−0.0331
LK*PDPNTLCDEFK (Lys 116)	1	1065.5890(+2)	0.0762
K*QTALVELLK (Lys 522)	1	848.7161(+2)	−0.0776
K*VPQVSTPTLVEVSR (Lys 412)	1	731.4685(+3)	0.0317
GLVLIAFSQYLQQC*PFDEHVK (Cyst 34)	0	1015.4614(+3)	−0.0309
LKPDPNTLCDEFK*ADEK (Lys 127)	2	858.2895(+3)	−0.0822
NECFLSHK*DDSPDLPK (Lys 106)	1	818.7910(+3)	0.0198
DAIPENLPPLTADFAEDK*DVCK (Lys 312)	1	1004.2688(+3)	−0.021
SLGK*VGTR (Lys 429)	1	685.2528(+2)	0.0073
EK*VLTSSAR (Lys 186)	1	765.8557(+2)	−0.01
ALK*AWSVAR (Lys 210)	1	777.3762(+2)	0.0036
DTHK*SEIAHR (Lys 4)	1	873.4264(+2)	0.0033
HKPK*ATEEQLK (Lys 535)	1	620.5352(+3)	0.0015
GACLLPK*IETMR (Lys 179)	1	942.6395(+2)	−0.0094

These glycated peptides were as follows: FK*DLGEEHFK (*C-N* glycation site is Lys 12) at *m/z* 902.1155(+2), GLVLIAFSQYLQQC*PFDEHVK (*C-S* glycation site is Cyst 34) at *m/z* 1015.4871(+3) and HKPK*ATEEQLK (*C-N* glycation site is Lys 535) at *m/z* 620.5411(+3). Additionally, 14 glycated peptides were also identified for the glycoconjugate prepared with an antigen-TF:BSA ratio of 8:1 respectively (Table 13.3b).

These glycated peptides were as follows: FK*DLGEEHFK (*C-N* glycation site is Lys 12) at *m/z* 902.1251(+2), LK*PDPNTLCDEFK (*C-N* glycation site is Lys 116) at *m/z* 1065.5890(+2), K*QTALVELLK (*C-N* glycation site is Lys 522) at *m/z* 848.7161(+2), K*VPQVSTPTLVEVSR (*C-N* glycation site is Lys412) at *m/z*

731.4685(+3), GLVLIAFSQYLQQC*PFDEHVK (*C-S* glycation site is Cyst34) at *m/z* 1015.3614(+3), LKPDPNTLCDEFK*ADEK (*C-N* glycation site is Lys 127) at *m/z* 858.2895(+3), NECFLSHK*DDSPDLPK (*C-N* glycation site is Lys 106) at *m/z* 818.7910(+3), DAIPENLPPLTADFAEDK*DVCK (*C-N* glycation site is Lys 312) *at m/z* 1004.2688(+3), SLGK*VGTR (*C-N* glycation site is Lys 429) at *m/z* 685.2528 EK*VLTSSAR (*C-N* glycation site is Lys 186) at *m/z* 765.8557(+2), ALK*AWSVAR (*C-N* glycation site is Lys 210) at *m/z* 777.3762(+2), DTHK*SEIAHR (*C-N* glycation site is Lys 4) at *m/z* 873.4264(+2), HKPK*ATEEQLK (*C-N* glycation site is Lys 535) at *m/z* 620.5352(+3) and GACLLPK*IETMR (*C-N* glycation site is Lys 179) at *m/z* 942.6395(+2). All aforementioned glycation sites were displayed in Fig. 13.7 [83].

As an example, the CID-fragmentation of the *S*-glycated peptide GLVLIAFSQYLQQC*PFDEHVK (*C-S* glycation site located at Cyst34) at *m/z* 1015.3614(+3) is shown in Table 13.4.

The MS/MS spectrum afforded the following series of products ions. First, we noted the product ions formed only by the fragmentation of the carbohydrate hapten: $[B_2 - Gal - CH_2CO - 2H_2O]^+$ at *m/z* 126.0578, $[B_2 - Gal- H2O]^+$ at *m/z* 186.0773, $[B_2 - Gal]^+$ at m/z 204.0845, B_2^+ at *m/z* 366.1487, Y_0^{2+} at 1341.1178 and Y_1^{2+} 1442.6371. Additionally, we also noted the peptide product ions resulting from the release of the carbohydrate portion from the glycopeptide $[-B_2^+]$ and these were the following: $[y_5 - B_2]^+$ at *m/z* 261.3074, $[b_7 - B_2]^+$ at *m/z* 348.4520, $[y_6 - B_2]^+$ at *m/z* 408.3712, $[b_8 - B_2]^+$ at *m/z* 435.4821, $[b_9 - B_2]^+$ at *m/z* 563.5424, $[y_8 - B_2]^+$ at *m/z* 665.4677, $[b_{10} - B_2]^+$ at *m/z* 726.6014, $[b_{11} - B_2]^+$ at *m/z* 839.7011, $[y_{10} - B_2]^+$ at *m/z*

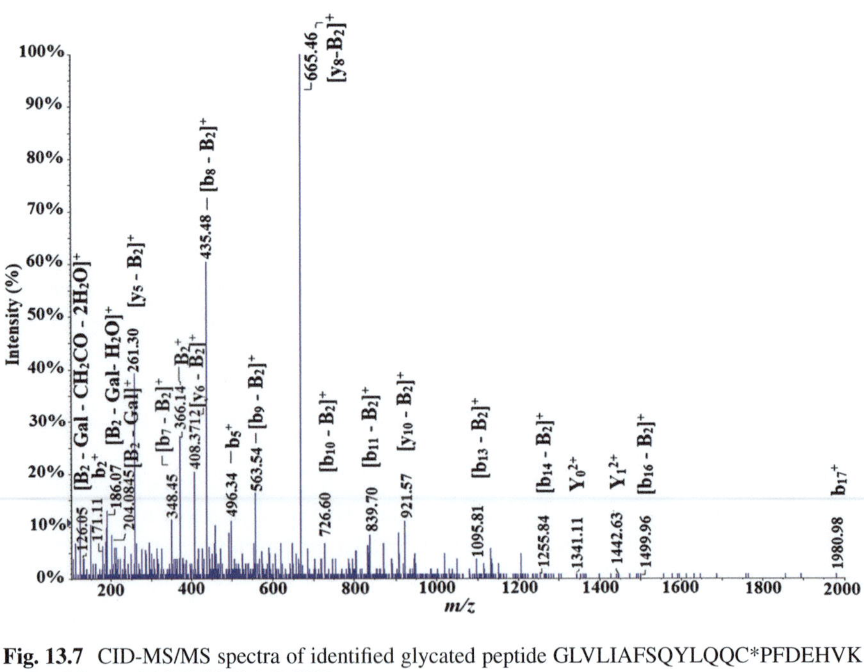

Fig. 13.7 CID-MS/MS spectra of identified glycated peptide GLVLIAFSQYLQQC*PFDEHVK 5 (Cys34) at *m/z* 1015.4614 (+3) [83]

Table 13.4 Identified product ions for the glycated peptide GLVLIAFSQYLQQC*PFDEHVK (Cyst34) at *m/z* 1015.4614(+3) [83]

Production	Calculated *m/z*	Observed *m/z*	Deviation (Da)
b_{17}^+	1980.9888	1980.9884	0.0004
$[b_{16} - B_2]^+$	1499.9619	1499.9621	−0.0002
Y_1^{2+}	1442.6635	1442.6371	0.0264
Y_0^{2+}	1341.1685	1341.1178	0.0507
$[b_{14} - B_2]^+$	1255.8407	1255.8424	−0.0017
$[b_{13} - B_2]^+$	1095.8100	1095.8145	−0.0045
$[y_{10} - B_2]^+$	921.5786	921.5777	0.0009
$[b_{11} - B_2]^+$	839.7000	839.7011	−0.0011
$[b_{10} - B_2]^+$	726.6014	726.6004	0.0010
$[y_8 - B_2]^+$	665.4615	665.4677	−0.0062
$[b_9 - B_2]^+$	563.5455	563.5424	0.0031
b_5^+	496.3493	496.3487	0.0006
$[b_8 - B_2]^+$	435.4869	435.4821	0.0048
$[y_6 - B_2]^+$	408.3781	408.3712	0.0069
B_2^+	366.1400	366.1487	−0.0087
$[b_7 - B_2]^+$	348.4549	348.4520	0.0029
$[y_5 - B_2]^+$	261.3097	261.3074	0.0023
$[B_2 - Gal]^+$	204.0866	204.0845	0.0021
$[B_2 - Gal- H_2O]^+$	186.0761	186.0773	−0.0012
b_2+	171.1128	171.1100	0.0028
$[B_2 - Gal - CH_2CO - 2H_2O]^+$	126.0550	126.0578	−0.0028

921.5777, $[b_{13} - B_2]^+$ at *m/z* 1095.8145, $[b_{14} - B_2]^+$ at *m/z* 1255.8424 and $[b_{16} - B_2]^+$ at *m/z* 1499.9621. Finally, we noted products ions created by the straight forward fragmentation of the peptide portion only and these are exemplified by the following product ions: b_2^+ at *m/z* 171.1100, b_5^+ at *m/z* 496.3487 and b_{17}^+ at *m/z* 1980.9884. It is important to mention that for the CID-MS/MS of this described precursor ion at *m/z* 1015.3614(+3) ($C_7H_{13}NO_2S^+$), no product ions containing the spacer linker arm were noted [83].

13.7 Conclusion

While it has been reported in several studies that MALDI-TOF-MS is a reliable technique for the average hapten: protein ratio determination, nevertheless, the digestion of the hapten-protein neoglycoconjugate vaccine followed by LC-MS/MS analysis of the digests is a powerful method for the discovery of the glycation sites.

Most of the time, the vaccines prepared using the squaric acid chemistry revealed the attachment of the hapten to lysine residues and leaded to the formation of a mixture of glycoforms. However, the coupling of haptens to protein carriers using the Michael addition showed that the reaction allowed the conjugation of hapten not only to lysine residues, but also on a cysteine residue.

Finally, in the two examples we discussed, most of the glycation sites were found to be located at the outer surface of the protein.

References

1. Plotkin SA (2008) Vaccines: correlates of vaccine-induced immunity. Clin Infect Dis 47:401–409
2. Heidelberger M, Avery OT (1923) The soluble specific substance of Pneumococcus. J Exp Med 38:73–79
3. Jones LH (2015) Recent advances in the molecular design of synthetic vaccines. Nat Chem 7:952–960
4. Pupo E, Aguila A, Santana H, Núnez JF, Castellanos-Serra L, Hardy E (1999) Mice immunization with gel electrophoresis-micropurified bacterial lipopolysaccharides. Electrophoresis 20:458–461
5. Davis MR Jr, Goldberg JB (2012) Purification and visualization of lipopolysaccharide from gram-negative bacteria by hot aqueous-phenol extraction. J Vis Exp 28:1–3
6. Reisser D, Pance A, Jeannin JF (2002) Mechanisms of the antitumoral effect of lipid A. BioEssays 24:284–289
7. Nagy G, Pál T (2008) Lipopolysaccharide: a tool and target in enterobacterial vaccine development. Biol Chem 389:513–520
8. Daum RS, Hogerman D, Rennels MB, Bewley K, Malinoski F, Rothstein E, Reisinger K, Block S, Keyserling H, Steinhoff M (1997) Infant immunization with Pneumococcal CRM197 vaccines: effect of saccharide size on immunogenicity and interactions with simultaneously administrated vaccines. J Infect Dis 176:445–455
9. Lefeber DJ, Kamerling JP, Vliegenthart JFG (2001) Synthesis of Streptococcus Pneumoniae type 3 neoglycoproteins varying in oligosaccharide chain length, loading and carrier protein. Chem Eur J 7:4411–4421
10. Paoletti LC, Kasper DL, Michon F, DiFabio J, Jennings HJ, Tosteson TD, Wessels MR (1992) Effects of chain length on the immunogenicity in rabbits of group B. Streptococcus type III oligosaccharide-Tetanus toxoid conjugates. J Clin Invest 89:203–209
11. Chernyak A, Kondo S, Wade TK, Meeks MD, Alzari PM, Fournier JM, Taylor RK, Kováč P, Wade WF (2002) Induction of protective immunity by synthetic *Vibrio cholerae* hexasaccharide derived from V. cholerae O1 Ogawa lipopolysaccharide bound to a protein carrier. J Infect Dis 185:950–962
12. Roy R (2004) New trends in carbohydrate-based vaccines. Drug Discov Today Technol 1:327–336
13. Ada G, Isaacs D (2003) Carbohydrate-protein conjugate vaccines. Clin Microbiol Infect 9:79–85
14. Landsteiner K (1945) The specificity of serological reactions. Harvard University Press, Cambridge
15. Avery OT, Goebel WF (1929) Chemo-immunological studies on conjugated carbohydrate-proteinsII. Immunological specificity of synthetic sugar-protein Ags. J Exp Med 50:533–550
16. Dick WE Jr, Beurret M (1989) In: Cruse JM, Lewis RE Jr (eds) Glycoconjugates of bacterial carbohydrate Ags, vol 10. Krager, Basel, pp 48–114
17. Tietze LF, Arlt M, Beller M, Glüsenkamp KH, Jähde E, Rajewsky MF (1991) Squaric acid diethyl ester: a new coupling reagent for the formation of drug biopolymer conjugates. Synthesis of squaric acid ester amides and diamides. Chem Ber 124:1215–1221

18. Glüsenkamp KH, Drosdziok W, Eberle G, Ja¨hde E, Rajewsky MFZ (1991) Naturforsch C. Biosci 46: 498-501
19. Tietze LF, Schröter C, Gabius S, Brinck U, Goerlach-Graw A, Gabius HJ (1991) Conjugation of p-aminophenyl glycosides with squaric acid diesters to a carrier protein and the use of the neoglycoprotein in the histochemical detection of lectines. Bioconjug Chem 2:148–153
20. Cohen S, Cohen SG (1966) Preparation and reactions of derivatives of squaric acid. Alkoxy-, hydroxy-, and aminocyclobutenediones. J Am Chem Soc 88:1533–1536
21. Grünefeld J, Bredhauer G, Zinner G (1985) Zur Reaktion von Quadratsäuredimethylester mit N,N-disubstituierten Hydrazin-Derivaten. Arch Pharm (Weinheim) 318:984–988
22. Bergh A, Magnusson BG, Ohlsson J, Wellmar U, Nilsson UJ (2001) Didecyl squarate – a practical amino-reactive cross-linking reagent for neoglycoconjugate synthesis. Glycoconj J 18:615–621
23. Kamath VP, Diedrich P, Hindsgaul O (1996) Use of diethyl squarate for the coupling of oligosaccharide amines to carrier proteins and characterization of the resulting neoglycoproteins by MALDI-TOF mass spectrometry. Glycoconj J 13:315–319
24. Hou SJ, Saksena R, Kováč P (2008) Preparation of glycoconjugates by dialkyl squarate chemistry revisited. Carbohydr Res 343:196–210
25. Saksena R, Adamo R, Kováč P (2007) Immunogens related to the synthetic tetrasaccharide side chain of the Bacillus anthracis exosporium. Bioorg Med Chem 15:4283–4310
26. Bongat AFG, Saksena R, Adamo R, Fujimoto Y, Shiokawa Z, Peterson DC, Fukase K, Vann WF, Kováč P (2010) Multimeric bivalent immunogens from recombinant Tetanus Toxin HC fragment, synthetic hexasaccharides and a glycopeptide adjuvant. Glycoconj J 27:69–77
27. Liebler DC (2008) Protein damage by reactive electrophiles: targets and consequences. Chem Res Toxicol 21:117–128
28. Ohe PC, Kuhne R, Ebert RU, Altenburger R (2005) Structural alerts - a new classification model to discriminate excess toxicity from narcotic effect levels of organic compounds in the acute daphnid assay. Chem Res Toxicol 18:536–555
29. Roberts DW, Schultz TW, Wolf EM, Aptula AO (2010) Experimental reactivity parameters for toxicity modeling: application to the acute aquatic toxicity of S2 electrophiles to *Tetrahymena pyriformis*. Chem Res Toxicol 23:228–234
30. Chipinda I, Hettick JM, Siegel PD (2011) Haptenation: chemical reactivity and protein binding. J Allergy 2011:1–11
31. Janatova JF, Fuller JK, Hunter M (1968) The heterogeneity of bovine albumin with respect to sulfhydryl and dimer content. J Biol Chem 243:3612–3622
32. Heredia KL, Bontempo D, Ly T, Byers JT, Halstenberg S, Maynard HD (2005) In situ preparation of protein-"smart" polymer conjugates with retention of bioactivity. J Am Chem Soc 127:16955–16960
33. Rappuoli R, De Gregorio E (2011) A sweet T cell response. Nat Med 17:1551
34. Roy R, Shiao TC (2012) Glycodendrimers as functional antigens and antitumorales vaccines. New J Chem 36:324–339
35. Huang YL, Hung JT, Cheung SKC, Lee HY, Chu KC, Li ST, Lin YC, Ren CT, Cheng TJR, Hsu TL, AL Y, CY W, Wong CH (2013) Carbohydrate-based vaccines with a glycolipid adjuvant for breast cancer. PNAS 110:2517–2522
36. Zhu J, Warren JD, Danishefsky SJ (2009) Synthetic carbohydrate-based anticancer vaccines: the memorial sloan-kettering experience. Expert Rev Vaccine 8:1399–1413
37. Keding SJ, Danishefsky SJ (2004) Prospects for total synthesis: a vision for a totally synthetic vaccine targeting epithelial tumors. Proc Natl Acad Sci U S A 101:11937–11942
38. Brown JR (1975) Structure of bovine serum slbumin. Fed Proc 34:591–600
39. Kandil AA, Chan N, Klein MH, Chong P (1997) Chemical synthesis of Haemophilus influenzae glycopeptide conjugates. Glycoconj J 14:13–17

40. Pozsgay V, Chu C, Pannell L, Wolfe J, Robbins JB, Schneerson R (1999) Protein conjugates of synthetic saccharides elicit higher levels of serum IgG lipopolysaccharide antibodies in mice than do those of the O-specific polysaccharide from shigella dysenteriae type 1. Proc Natl Acad Sci U S A 96:5194–5197
41. Peeters CC, Evenberg D, Hoogerhout P, Kayhty H, Saarinen L, van Boeckel CA, van der Marel GA, van Boom JH, Poolman JT (1992) Synthetic trimer and tetramer of 3- b-D-ribose-(1-1)-D-ribitol-5-phosphate conjugated to protein induce antibody responses to Haemophilus influenzae type b capsular polysaccharide in mice and monkeys. Infect Immun 60:1826–1833
42. Snippe H, Zigterman JWJ, van Dam JEG, Kamerling JP (1988) Oligosaccharide-haptenated liposomes used as a vaccine to Streptococcus pneumoniae. In: Gregoriadis G (ed) Liposomes as drug carriers in: trends and progress. Wiley, New York, pp 183–196
43. Boas U, Heegaard PMH (2004) Dendrimers in drug research. Chem Soc Rev 33:43–63
44. Benaissa-Trouw B, Lefeber DJ, Kamerling JP, Vliegenthart JF, Kraaijeveld K, Snippe H (2001) Synthetic polysaccharide Type 3-related Di-, Tri-, and Tetrasaccharide-CRM(197) conjugates induce protection against Streptococcus pneumoniae Type 3 in mice. Infect Immun 69:4698–4701
45. Snippe H, van Houte AJ, van Dam JE, De Reuver MJ, Jansze M, Willers JM (1983) Immunogenic properties in mice of hexasaccharide from the capsular polysaccharide of Streptococcus pneumoniae Type 3. Infect Immun 40:856–861
46. Chabot DJ, Scorpio A, Tobery SA, Little SF, Norris SL, Friedlander AM (2004) Anthrax capsule vaccine protects against experimental infection. Vaccine 23:43–47
47. Daubenspeck JM, Zeng H, Chen P, Dong S, Steichen CT, Krishna NR, Pritchard DG, Turnbough CL Jr (2004) Novel oligosaccharide side chains of the collagen-like region of BclA, the major glycoprotein of the Bacillus anthracis exosporium. J Biol Chem 279:30945–30953
48. Jahouh F, Xu P, Vann WF, Kováč P, Banoub JH (2013) Mapping the glycation sites in the neoglycoconjugate from hexasaccharide antigen of *Vibrio cholerae* serotype ogawa and the recombinant Tetanus Toxin C-fragment carrier. J Mass Spectrom 48:1083–1090
49. Jahouh F, Saksena R, Aiello D, Napoli A, Sindona G, Kovác P, Banoub JH (2010) Glycation sites in neoglycoconjugates from the terminal monosaccharide antigen of the O-PS of *Vibrio cholerae* O1, serotype ogawa, and BSA revealed by matrix-assisted laser desorption-ionization tandem mass spectrometry. J Mass Spectrom (10):1148–1159
50. Jahouh F, Hou S, Kovác P, Banoub JH (2012) Determination of glycation sites by tandem mass spectrometry in a synthetic lactose-bovine serum albumin conjugate, a vaccine model prepared by dialkyl squarate chemistry. Rapid Commun Mass Spectrom 26:749–758
51. Jahouh F, Hou SJ, Kováč P, Banoub JH (2011) Determination of the glycation sites of *Bacillus anthracis* neoglycoconjugate vaccine by MALDI-TOF/TOF-CID-MS/MS and LC-ESI-QqTOF-tandem mass spectrometry. J Mass Spectrom 46:993–1033
52. Roy R, Shiao TC (2011) Organic chemistry and immunochemical strategies in the design of potent carbohydrate-based vaccines. Chimia (Aarau) 65:24–29
53. Icart LP, Fernandez-Santana V, Veloso RC, Carmenate T, Sirois R, Roy R, Verez Bencomo V (2007) T-cell immunity of carbohydrates. In: Roy R (ed) Carbohydrate-based vaccines. ACS Symp Ser 989: 1-20
54. Gao GF, Jakobsen BK (2000) Molecular interactions of coreceptor CD+8 and MHC Class I: the molecular basis for functional coordination with the T-cell receptor. Immunol Today 21:630–636
55. Springer GF (1997) Immunoreactive T and Tn epitopes in cancer diagnosis, prognosis, and immunotherapy. J Mol Med 75:594–602
56. Campbell BJ, Finnie IA, Hounsell EF, Rhodes JM (1995) Direct demonstration of increased expression of Thomsen-Friedenreich (TF) antigen in colonic adenocarcinoma and ulcerative colitis mucin and its concealment in normal mucin. J Clin Invest 95:571–576
57. Springer GF (1984) T and Tn, general carcinoma autoantigens. Science 224:1198–1207
58. Dippold W, Steinborn A, Büschenfelde KHM (1990) The role of the Thomsen-Friedenreich antigens a tumor-associated molecule. Environ Health Perspect 88:255–257

59. Laštovičková M, Chmelik J, Bobalova J (2009) The combination of simple MALDI matrices for the improvement of intact glycoproteins and glycans analysis. Int J Mass Spectrom 281:82–88
60. Issaq HJ, Conrad's TP, Prieto DA, Tirumalai R, Veenstra TD (2003) SELDI-TOF MS for diagnostic proteomics. Anal Chem 75:148A–155A
61. Liu C (2011) The application of SELDI-TOF-MS in clinical diagnosis of cancers. J Biomed Biotechnol 2011:1–6
62. Morelle W, Michalski JC (2007) Analysis of protein glycosylation by mass spectrometry. Nat Protoc 2:1585–1602
63. Roepstorff P, Fohlman J (1984) Proposal for a common nomenclature for sequence ions in mass spectra of peptides. Biol Mass Spectrom 11(11):601
64. Johnson RS, Martin SA, Biemann K, Stults JT, Watson JT (1987) Novel fragmentation process of peptides by collision-induced decomposition in a tandem mass spectrometer: differentiation of leucine and isoleucine. Anal Chem 59:2621–2625
65. Domon B, Costello C (1988) A systematic nomenclature of carbohydrate fragmentation in FAB-MS/MS spectra of glycoconjugates. Glycoconj J 5:397–409
66. Mock M, Fouet A (2001) Anthrax. Annu Rev Microbiol 55:647–671
67. Priest FG (1993). In: Sonenshein AL, Hoch JA, Losick R (Eds) Bacillus subtilis and other gram-positive vacteria: biochemistry, physiology, and molecular biology. American society for microbiology, Washington
68. Nicholson WL, Munakata N, Horneck G, Melosh HJ, Setlow P (2000) Resistance of bacillus endospores to extreme terrestrial and extraterrestrial environments. Microbiol Mol Biol Rev 64:548–572
69. Boutiba-Ben Boubaker I, Ben Redjeb S (2001) Bacillus anthracis: causative agent of anthrax. Tunis Med 79:642–646
70. Read TD, Salzberg SL, Pop M, Shumway M, Umayam L, Jiang L, Holtzapple E, Busch JD, Smith KL, Schupp JM, Solomon D, Keim P, Fraser CM (2002) Comparative genome sequencing for discovery of novel polymorphisms in bacillus anthracis. Science 296:2028–2033
71. Turnbull PCB (1999) Definitive identification of bacillus anthracis-a review. J Appl Microbiol 87:237–240
72. Reed LJ, Muench H (1938) A simple method for estimating fifty percent endpoints. Am J Hyg 27:493–497
73. Hoffmaster AR, Fitzgerald CC, Ribot E, Mayer LW, Popovic T (2002) Molecular subtyping of bacillus anthracis and the 2001 bioterrorism-associated anthrax outbreak, United States. Emerg Infect Dis 8:1111–1116
74. Chun JH, Hong KJ, Cha SH, Cho MH, Lee KJ, Jeong DH, Yoo CK, Rhie GE (2012) Complete genome sequence of *Bacillus anthracis* H9401, an isolate from a Korean patient with anthrax. J Bacteriol 194:4116–4117
75. Williams DD, Benedek O, Turnbough CL Jr (2003) Species-specific peptide ligands for the detection of bacillus anthracis spores. Appl Environ Microbiol 69:6288–6293
76. Burkitt WI, Giannakopulos AE, Sideridou F, Bashir S, Derrick PJ (2003) Discrimination effects in MALDI-MS of mixtures of peptides-analysis of the proteome. Austr J Chem 56:369–377
77. Kratzer R, Eckerskorn C, Karas M, Lottspeich F (1998) Suppression effects in enzymatic peptide ladder sequencing using ultraviolet matrix assisted laser desorption/ionization - mass spectrometry. Electrophoresis 19:1910–1919
78. Shamsuddin AM, Tyner GT, Yang GY (1995) Common expression of the tumor marker D-galactose-ß-[1→3]-N-acetyl-D-galactosamine by different adenocarcinomas: evidence of field effect phenomenon. Cancer Res 55:149–152
79. Roy R, Baek MG, Rittenhouse-Olson K (2001) Synthesis of N,N'-bis(acrylamide) acetic acid base-T antigen glycodendrimers and their mouse monoclonal IgG antibody binding properties. J Am Chem Soc 123:1809–1816
80. Baek MG, Rittenhouse-Olson K, Roy R (2001) Synthesis and antibody binding properties of glycodendrimers bearing the tumor related T antigen. Chem Commun 2001:257–258

81. Baek MG, Roy R (2002) Glycodendrimers: novel glycotope isosteres unmasking sugar coding. Case study with T-Ag markers from breast cancer MUC1 glycoprotein. Rev Mol Biotechnol 90:291–309
82. Dubois M, Gilles KA, Hamilton JK, Rebers PA, Smith F (1956) Colorimetric method for determination of sugars and related substances. Anal Chem 28:350–356
83. Demian WL, Kottari N, Shiao TC, Randell E, Roy R, Banoub JH (2014) Direct targeted glycation of the free sulfhydryl group of cysteine residue (Cys-34) of BSA. Mapping of the glycation sites of the anti-tumor Thomsen-Friedenreich neoglycoconjugate vaccine prepared by Michael addition reaction. J Mass Spectrom 49:1223–1233

Chapter 14
Explosive Detection Strategies for Security Screening at Airports

Peter Pallister, Terri D'Souza, Chelsea Black, Nigel Hearns, and Jeffrey C. Smith

Abstract Commercial airline traffic is at an all-time high during a period in history when security threats are real and the access to explosive materials is relatively easy. Governments, in cooperation with travel and security industries have established numerous strategies and technologies to allow flying to remain one of the safest modes of transportation today. This short chapter provides a brief outline of some of the main technological advances that are used to detect explosives and ensure the safety of passengers when flying. This chapter describes technologies that are used to scan for explosives within other objects as well as those that are able to detect explosives or explosive residues on the surface of objects or persons with a focus on those that employ mass spectrometry or mass spectrometry-based tools.

14.1 Introduction

Modern security screening methods for detecting explosives cover a wide variety of technologies and are aimed at targeting the explosive chemical compound or the overall explosive device. In recent years there have been significant advances to both fronts. Although the intentional harmful use of explosives is a risk anywhere in the world, stringent security measures have been established for air travel due to the vulnerability of this form of travel to devastating loss should an explosion occur mid-flight. The World Bank has published data indicating that by the year 2015, over 3.4 billion human passengers have travelled by air and that this number is set to increase exponentially over time [1]. Similarly, tools to detect explosives for air

P. Pallister • T. D'Souza • C. Black • J.C. Smith (✉)
Department of Chemistry, Carleton University, Ottawa, ON K1S 5B6, Canada
e-mail: JeffCSmith@cunet.carleton.ca

N. Hearns
National Forensic Laboratory Services, Royal Canadian Mounted Police, Ottawa, ON K1A 0R2, Canada

J.H. Banoub, R.M. Caprioli (eds.), *Molecular Technologies for Detection of Chemical and Biological Agents*, NATO Science for Peace and Security Series A: Chemistry and Biology, DOI 10.1007/978-94-024-1113-3_14

travel have emerged over the years and the literature indicates that new developments in security technology are continually being made. This short chapter will briefly outline some of the main security technologies that currently exist to ensure that flying remains one of the safest modes of transportation available today with a focus on the use of mass spectrometry-related tools to accomplish this task.

14.2 Security Technologies to Detect Explosives Contained Within Objects

Explosive compounds are typically contained within a hard (e.g. metal) container until detonation. Several technologies exist that identify suspicious looking containers or objects, rather than the explosive chemical itself, employing strategies that permeate surfaces or objects to detect them. Metal detectors, X-ray scanners, and more recently radio frequency (RF) scanners or millimetre-wave scanners are being used routinely as screening technologies. Metal detectors and X-ray scanners are not new technologies but are still used routinely at most portals or points of entry for detecting potential threats or explosive devices. Metal detectors are relatively simple devices that can be stationary or held as a portable "wand" and consist of an electronic "transmitter coil" that sets up a magnetic field as well as a "receiver coil" that has current induced in it by the first coil. When electrically conductive objects come in close proximity to the magnetic field, eddy currents are induced in the metal, which produces a magnetic field of its own, which is detected by the receiver coil, Fig. 14.1. Metal detectors enable a screening operation to determine whether an object or person contains electrically conductive materials that may comprise the

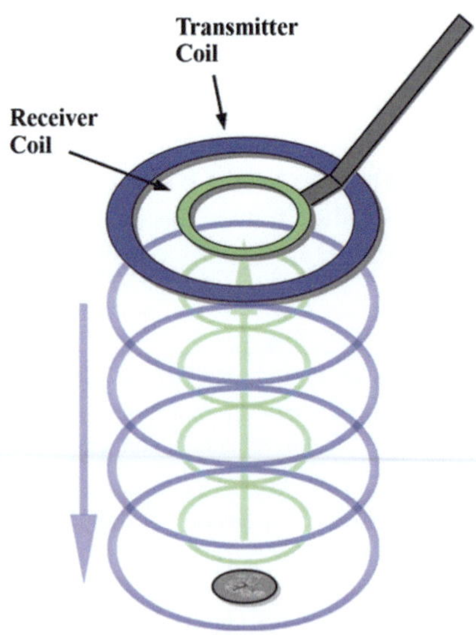

Fig. 14.1 Simple schematic of a metal detector. A transmitter coil creates a magnetic field that is detected in a receiver coil

components of an explosive device. Similarly, X-ray scanners allow a security checkpoint to gain insight into the contents of a package. Due to the relative energy of X-rays, these scanners are not used on humans directly but are used to scan both carry-on and checked luggage at airport terminals. X-ray scanners irradiate an object with electromagnetic radiation in the 0.01–10 nm wavelength range. At this range, the energy of the waves ranges between 100 eV and 100 keV and has frequencies between 3×10^{16} and 3×10^{19} Hz [2]. The short wavelength of X-ray electromagnetic radiation results in it passing through most soft objects without being absorbed while harder or denser objects can absorb these high-energy photons. Detection of the radiation creates an image based on what was absorbed which allows the expert user to view the hard objects within a closed container. X-ray scanners are particularly useful to detect weapons such as firearms or knives and may detect the hard physical structures, such as electronics, of an explosive device. X-ray scanners are also able to image liquids that are concealed within packages that may be accelerants or explosive compounds. An x-ray scanner is of relatively simple design with a conveyor belt that allows packages to travel past an enclosed beam of X-rays for imaging, Fig. 14.2.

Millimetre-wave scanners were developed in the early 1990's [3, 4] but only in the last decade have they seen significant adoption as commercial threat screening devices. These systems use millimetre-wave (24–60 GHz frequency range) electromagnetic radiation to produce 3D images of the target, Fig. 14.3.

RF-scanners are particularly useful as they penetrate clothing easily and are sensitive to concealed weapons and explosives. They are advantageous over X-ray technology as there is a reduced risk to human health using lower energy radiation.

Fig. 14.2 An X-ray scanner. Packages pass through the device on the conveyor belt where they are irradiated with X-ray electromagnetic radiation to permit imaging of their contents

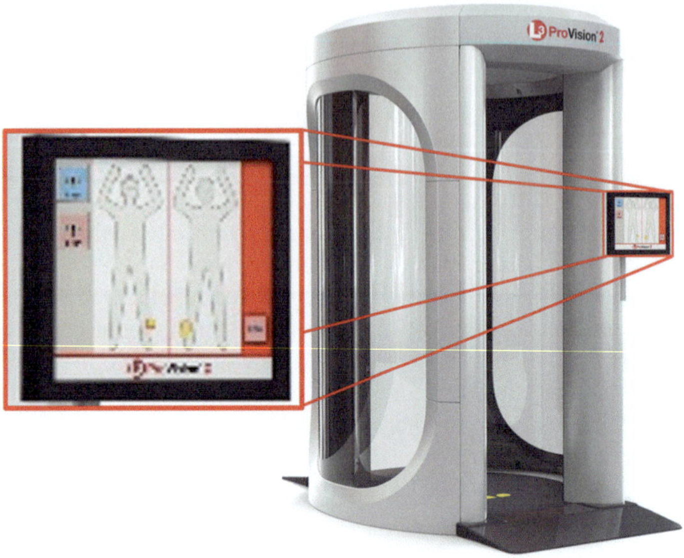

Fig. 14.3 A radio-frequency (RF) scanner. Millimetre-wave electromagnetic radiation passes through clothes but reflects off of skin and tissue. Hard objects that are concealed under clothing on a passenger are imaged, depicted by a yellow region in the example above.

Other advantages include the speed and convenience at which passengers may move through a security checkpoint as metal objects in clothing (e.g. belts, jewelry, and shoes) do not need to be removed prior to screening, unlike metal detectors.

There are also a few methods that are currently employed that detect individual chemical compounds in an explosive device or in residue from explosives handling from within objects. These stand-off or long-range techniques to detect explosives are not as common as those mentioned above in regards to passenger air travel but are more commonly employed at customs checkpoints for the shipping of packages. A less widely used technology uses neutron radiation to determine chemical composition [5]. However, due to the shielding requirements and poor public appeal this technology has not seen significant adoption by commercial security. Another stand-off technology used for explosives detection is a spectroscopic method that uses Raman lasers to detect explosive composition [6]. While this technology is currently not used extensively in commercial security, research is ongoing in this area [7]. In addition to stand-off techniques, there are a number of methods that rely detecting the explosive compound in the vapour phase. Gas chromatography with chemiluminescence and gas chromatography with electron capture detector are two methods that provide trace explosives detection particularly as hand-held devices [8]. However, arguably the most widely used technology currently for trace explosives detection is ion mobility spectrometry (IMS).

14.3 Security Technologies to Detect Explosives Contained on the Surface of Objects

IMS is not a new technique and the first discoveries related to ion mobility date back to the early 1900's. However, the first commercial, widely adopted benchtop IMS instrument used for detecting explosives was not released until 1990, the IONSCAN from Barringer Research Ltd., now Smiths Detection. There have been many improvements made to these devices since then [9] but the working principle remains the same, as illustrated in Fig. 14.4.

A sample of analyte is volatilized into the gas phase and ionized by some means, traditionally via a sealed radioactive ^{63}Ni source. Small bursts of ions are released into a drift tube that has a uniform electric field gradient along the length of the tube. The ions are accelerated through the tube and are separated based on their "mobility" constant, K. The mobility of a given ion is a function of the ion's charge, mass, collision cross section, and temperature. So, in general, heavier, larger analytes have a smaller mobility and arrive at the detector at a later time than lighter, smaller analytes. Given this working principle, IMS actually acts like a chromatography technique, and indeed when the technique was beginning to be commercialized it was initially called "plasma chromatography" [10]. IMS has been shown to work very well for trace detection of explosives but does not work for extremely low vapour pressure compounds and is, in general, limited to nanogram or subnanogram detection limits [9, 11]. As well, given that IMS is essentially a form of chromatography, positive identification of compounds must be made through the use of comparison with standards on a given instrument. Given the limitations of IMS, the next logical step to improving trace explosives detection and explosives screening would be to use mass spectrometry (MS), which has an inherently higher sensitivity and provides accurate mass-to-charge information which reduces the potential for false positives. While MS would provide significantly improved detection capabilities,

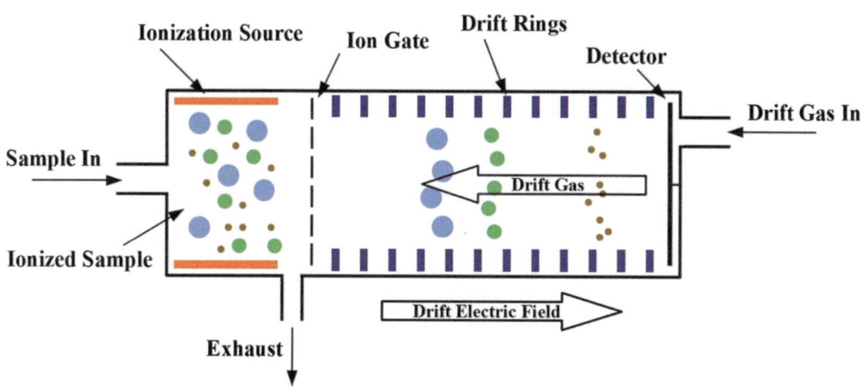

Fig. 14.4 An ion mobility spectrometer. Ions are separated in the gas phase as they travel under the influence of an electric field against a drift gas flowing in the opposite direction

there are a number of factors that prevent MS from being widely adopted for commercial security applications. MS is performed under reduced pressure requiring potentially sophisticated vacuum systems.

In either case, for both IMS and MS, one of the limiting factors for analysis of explosives is the ability to ionize the analyte efficiently in a potentially complex matrix. As mentioned, IMS typically uses a radioactive ^{63}Ni source but it requires efficient volatilization of the analyte before being ionized by the radioactive source. MS uses a wide variety of ionization methods [12] but most either require the analyte to be in the gas phase first (like with ^{63}Ni) or they require the analyte to be in solution first, which increases sample preparation time. However, in the last decade, novel ionization sources have been introduced that overcome the potential cost and time constraints needed for commercial security screening applications which will be discussed in the next section.

14.4 Novel Ionization Methods

Desorption electrospray ionization (DESI) was first introduced in 2004 and is a very simple modification of a more traditional electrospray ionization (ESI) source, Fig. 14.5 [13].

A DESI source consists of a spray capillary and a surrounding nebulizer capillary. A high voltage (around 4 kV) is placed on the spray capillary while a solution is directed through it. A nebulizing gas is directed through the nebulizer capillary to

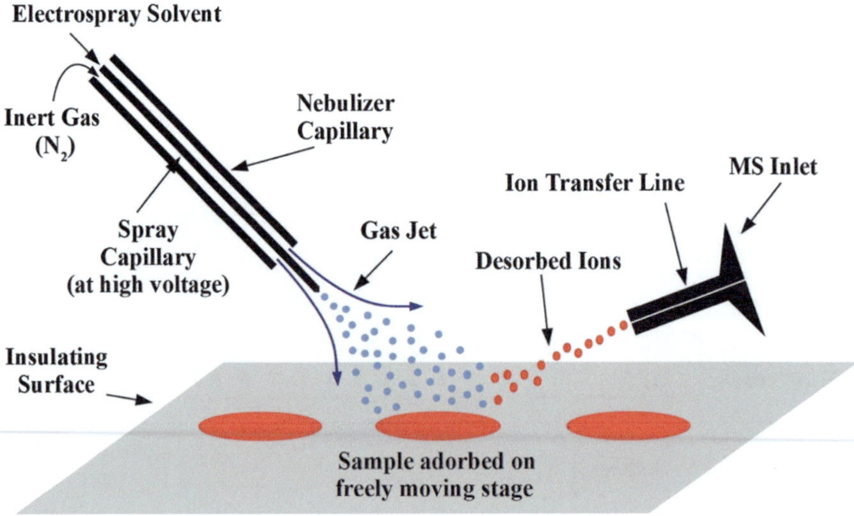

Fig. 14.5 Desorption electrospray ionization (DESI) produces analyte ions for MS detection via desorbing analyte from a surface using positively charged solvent ions

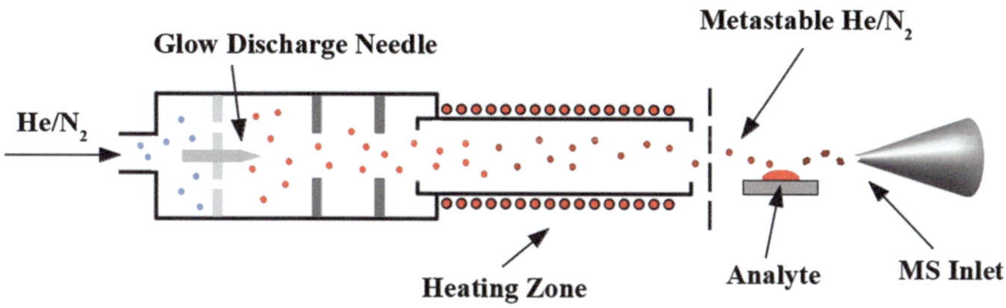

Fig. 14.6 Direct analysis in real time (DART) ionization source. Samples are ionized via Penning ionization from an excited stream of helium or nitrogen

aid in the production of charged microdroplets of solution exiting the source. The electrospray is then directed at an insulating sample. The spray interacts with the sample at the surface causing desorption of ionized analyte which then enters the orifice to the mass spectrometer. The major advantages of DESI are that the target can be almost any surface (e.g. dollar bill, luggage, etc.) and analysis time can be extremely short. This significantly enhances the ease of use and reduces sample preparation time. One of the early applications of DESI was analysis of trace levels of explosives on paper, plastic, and metal [14]

Direct analysis in real time (DART) is another ionization method introduced recently [15] that is extremely fast and easy to use, Fig. 14.6.

DART is similar to DESI in that ionization of desired analyte occurs by directing an energetic spray of particles at the target producing ions. However, instead of a solution spray-based approach, DART creates a plasma from a source gas, typically helium or nitrogen. The plasma is created by directing the source gas through a chamber with an applied voltage followed by a heated tube producing a stream of metastable excited-state atoms and ions. This plasma stream is directed at a surface to produce ions which then enter the orifice of a mass spectrometer [16]. As with DESI the target can be almost any material and analysis time is very short. As well, positioning of the source is not critical for ion production making it suitable for fast analysis of complex target. Indeed, in the original paper by Cody et al. describing the DART source, trace levels of explosives (TNT, HMX, Tetryl, etc.) were detected on targets such as a man's necktie and water from a muddy pond [15].

14.5 Conclusion and Future Developments for Explosives Detection

In light of the global security climate, the technologies described above will continue to be used for security screening for the foreseeable future. Technologies to detect concealed explosive devices remain the most predominant tools at checkpoints; however, advances in MS-based technologies may give rise to their increased prevalence in passenger and cargo screening. The advent of novel ambient pressure ionization sources allows for analysis of samples and sample matrices of greatly increased complexity. This clearly increases the accessibility of MS as a tool for analyzing improvised explosive devices (IEDs) and concealed explosives. Given the somewhat standoff nature of DART and DESI, as well as the fast analysis time, threats can be easily detected before reaching their intended targets. Additionally, these techniques lend themselves to highly sensitive, fast analysis of post-blast explosive fragments. Since DART and DESI effectively ionize analyte molecules from the surfaces of a wide variety of substrates, these techniques can be easily implemented to analyze explosive compound residues as well as fabrication materials used in a pre-blast, or post-blast, IEDs. The analysis of post-blast fragments would be extremely useful for forensic analysis of blast sites. Indeed, DESI-MS has been used recently for the study of explosives in soil samples in post-blast scenarios [17]. It is interesting to note that in this work the chloride adduct of RDX, PETN, Tetryl, and HMX was used for quantitation instead of the molecular ion. Additionally, the MS response was very sensitive to the solvent spray composition. While it is possible to obtain ultra-trace detection of explosives with reasonable sensitivity (sub-μg/kg) there is still room for improvement. It is possible that the controlled use of a dopant to force adduct formation could improve sensitivity greatly.

Significant technological improvements have been made to detect explosives over the past few decades. It is now possible to accurately determine with high sensitivity the presence of explosive compounds with novel ionization sources and mass spectrometry. Continued research will lead to future developments in this field that will provide rapid analysis with easy-to-use instrumentation, ultimately reducing the threats from explosives globally.

References

1. http://data.worldbank.org/indicator/IS.AIR.PSGR. Accessed 28 Feb 2017
2. https://en.wikipedia.org/wiki/X-ray. Accessed 28 Feb 2017
3. Sheen DM, McMakin DL, Collins HD, Hall TE (1993) Near-field millimeter-wave imaging for weapons detection. Proc SPIE 1824:223–233
4. Sheen DM, McMakin DL, Collins HD (1995) Circular scanned millimeter-wave imaging system for weapon detection. Proc SPIE 2511:122–130
5. Grodzins L (1991) Nuclear techniques for finding chemical explosives in airport luggage. Nucl Instr Meth Phys Res B 56:829–833
6. Gares KL, Hufziger KT, Bykov SV, Asher SA (2016) Review of explosive detection methodologies and the emergence of standoff deep UV resonance Raman. J Raman Spectrosc 47:124–141
7. Zachhuber B, Ostmark H, Carlsson T (2014) Spatially offset hyperspectral standoff Raman imaging for explosive detection inside containers. Proc SPIE 9073:90730J
8. Singh S, Singh M (2003) Explosives detection systems (EDS) for aviation security. Signal Proc 83:31–55
9. Eiceman GA, Karpas Z, Hill HH Jr (2014) Ion mobility spectrometry, 3rd edn. CRC Press, Boca Raton
10. Cohen MJ, Karasek FW (1970) Plasma chromatography – a new dimension for gas chromatography and mass spectrometry. J Chromatogr Sci 8:330–337

11. Ewing RG, Atkinson DA, Eiceman GA, Ewing GJ (2001) A critical review of ion mobility spectrometry for the detection of explosives and explosive related compounds. Talanta 54:515–529
12. Ekman R, Silberring J, Westman-Brinkmalm A, Kraj A (2009) Mass Spectrometry: Instrumentation, Interpretation, and Applications. Wiley, Hoboken
13. Takats Z, Wiseman JM, Gologan B, Cooks RG (2004) Mass spectrometry sampling under ambient conditions with desorption electrospray ionization. Science 306:471–473
14. Takats Z, Cotte-Rodriguez I, Talaty N, Chen H, Cooks RG, (2005) Direct, trace level detection of explosives by desorption electrospray ionization mass spectrometry. Chem Commun 1950–1952
15. Cody RB, Laramee JA, Durst HD (2005) Versatile new ion source for the analysis of materials in open air under ambient conditions. Anal Chem 77:2297–2302
16. Cooks RG, Ouyang Z, Takats Z, Wiseman JM (2006) Ambient mass spectrometry. Science 311:1566–1570
17. Bianchi F, Gregori A, Braun G, Crescenzi C, Careri M (2015) Micro-solid-phase extraction coupled to desorption electrospray ionization-high-resolution mass spectrometry for the analysis of explosives in soil. Anal Bioanal Chem 407:931–938

Chapter 15
Synthesis and ESI-MS/MS Fragmentation Study of Two New Isomeric Oxazolidin-2-One Derivatives

Ahoya Anothane Caleb, Youssef Ramli, Hicham Benabdelkamel, Rachid Bouhfid, Nour-eddine Es-Safi, El Mokhtar Essassi, and Joseph H. Banoub

Abstract Two new isomeric oxazolidin-2-one derivatives have been synthesized through reaction of (*E*)-3-styrylquinoxalin-2-one with bis-dichloroethylamine hydrochloride in presence of potassium carbonate. The structures of the formed products were established by NMR and mass spectroscopy. The fragmentation patterns of the two novel synthesized oxazolidin-2-ones (**1**, **2**), possessing the same backbone structure, were investigated using electrospray ionization mass spectrometry (ESI-MS) and tandem mass spectrometry (MS/MS) techniques. A simple methodology, based on the use of ESI (positive ion mode) and by increasing the declustering potential in the atmospheric pressure/vacuum interface, collision induced dissociation (CID), was used to enhance the formation of the fragment ions. In general, the novel synthetic oxazolidin-2-one derivatives afforded, in the gas phase, protonated molecules leading to the confirmation of the molecular masses and chemical structures of the studied compounds. The breakdown routes of the protonated molecules were rationalized by conducting low-energy collision CID-MS/MS analyses (product ion scans) using. The structural similarity between

---

*This work is dedicated to the memory of Dr. Ahoya Anothane Caleb.

A.A. Caleb
Laboratoire de Chimie Organique Hétérocyclique, Mohammed V University in Rabat, Rabat, Morocco

Y. Ramli
Laboratoire de Chimie Thérapeutisue, Faculté de Médecine et de Pharmacie de Rabat, Mohammed VFUniversity in Rabat, Rabat, Morocco

H. Benabdelkamel
King Saud University College of Medecine, 1141 Riyadh, Saudi Arabia

R. Bouhfid
Moroccan Foundation for Advanced Science, Innovation and Research (MASCIR), Rabat Design Center, Rue Mohamed Al Jazouli, Madinat El Irfane, Rabat, Morocco

J.H. Banoub, R.M. Caprioli (eds.), *Molecular Technologies for Detection of Chemical and Biological Agents*, NATO Science for Peace and Security Series A: Chemistry and Biology, DOI 10.1007/978-94-024-1113-3_15

the two studied oxazolidin-2-ones explains the similarity observed in their CID-MS/MS spectra and their fragmentation pathways. ESI-MS/MS recorded with increased collision energy values permits us to observe different product ions and allow us to differentiate the two studied isomers. ESI-MS and CID-MS/MS analyses have thus proven to be a specific and very sensitive method for the structural investigation of the two novel synthesized oxazolidin-2-one derivatives.

Keywords Quinoxaline • Oxazolidin-2-one • Potassium carbonate • ESI-MS • Tandem MS • MS-MS • Fragmentation pathways

15.1 Introduction

Oxazolidin-2-ones derivatives are a very important class of heterocyclic compounds which have attracted attention in various areas of drug development for antibacterial activity, fine chemicals, pesticides, herbicides and have found applications as drugs and agrochemicals [3–9]. Oxazolidin-2-ones have also been widely used as excellent intermediates in the synthesis of natural products and pharmacologically active compounds [10] such as fine chemicals, pharmaceuticals, pesticides, and herbicides [11]. Thus, oxazolidinone skeleton has been incorporated into a wide variety of therapeutically interesting compounds that have potent biological activities [12–16].

Owing to their importance in such important biological and synthetic research areas, the chemistry of oxazolidinones derivatives has received considerable attention [17]. From a synthetic point of view, a wide choice of synthetic methods are available in the literature to prepare oxazolidin-2-ones such as base-mediated cyclization of N-boc-β-amino alcohols [18–20], reduction followed by carbonyl insertion of β-azido alcohols [18, 19], carbonyl insertion in β-amino alcohols using

N.-e. Es-Safi
Laboratoire de Chimie Organique et d'Etudes Physico-chimiques, Ecole Normale Supérieure de Takaddoum, Mohammed V University in Rabat, Rabat, Morocco

E.M. Essassi (✉)
Laboratoire de Chimie Organique Hétérocyclique, Mohammed V University in Rabat, Rabat, Morocco
e-mail: emessassi@yahoo.fr

J.H. Banoub
Fisheries and Oceans Canada, Science Branch, Special Projects, Chemistry Department, Memorial University, St John's, NL, Canada
e-mail: banoubjo@dfo-mpo.gc.ca

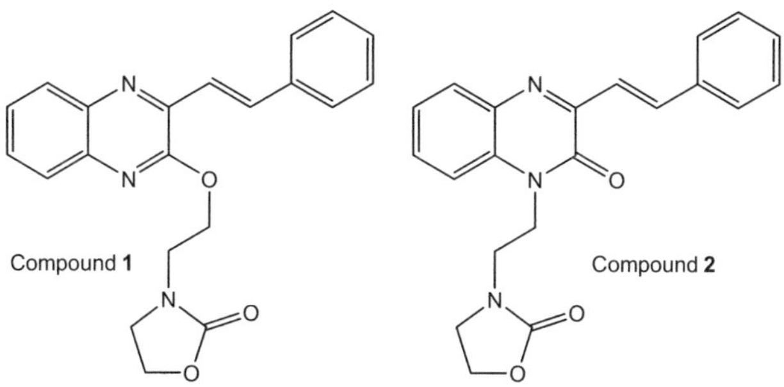

Fig. 15.1 Structure of the isobaric/diastereomeric oxazolidin-2-one derivatives **1**, **2**

either phosgene derivatives [18, 19], carbon dioxide pressure [20] or 1,10-carbonyldiimidazole [21, 22]. Reaction of amino alcohols with phosgene and related compounds furnishes oxazolidinones [23, 24]. Also, treatment of amino alcohols with ethyl chloroformate produced carbamates which in turn can be cyclized to 2-oxazolidinones [24–26]. However, most of these methods suffer from either lower yields or use of hazardous reagents.

In continuation of our work on the synthesis of new heterosyclic systems containing quinoxaline and oxazolidinone moieties [11–26], we herein report the synthesis and characterization of two new isomeric oxazolidin-2-one derivatives which have been obtained from (*E*)-3-styrylquinoxaline-2-one. Their characterization was achieved through NMR and electrospray ionization mass spectrometry ESI-MS (+ ion mode) analysis. In addition, and in a continuation of our research work on the structural elucidation and MS fragmentation breakdown of novel biomolecules, we present herein the single-stage ESI-MS of the two new synthesized oxazolidin-2-ones measured with a hybrid QqTOF-MS instrument. We also report the low-energy collision-induced dissociation (CID-MS/MS) analyses of the selected protonated molecules.

The two novel isobaric/diastereomeric oxazolidin-2-ones derivatives **1**, **2** are shown in Fig. 15.1. To the best of our knowledge, there have been no previous reported studies on the ESI-MS characterization and/or Cid-MS/MS fragmentation routes of this type of compounds.

15.2 Experimental

15.2.1 General

Melting points were determined on a Koffler melting point apparatus and are uncorrected. ^{1}H, ^{13}C and DEPT NMR spectra were recorded with an AVANCE 300 Bruker instrument in Me_2SO-d_6 using Me_4Si as an internal standard. Mass spectra were recorded on a QqTOF-MS. Positive ESI was used as ionization mode.

15.2.2 Synthesis

The two studied isobaric oxazolidin-2-one derivatives were synthesized from (*E*)-3-styrylquinoxaline-2-one (compound A) as described in Scheme 15.1. The isolated compounds were structurally elucidated by ESI-MS (+ ion mode) and low-energy collision dissociation tandem mass spectrometry (CID-MS/MS) and by 1D ^{1}H- and ^{13}C–NMR analysis.

In a 100 mL reactor flask, 3-styrylquinoxaline-2-one (0.0125 mol) was mixed with bis(2-chloroethyl)amine hydrochloride (2.66 mol) in dimethylformamide (40 mL) and in presence of potassium carbonate (2.87 mmol) and tetra-n-butylammonium bromide (BTBA) (1 mmol). The mixture was refluxed in a sand bath, under stirring and the reaction was monitored through TLC. After disappearance of the starting material, the solvent was evaporated under reduced pressure. Separation of the two compounds was performed through silica gel column chromatography using the mixture hexane/ethyl actetae (4/6) as eluent. The pure compounds **1** (168–170 °C) and **2** (184–186 °C) were obtained with yields of 37 and 57% respectively after recristallization from the same mixture.

In a 100 mL reactor flask, 3-styrylquinoxaline-2-one (0.0125 mol) was mixed with bis(2-chloroethyl)amine hydrochloride (2.66 mol) in dimethylformamide (40 mL) and in presence of potassium carbonate (2.87 mmol) and tetra-n-butylammonium bromide (BTBA) (1 mmol). The mixture was refluxed in a sand bath, under stirring and the reaction was monitored through TLC. After disappearance of the starting material, the solvent was evaporated under reduced pressure.

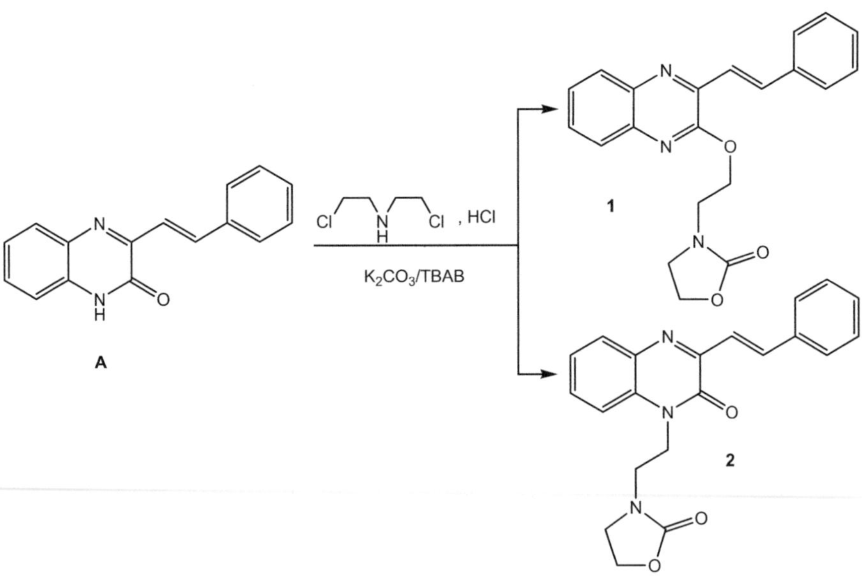

Scheme 15.1 Synthesis of the two isobaric compounds **1**, **2**

Separation of the two compounds was performed through silica gel column chromatography using the mixture hexane/ethyl actetae (4/6) as eluent. The pure compounds **1** (168–170 °C) and **2** (184–186 °C) were obtained with yields of 37 and 57% respectively after recrystallization from the same mixture.

(*E*)-3-(2-(3-styrylquinoxalin-2-yloxy) ethyl oxazolidine-2-one: 1
^{1}H–NMR (Me_2SO-d_6, 300 MHz), δ ppm: 3.7 (2H, t, $J = 7.8$ Hz, α-CH_2 NC = O); 3.8 (2H, t, $J = 5.4$ Hz, β-CH_2 N(=O)); 4.3 (2H, t, J = 7.8 Hz, −CH_2 OC); 4.7 (2H, t, $J = 5.4$ Hz, −CH_2OC = O); 7.0–8.0 (9H, m, CHar).

^{13}C NMR (Me_2SO-d_6, 75 MHz), δ ppm: 43.2 (CH_2); 44.8 (CH_2); 61.8 (CH_2); 63.5 (CH_2); 121.6 (CH); 137.1 (CH); 125.9–136.44 (CHar); 138.9 (Cq); 139.2 (Cq); 140.1 (Cq); 143.6 (Cq); 149.2 (Cq); 154.9 (Cq).

(*E*)-3-(2-(2-oxo-3-styrylquinoxalin-1(2H)-yl) ethyl) oxazolidin-2-one: 2
^{1}H–NMR (Me_2SO-d_6, 300 MHz), δ ppm: 3.6 (2H, t, J = 6.0 Hz, α-CH_2 N(=O)); 3.6 (2H, t, J = 6.6 Hz, β-CH_2 N(=O)); 4.2 (2H, t, J = 6.6 Hz, −CH_2 OC); 4.4 (2H, t, J = 6.0 Hz, −CH_2 OC(=O)); 7.4–8.0 (9H, m, CHar).

^{13}C NMR (Me_2SO-d_6, 75 MHz), δ ppm: 39.8 (CH_2); 40.1 (CH_2); 40.1 (CH_2); 61.9 (CH_2); 121,9 (CH); 137,1 (CH); 114.2 (CH); 123.7 (CH); 127.7 (CH); 128.9 (CH); 129.5 (CH); 130.1 (CH); 132.1 (Cq); 132.1 (Cq); 135.8 (Cq); 151.3 (Cq); 154.4 (Cq); 157.9.

15.3 Results and Discussion

The two synthesized isobaric heterocyclic compounds, $C_{21}H_{19}N_3O_3$, are new synthetic molecules containing oxazolidine and quinoxaline rings with a styryl substituent linked in the position 3 of the quinoxaline moiety. The structure of both compounds is built up from two fused six-membered rings linked to a five-membered oxazolidin-2-one ring by a C-2 chain. The two isomeric compounds differed by the position of the C-2 chain on the quinoxaline skeleton being on the 3 position oxygen atom of the quinoxaline moiety for compound 1 and on the position 2 nitrogen atom of the quinoxaline moiety.

The molecular structures of compounds **1** and **2** were elucidated on the basis of their ESI-MS, ^{1}H and ^{13}C NMR spectral data. The NMR data were in agreement with the proposed structures.

In order to characterize the fragmentation pathways of these oxazolidin-2-ones, the ESI- MS spectra of compounds **1** and **2** were recorded in the positive ion mode. In agreement with their chemical structures the ESI-MS indicated the formation of the protonated molecules $[M + H]^+$ respectively at *m/z* 362 for both compounds. The low-energy CID-MS-MS analyses conducted on the precursor protonated molecules generated from both oxazolidin-2-one derivatives resulted in the formation of a series of the diagnostic product ions and the obtained results are discussed below.

15.3.1 Product Ion Scan of the Protonated Molecule $[M + H]^+$ 1a at m/z 362

In order to investigate the fragmentations pathways observed in compound **1**, low-energy collision CID-MS/MS analyses were conducted in order to determine the fragmentation routes leading to the formation of the various product ions observed in the conventional mass spectra obtained at different fragmentation voltages. The product ion scan of the protonated molecule $[M + H]^+$**1a** at m/z 362 was recorded with a CE = 30 eV and the obtained spectrum is showed in Fig. 15.2. The CID-spectrum showed four major product ion signals located at m/z 70, 114, 249 and 275.

A tentative breakdown pattern of some of this major series of diagnostic product ions is presented in Scheme 15.2 showing three major fragmentation routes. The protonated molecule **1a** eliminates an oxazolidinone molecule giving the product ion **1b** (m/z: 275) which possesses a 2-vinyloxy-3-styryl-quinoxalinium structure. The protonated molecule **1a** at *m/z* 362 can also lose a 3-styryl-quinoxalin-2-one molecule yielding the product ion **1c** at *m/z 114* which represents the base peak of the obtained MS/MS spectrum. This last formed product ion could also eliminate a carbon dioxide molecule to give the ion **1d** at *m/z* 70. Finally the precursor protonated molecule **1a** could eliminate a 1-vinyl-oxazolin-2one molecule giving the product ion **1e** at *m/z* 249 which possess the 3-styryl-quinoxalin-2one structure.

A mechanism for the formation of the ions **1b, 1c** and **1e** from **1a** is tentatively proposed in Scheme 15.3.

In order to have more information on the gas phase fragmentations of compound **1**, *quasi*-MS^3 analysis was conducted on one of the primary product ion **1b** generated from the precursor protonated molecule **1a** [27]. The quasi-MS^3 spectra of the

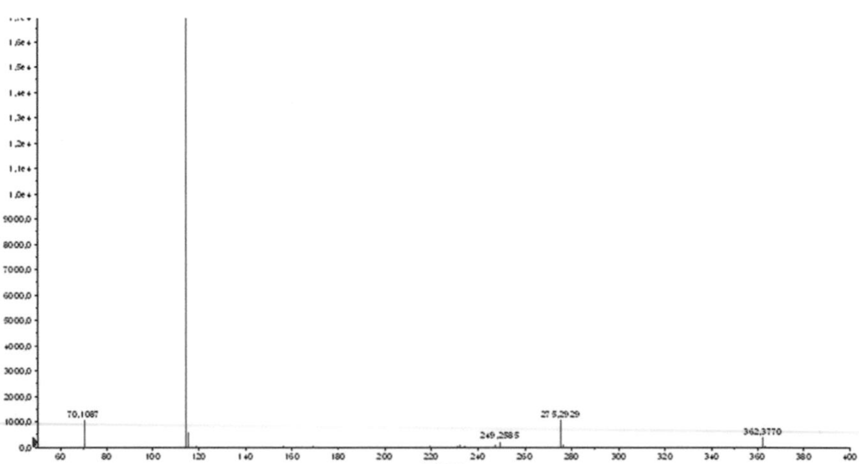

Fig. 15.2 Low-energy CID-MS/MS spectra of the protonated molecule **1a** (m/z 362) recorded with a CE of 30 eV

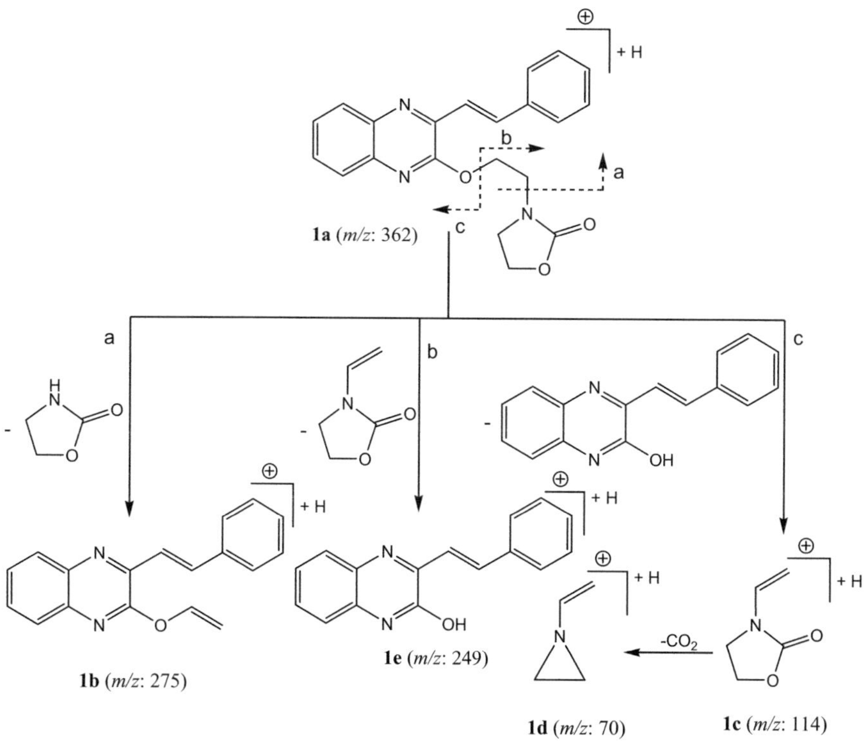

Scheme 15.2 Proposed fragmentation pathways observed for the CID-MS/MS of the precursor protonated molecule **1a** at m/z 362

product ion **1b** at *m/z* 275 recorded with a CE = 20 eV showed three secondary generated product ions at *m/z* 247, 232 and 169 (Fig. 15.3). The formation of these second generation product ions could be tentatively explained as indicated in Scheme 15.4. The product ion **1f** at *m/z* 247 could be formed through a rearrangement involving the loss of an neutral ethyne and a hydrogen molecules giving the furoquinoxaline structure. This latter could yield the product ion 1 g at *m/z* 169 through the loss of neutral benzene molecule. The formation of product ion **1 h** at *m/z* 232 occurred from the precursor at *m/z* 275 by loss of a radical vinyloxy giving a structure 2-styryl-quinoxaline with an odd electron number (Scheme 15.4).

15.3.2 Product Ion Scan of the Protonated Molecule $[M + H]^+$ 2a at m/z 362

After having investigated the low-energy CID-MS/MS behaviour of compound **1**, we were interested to examine the gas-phase behaviour of its isobaric diastereomer **2** under the same conditions. Low-energy CID-MS/MS of the protonated molecule

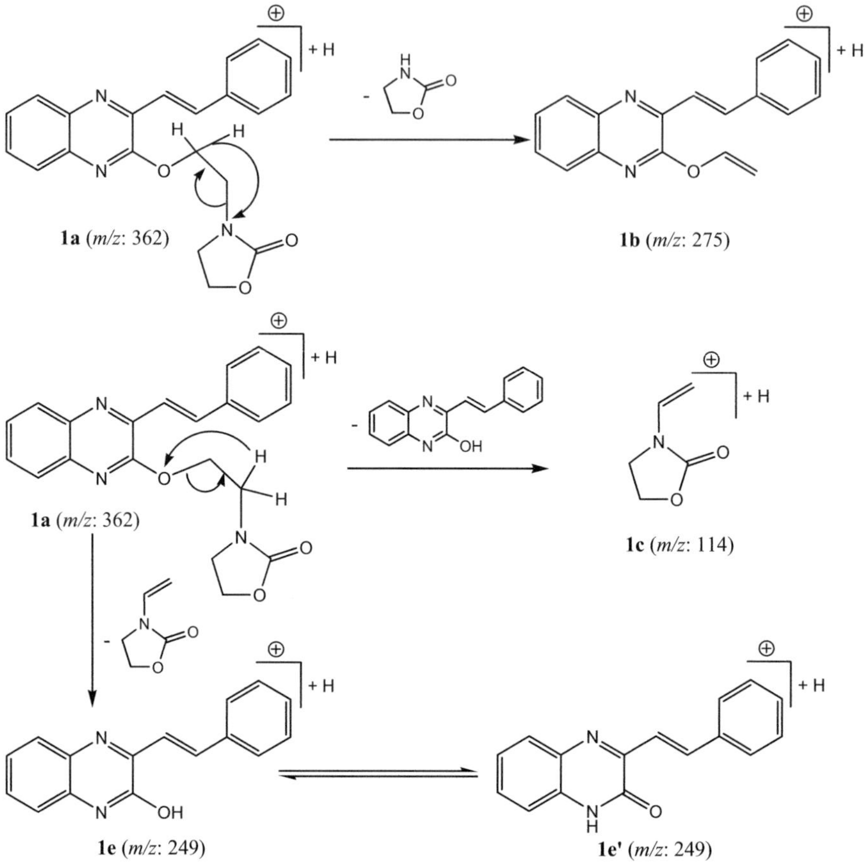

Scheme 15.3 Proposed mechanisms for the formation of the product ions **1b** at *m/z* 275, **1c** at *m/z* 114 and **1e** at m/z 249 obtained from the precursor protonated molecule **1a** at *m/z* 362

2a at m/z 362 afforded a series of the product ions as shown in Fig. 15.4. The CID-spectra showed four major product ion signals at *m/z* 70, 114, 249 and 275.

The gas-phase CID-fragmentation routes of this precursor protonated molecule **2a** at *m/z* 362 are tentatively described to occur by three major breakdown patterns presented in Scheme 15.5. The precursor protonated molecule **2a** at *m/z* 362 eliminates an oxazolidinone molecule to afford the product ion **2b** at *m/z* 275 (base peak). The product **2a** at *m/z* 362 can also yield the 3-styryl-quinoxaline product ion **2c** at *m/z* 249 by elimination of 1-vinyl-oxazolidin-2-one molecule. This latter product ion yields the furoquinoxalinium ion **2d** at *m/z* 247 through an intra-molecular cyclization followed by a loss of a hydrogen molecule. Finally the precursor ion **2a** at *m/z* 362 can also eliminate a 3-styryl quinoxalin-2-one molecule to give the 1-vinyloxazolidin-2-one product ion **2e** at *m/z* 114. This latter lose a CO_2 molecule to give the 1-vinylaziridine product ion **2f** at *m/z* 70.

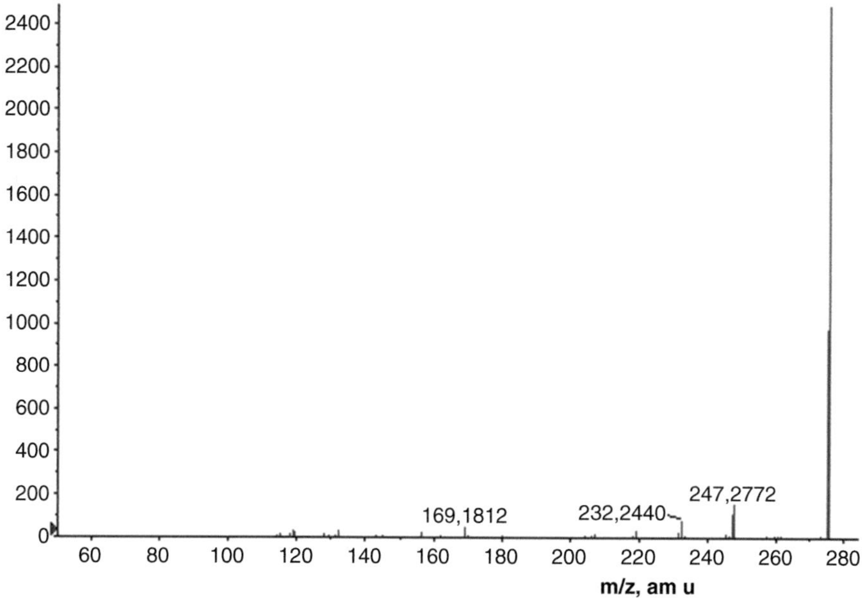

Fig. 15.3 Low-energy *quasi*-MS^3 analysis of the primary product ion **1b** at *m/z* 275 recorded with a CE of 30 eV

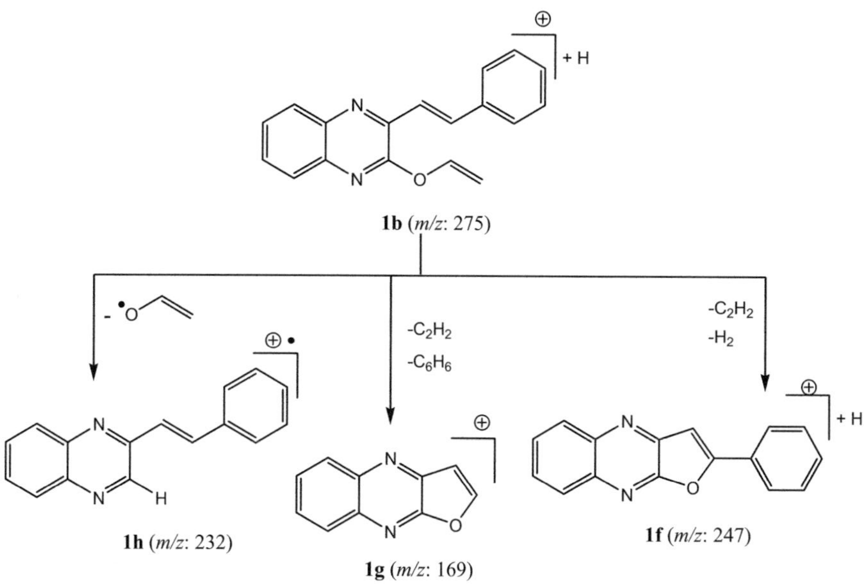

Scheme 15.4 Proposed fragmentation pathways observed for the *quaisi*-MS3 analysis of the precursor primary product ion **1b** at m/z 275

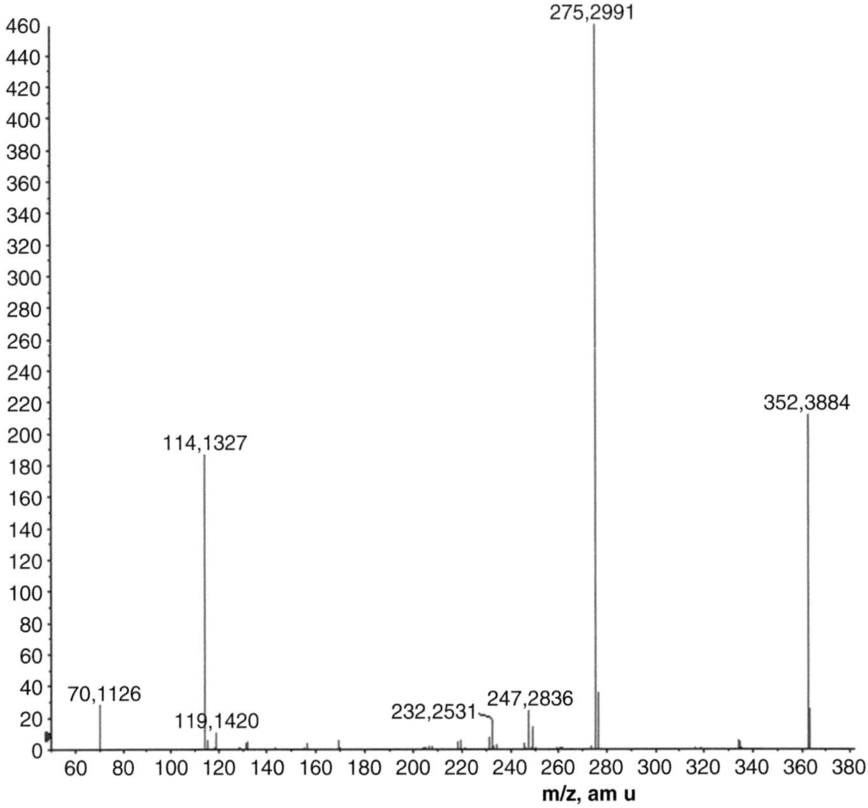

Fig. 15.4 Low-energy CID-MS/MS spectra of the protonated molecule **2a** at *m/z* 362 recorded with a CE of 30 eV

It is interesting to mention that the formation of the product ion **2b** at *m/z* 275 could be explained through a fragmentation involving the migration of the methylene hydrogen linked to the lactam nitrogen atom with concomitant elimination of the oxazolidinone molecule. A mechanism for the formation of these ions is summarized in Scheme 15.6.

In addition, we proposed that the product ion **2d** at *m/z* 247 could be formed from the tautomeric form of the 3-styryl quinoxalinium ion through a rearrangement involving the hydroxyl group giving the dihydrofuroquinoxalinium ion **2c'** at m/z 249 (Scheme 15.7). This latter aromatizes to produce, the furoquinoxalinium secondary product ion **2d** at m/z 247. Finally, the ion product **2e** at *m/z* 114 is formed through the rearrangement of the precursor protonated molecule at *m/z* 362 as indicated in Scheme 15.7 involving the methylene hydrogen linked to the oxazolidinone nitrogen atom.

The CID-fragmentation pathways of the second generation product ion **2b** at *m/z* 275 created from the protonated ion **2a** at *m/z* 362 was investigated through a quasi

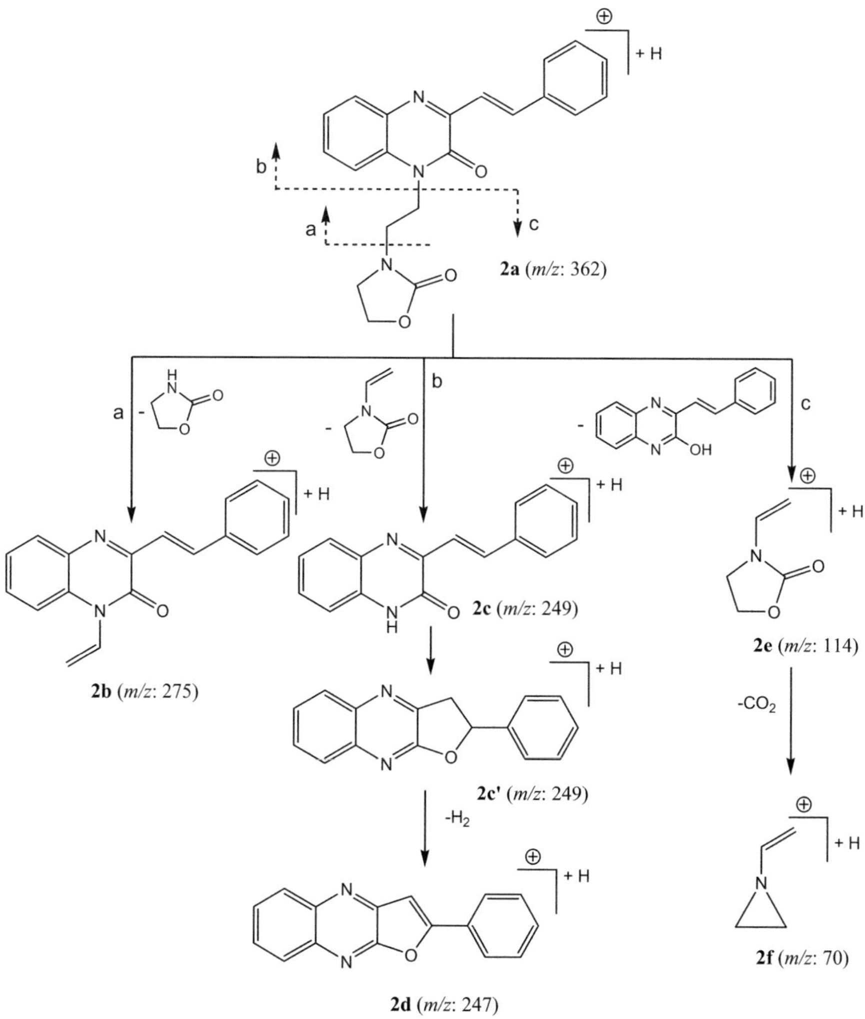

Scheme 15.5 Proposed fragmentation pathways observed for the CID-MS/MS of the precursor protonated molecule **2a** at m/z 362

MS^3 analysis performed with a CE = 20 eV (Fig. 15.5). The obtained CID-spectrum showed the appearance of two major product ions at *m/z* 247 and 232, which corresponded respectively to 1-vinyl-2-styryl benzimidazole and 3-styryl quinoxaline structures. The formation of the product ion **2 g** at *m/z* 247 was obtained from the precursor product ion through the loss of carbon monoxide molecule. The product ion **2 h** at *m/z* 232 was formed through the migration of a vinyl group from the lactam oxygen atom yielding the 2-vinyloxy-3-styryl quinoxalinium product ion **2 h'**. This latter eliminates a vinyloxy radical to afford the product ion **2 h** at *m/z* 232 (Scheme 15.8).

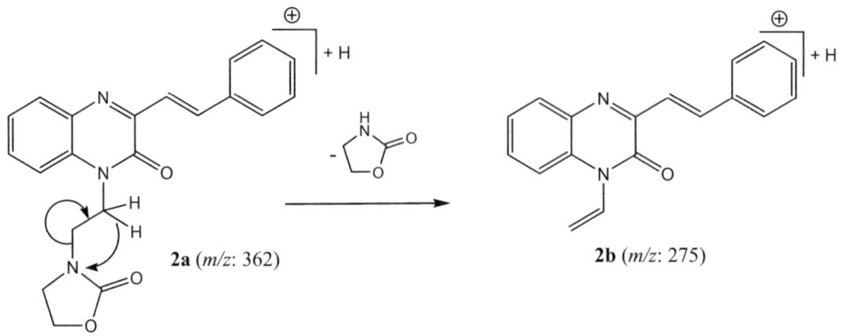

Scheme 15.6 Proposed mechanism for the formation of product ion **2b** at *m/z* 275 from the protonated molecule **2a** at *m/z 362*

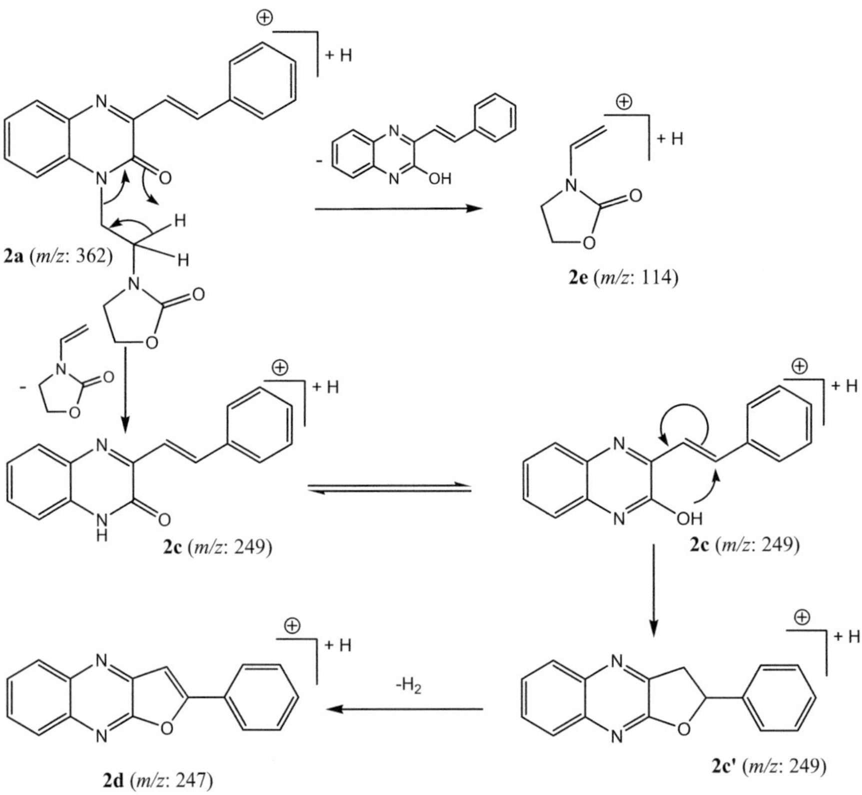

Scheme 15.7 Proposed mechanism for the product ions **2d** (m/z 275) and **2e** (m/z 114) from the protonated molecule **2a** (m/z 362)

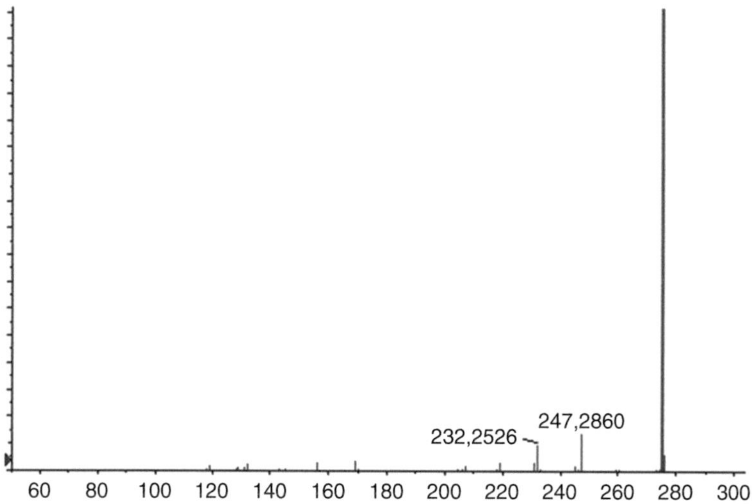

Fig. 15.5 *Quasi*-MS^3 of the primary product ion **2b** at *m/z* 275 recorded with a CE of 20 eV

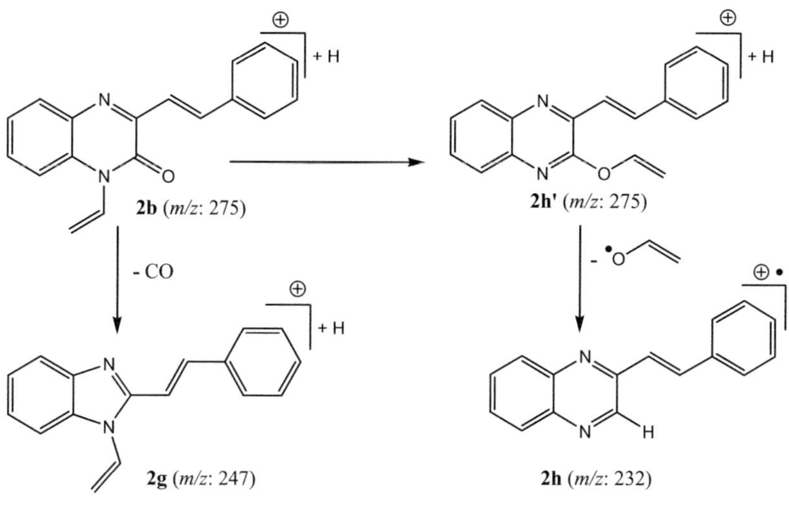

Scheme 15.8 Proposed fragmentation pathways observed for the quasi-MS^3 of the selected product ion **2b** at *m/z* 275

It should be noted that most of the CID-fragmentation routes discussed herein this manuscript involved the cleavage of the ethyl oxazolidin-2-one moiety. For this reason, it became very difficult to differentiate between the two isobaric diastereomers. Therefore, we considered to ensure this potential differentiation, to induce the fragmentation involving the quinoxaline heterocycle. In order to produce such fragmentations, CID-MS/MS analyses were recorded with increasing collision energy values. Accordingly, the CID-MS/MS analysis of the protonated molecule **2a** was

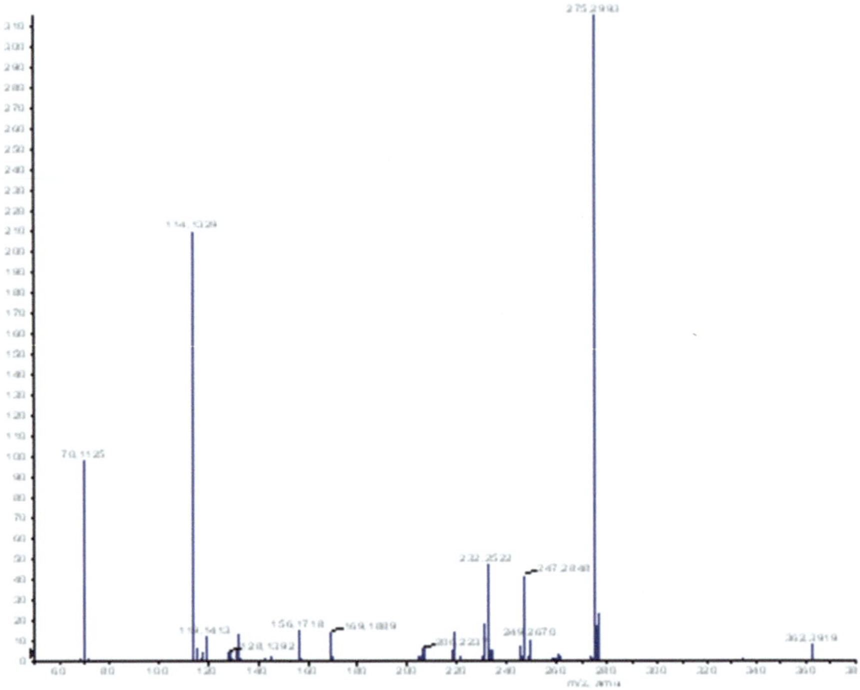

Fig. 15.6 Low-energy CID-MS/MS of the protonated molecule $[M + H]^+$ **2a** at *m/z* 362 recorded with a CE of 40 eV

Scheme 15.9 Fragmentation observed in the MS/MS of the ion **2a** with a CE of 40 eV

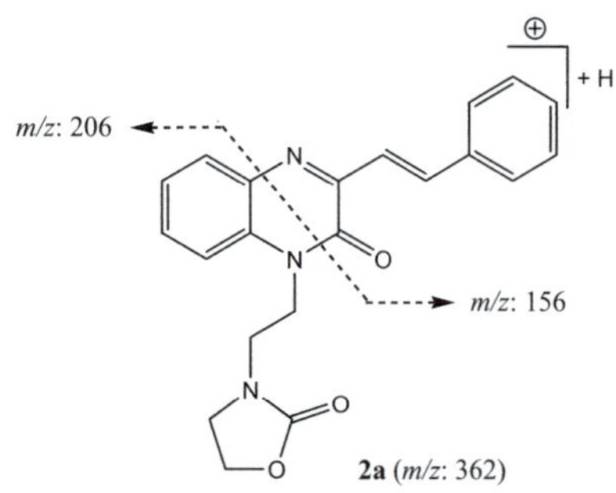

recorded with a higher collision energy of 40 eV; and it showed a series of product ions which included two major ones at *m/z* 156 and 206 (Fig. 15.6.).

The genesis of this series of product ions could only be rationalized through the fragmentation involving the cleavage of the two bonds of the quinoxaline heterocycle ring (Scheme 15.9). The obtained product ion at *m/z* 206 contained the ethyl oxazolidin-2-one moiety and confirmed the existence of the corresponding isobaric diastereomer **2**. This example shows the practicality of low-energy CID-MS/MS technique for the analysis and the differentiation of such isobaric oxazolidin-2-one derivatives.

15.4 Conclusion

In summary, the present work present results the synthesis of two new isobaric oxazolidin-2-one diastereomers from a quinoxalin-2-one derivative bearing a styryl substituent at the position 3. The molecular structures of the two synthesized compounds were elucidated on the basis of their NMR and MS data. The obtained results showed that the two isomers differed by the position of the linkage point of the ethyl-oxazolidin-2-one moiety on the quinoxalin-2-one skeleton, being the oxygen for compound **1** and the N-1for compound **2**.

The study of the ESI-MS and low-energy CID-MS/MS fragmentation routes of two novel isobaric compounds allowed us to propose plausible fragmentation pathways of the observed product ions and propose mechanisms of their formation. The structural similarity between the two oxazolidin-2-ones **1** and **2** explains the similarity observed in their CID-MS/MS spectra and their fragmentation pathways. However, low-energy CID-MS/MS recorded with increased collision energy value permited us to observe different product ions and allow us to differentiate the two studied novel compounds. This is a very interesting finding, which clearly demonstrate the potential of mass spectrometry for the structural elucidation of oxazolidin-2-ones.

References

1. Ford CW, Zurenko GE, Barbachyn MR (2001) The discovery of linezolid, the first oxazolidinone antibacterial agent. Curr Drug Targ Infect Dis 1:181–199
2. Kim SH, Jung MH, Yoo KH, Cho JH, Oh CH (2008) Synthesis and antibacterial activities of novel oxazolidinones having cyclic sulfonamide moieties. Bioorg Med Chem Let 18:5815–5818
3. Gregory WA WA, Brittelli DR DR, Wang CLJ CLJ, Wuonola MA MA, Mc Ripley RJ RJ, Eustice DC DC, Eberly VS VS, Bartholomew PT PT, Slee AM, PT FM (1989) Antibacterials. Synthesis and structure-activity studies of 3-aryl-2-oxo-oxazolidines. J Med Chem 32:1673–1681
4. Hideaki K, Masayuki M, Hiroyki K, Tetsu-ichiro KMK, Hiroyuki O (1999) Cytoxazone: a novel cytokine modulator containing a 2-oxazolidinone ring produced by Streptomyces sp. J Org Chem 64:1052–1053
5. Wouters VJ (1998) Use of chiral glycerol 2,3-carbonate in the synthesis of 3-aryl-2-oxazolidinone. Curr Med Chem 5:137–162
6. Gottschlich C, Ruecker H (1992) (5S)-3-aryl-5-(1-piperidinylmethyl)-2-oxazolidinones, a new class of potential neuroleptics with a high affinity for sigma receptors. Bioorg Med Chem Lett 2:165–116
7. Jeon H, Jo N, Cho J, Oh CH (2007) Synthesis and antibacterial evaluation of 1β-methyl-2-[5-(1-methoxyimino-2-substituted sulfonamide ethyl)pyrrolidin-3-ylthio] carbapenems and their related compounds. Eur J Med Chem 42:358–366
8. Evans DA, Ny HP, Rieger DL (1993) Total synthesis of the macrolide antibiotic Rutamycin B. J Am Chem Soc 115:11446–11459

9. Jones TK, Reamer RA, Desmond R, Mills SG (1990) Chemistry of tricarbonyl hemiketals and application of Evans technology to the total synthesis of the immunosuppressant (−)-FK-506. J Am Chem Soc 112:2998–3017
10. Bouzina A, Grib I, Bechlem K, Belhani B, Aouf N-E, Berredjem M (2016) Efficient synthesis of novel *N*-acylsulfonamide oxazolidin-2-ones derivatives. Karbala Int J Mod Sci 2:98–103
11. Sebbar NK, Mekhzoum MEM, Essassi EM, Zerzouf A, Talbaoui A, Bakri Y, Saadi M, LEI A (2016) Novel 1,4-benzothiazine derivatives: synthesis, crystal structure, and anti-bacterial properties. Res Che Intermed 42:6845–6862
12. Ramli Y, Essassi EM (2015) Advances in synthetic approaches, functionnalization and biological properties of Quinoxaline derivatives. In: Taylor J C (ed) Advances in chemistry research, Volume 27, Nova Science Publishers, Inc, New York, pp 109–160. ISBN:978-1-63482-545-00
13. Alsubar A, Bouhfid R, Essassi EM (2009) Synthesis of new oxindole derivatives containing an oxazolidin-2-one. ARKIVOC 12:337–346
14. Caleb AA, Bouhfid R, Essassi EM, El Ammari L (2009) 3-[2-(3-methyl-2-oxo-1,2-dihydroquinoxalin-1-yl)ethyl]oxazolidin-2-one. Acta Cryst E65:o2024–o2025
15. Bouayad K, Kandri Rodi Y, Ouzidan Y, Essassi EM, Saadi M, El Ammari L (2015) Crystal structure of 5-chloro-1,3-bis[2-(2- oxo-1,3-oxazolidin-3-yl)ethyl]-1H-benzimidazol 2(3H)-one. Acta Cryst E71:o735–o736
16. Daouda B, Doumbia ML, Essassi EM, Saadi M, El Ammari L (2013) 3-[2-(3-phenyl-2-oxo-1,2-dihydroquinoxalin-1-yl)ethyl]-1,3-oxazolidin-2-one. Acta Cryst E69:o662
17. Ramli Y, Zouihri H, Essassi E, Ng SW (2012) 3-{2-[(3-{(E)-2-[4-(Dimethylamino)- phenyl] ethenyl}quinoxalin-2-yl)oxy]- ethyl}-1,3-oxazolidin-2-one monohydrate. Acta Cryst E68:o241
18. Ouzidan Y, Kandri Rodi Y, Fronczek FR, Venkatraman R, El Ammari L, Essassi EM (2011) 1,3-Bis[2-(2-oxo-1,3-oxazolidin-3-yl)- ethyl]-1H-benzimidazol-2(3H)-one. Acta Cryst E67:o362–o363
19. Ouzidan Y, Jasinski JP, Butcher RJ, Golen JA, Essassi EM, El Ammari L (2011) 3-[2-(6-Bromo-2-phenyl-3H-imidazo- [4,5-b]pyridin-3-yl)ethyl]-1,3-oxazolidin- 2-one. Acta Cryst E67:o1095
20. Daouda B, Brelot L, Doumbia ML, Essassi EM, Ng SW (2011) 3-{2-[(3-Phenylquinoxalin-2-yl)oxy]- ethyl}-1,3-oxazolidin-2-one>. Acta Cryst E67:o1235
21. Ahmed Moussaif A, Essassi EM, Lazar S, Zouihri H, Leger JM (2010) 3-[2-(1H-Benzimidazol-2-ylsulfanyl)- ethyl]-1,3-oxazolidin-2-one. Acta Cryst E66:o3137
22. Daouda B, Ahabchane NH, Zouihri H, Essassi EM, Ng SW (2010) 1-[2-(2-Oxo-1,3-oxazolidin-3-yl)ethyl]-4- phenyl-1H-1,5-benzodiazepin-2(3H)-one. Acta Cryst E66:o2080
23. Bel-Ghacham H, Kandri Rodi Y, Saffon N, Essassi EM, Ng SW (2010) 6-Bromo-1-[2-(2-oxo-1,3-oxazolidin-3- yl)ethyl]-1H-imidazo[4,5-b]pyridin- 2(3H)-one. Acta Cryst E66:o456
24. Abdussalam Alsubari A, Bouhfi R, Zouihri H, Essassi EM, Seik Weng Ng SW (2010) 3-[2-(5H-Indolo[2,3-b]quinoxalin-5-yl)- ethyl]-1,3-oxazolidin-2-one. Acta Cryst E66:o2461
25. Ahoya CA, Rachid Bouhfid R, Ballo Daouda B, Essassi EM, El Ammari L (2010) 3-[2-(3-Methylquinoxalin-2-yloxy)ethyl]-1,3-oxazolidin-2-one. Acta Cryst E66:o1050
26. Banoub J. (ed) (2011) Detection of biological agents for the prevention of terrorism. NATO science for peace and securuty Series A.: Chemistry and biology, DOI 10. 1007/978-90-481-9815-3_1, Springer Science+Business Media B. V 319–360.
27. Joly N, El Aneed A, Martin P, Banoub J (2005) Structural determination of the novel fragmentation routes of morphine opiate receptor antagonists using electrospray ionization quadrupole time-of-flight tandem mass spectrometry. Rapid Commun Mass spectrom 19:3119–3130

Chapter 16
The Use of Multiple Fragmentation Events in a Single Laser Shot for Improved Drug Quantification by MALDI TOF/TOF Mass Spectrometry

Boone M. Prentice, Chad W. Chumbley, Brian C. Hachey, Jeremy L. Norris, and Richard M. Caprioli

Abstract Matrix-assisted laser desorption/ionization time of flight (MALDI-TOF) mass spectrometry is an important molecular detection technology due to its extensive mass range, sensitivity, rapid analysis time, and minimal sample preparation. However, quantitative MALDI-TOF approaches have historically suffered from poor accuracy and precision, mainly due to the variability in analyte distribution and a non-uniform sample surface and analyte suppression [1-5]. In order to alleviate spectral complexity at low m/z and ensure chemical specificity, tandem mass spectrometry (MS/MS) is often employed. Herein, we demonstrate that multiple MS/MS events can be performed in a single laser shot [8,9]. This allows the reference of analyte instensity to that of an internal standard in each laser shot, thereby improving accuracy, precision, and throughput.

B.M. Prentice (✉) • B.C. Hachey • J.L. Norris
Mass Spectrometry Research Center, Department of Biochemistry,
Vanderbilt University, Nashville, TN, USA
e-mail: boone.m.prentice@vanderbilt.edu

C.W. Chumbley
Mass Spectrometry Research Center, Department of Biochemistry,
Vanderbilt University, Nashville, TN, USA

Department of Chemistry, Vanderbilt University, Nashville, TN, USA

R.M. Caprioli
Department of Biochemistry, Mass Spectrometry Research Center,
Vanderbilt University School of Medicine, Nashville, TN, USA

J.H. Banoub, R.M. Caprioli (eds.), *Molecular Technologies for Detection of Chemical and Biological Agents*, NATO Science for Peace and Security Series A: Chemistry and Biology, DOI 10.1007/978-94-024-1113-3_16

Matrix-assisted laser desorption/ionization time of flight (MALDI-TOF) mass spectrometry has become a popular qualitative analysis tool due to its extensive mass range, sensitivity, rapid analysis time, and minimal sample preparation. These types of molecular detection strategies can play important roles in the analyses of many different chemical and biological agents. However, quantitative MALDI-TOF approaches have historically suffered from poor accuracy and precision, mainly due to the variability in analyte distribution and a non-uniform sample surface [1, 2]. The accurate quantification of small molecules and metabolites is further complicated by the vast excess of matrix signal that can dominate ion signal in the low *m/z* region of the mass spectrum and often necessitates the use of an internal standard [3–5]. In order to alleviate this spectral complexity at low *m/z* and ensure chemical specificity, tandem mass spectrometry (MS/MS) is often employed.

Conventional MALDI-MS/MS modalities, with some exceptions, typically scan for only a single MS/MS transition per laser shot [6, 7]. Herein, we demonstrate that multiple MS/MS events can be performed in a single laser shot [8, 9]. For the purposes of this discussion, an MS/MS event refers to the isolation and fragmentation of a single *m/z* window. As such, this technology allows multiple MS/MS isolation and fragmentation events to be performed in each single laser shot. This allows the reference of analyte instensity to that of an internal standard in each laser shot, thereby improving accuracy, precision, and throughput.

Experiments were performed on a MALDI-TOF/TOF mass spectrometer (300 Tandem, SimulTOF Systems, Sudbury, MA) [10, 11]. The instrument was operated in positive ion MS/MS mode at 4 kV and is equipped with a 349 nm frequency-tripled Nd:YLF MALDI laser capable of repetition rates of up to 5 kHz (Spectra-Physics, Santa Clara, CA). Figure 16.1 illustrates the process of individually selecting and fragmenting two distinct precursor ions in a single laser shot using this instrument [8]. Ions are initially extracted from the first source region of the instrument and enter the first field-free drift region (TOF-1). A timed ion selector (TIS) is then used to isolate a precursor ion of interest (t = 30 μs for the blue ion, t = 38 μs for the red ion). The TIS is opened and closed multiple times within the same mass spectrum, allowing multiple ions of increasing *m/z* to be isolated. After isolation, the ions enter a collision cell and second source region. Similar to TIS isolation, the second source region is ramped multiple times within the same mass spectrum to reaccelerate ions of increasing *m/z*. Ions then enter the TOF-2 region of the instrument prior to striking the multichannel plate detector.

In quantification experiments, multiple TOF/TOF events can be useful in experiments where the internal standard is quite different in precursor *m/z* compared to the analyte (*i.e.*, when simply broadening the MS/MS window is not feasible because it would transmit an excess of chemical noise). This is demonstrated in Fig. 16.2 for the quantification of the drug enalapril, an angiotensin-converting

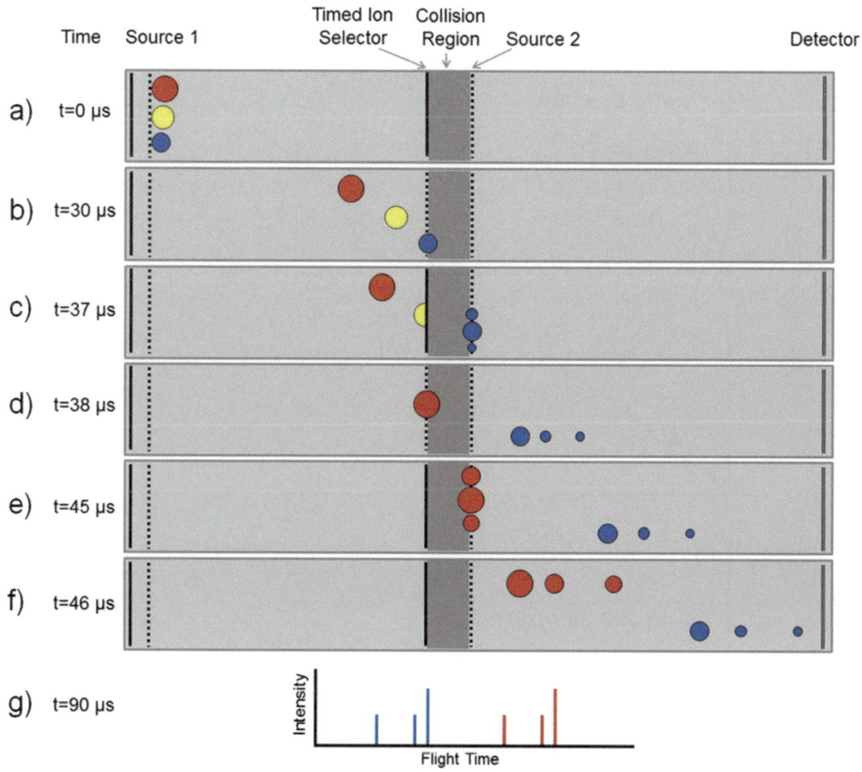

Fig. 16.1 Scheme showing the individual selection and reacceleration of two distinct precursor ions in a single laser shot. (a) Ions are extracted from the source 1 region of the instrument and separated according to *m/z* in TOF-1. (b) The TIS opens to allow the first (blue) ion through. (c) As the first ion is fragmented and passes through the collision cell to the source 2 region, the TIS closes again to prevent the second (yellow) ion from passing. (d) While the first (blue) ion is reaccelerated into the TOF-2 region, the TIS reopens to allow the third (red) ion to pass into the collision cell. (e) The second (blue) ion is traversing the TOF-2 region while the third (red) ion has reached the source 2 region and the TIS has closed again. (f) The first (blue) ion is still traversing the TOF-2 region and the third (red) ion has been reaccelerated. (g) The fragment and precursor ions of both the blue and red ions have reached the detector and are visible in the resulting spectrum. Note: fragment ions are shown here to arise from CID whereas actual experiments employed PSD. The reflectron in the TOF-2 region has been omitted for simplicity. Ion intensities and fragment ion sizes are not shown to scale (Reprinted with permission from Ref. [8])

enzyme (ACE) inhibitor used to treat hypertension, in 45% pooled human plasma. This drug was analyzed in a concentration range of 0.03–10.0 μM using 0.50 μM of the drug ramipril as an internal standard [8]. These samples were prepared using an acetonitrile protein precipitation protocol and the extract was manually spotted (1 μL) on a stainless steel MALDI target in a 1:1 mixture with 50 mg/mL of dihy-

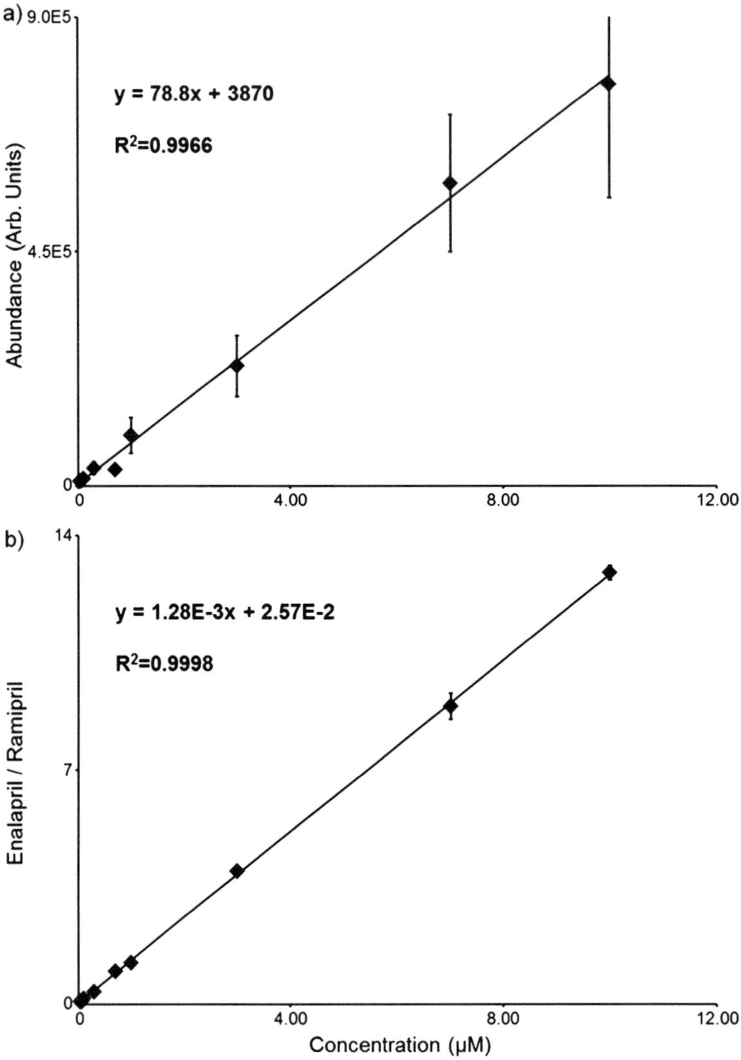

Fig. 16.2 Linearity and relative standard devations are improved for enalapril quantification from (**a**) without normalization to (**b**) upon normalization to the ramipril internal standard (n = 5) (Reprinted with permission from Ref. [8])

droxybenzoic acid (DHB, the MALDI matrix) in 20% methanol. Enalapril (*m/z* 377) and ramipril (*m/z* 417) are sufficiently different in their *m/z* values such that broadening the MS/MS window to ~40 Da to accommodate both ions would transmit an excess of chemical noise, diminishing sensitivity. However, using multiple discrete fragmentation events allows enalapril to be normalized to the ramipril internal standard, resulting in dramatically improved accuracies (data not shown), improved linearity and > four-fold improvements in relative standard deviation (Fig. 16.2). This assay resulted in a limit of quantification (LOQ) of ~11 ng/mL, a value well below the expected serum level following a single 10 mg dose (50–250 ng/mL) [12, 13].

Finite ion transit times through the device as well as power supply limitations require precursor ions to differ by ~6% in *m/z* in order to discretely isolate and reaccelerate ions using this technology, preventing the use of multiple MS/MS windows in the discrete analysis of a small molecule and a stable isotope analogue. However, when using isotopically labeled compounds as internal standards, the MS/MS window can be broadened to accommodate both the analyte and internal standard, and multiple MS/MS events can be used to analyze multiple analyte/internal standard pairs, thereby increasing assay throughput. This is demonstrated in Fig. 16.3. for the simultaneous quantification of three drugs, promethazine (PMZ), enalapril (ENP), and verapamil (VPM), using deuterated analogues as internal standards. [8] Normalization to the internal standards resulted in improvements in accuracies, linarites, and precisions (data not shown). This assay provided accurate quantification within toxic ranges (LOQs: ~199 ng/mL for PMZ, ~264 ng/mL for ENP, ~318 ng/mL for VPM) while increasing throughput by performing three assays in a single experiment.

These results demonstrate the individual MS/MS analysis of multiple precursor ions in a single laser shot on a MALDI-TOF/TOF mass spectrometer [8, 9]. Tandem mass spectrometry affords chemical specificity and improves sensitivity by eliminating chemical noise. This novel methodology allows the analyte intensity to be referenced to that of the internal standard in each single laser shot, improving accuracy, precision, and throughput. A unique benefit to this system is that, because precursor ions are separated in the TOF-1 region of the instrument, isomeric fragment ions (*e.g.*, the fragment ions of enalapril and ramipril, *m/z* 235) are easily resolved (data not shown).

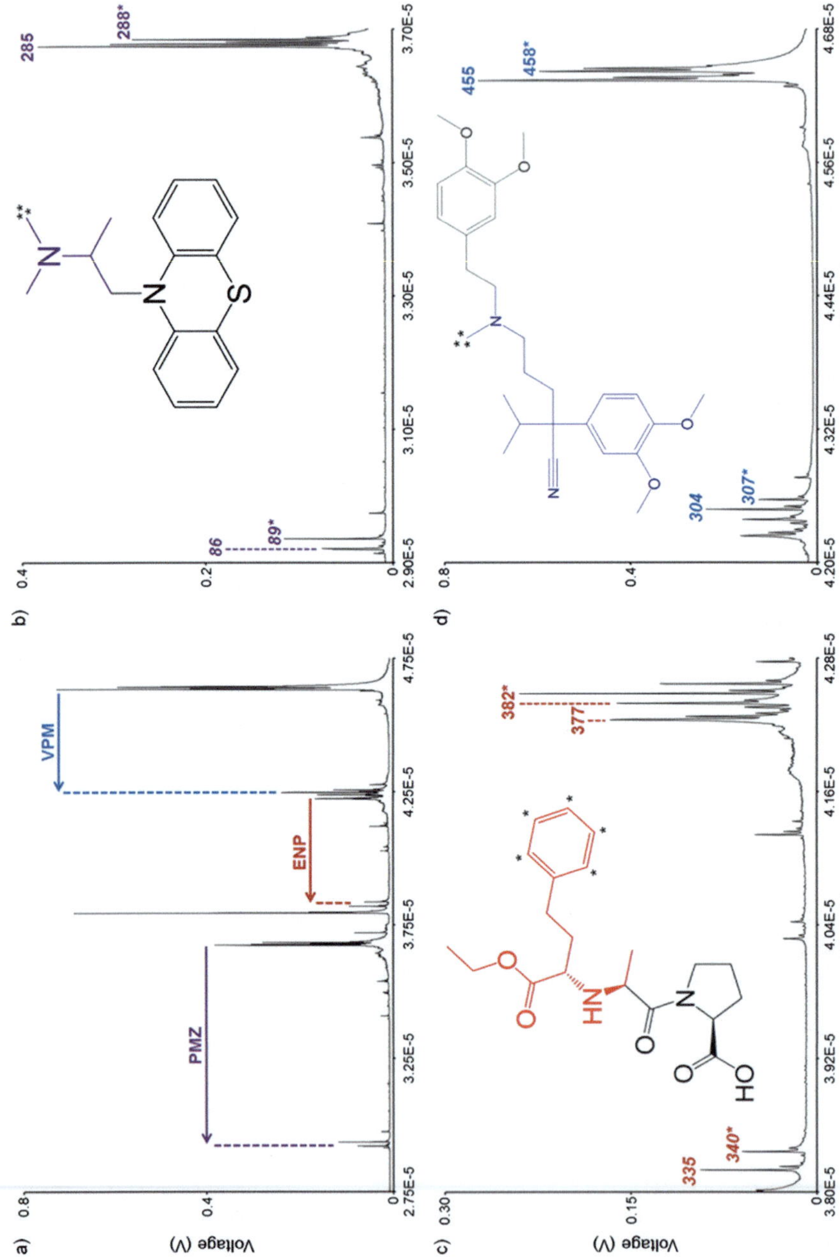
a)
b)
c)
d)
Voltage (V)
Voltage (V)
Time (s)
Time (s)
PMZ
ENP
VPM
285
288*
86
89*
455
458*
304
307*
335
340*
377
382*
0.8
0.4
0
0.4
0.2
0
0.30
0.15
0
0.8
0.4
0
2.75E-5
3.25E-5
3.75E-5
4.25E-5
4.75E-5
2.90E-5
3.10E-5
3.30E-5
3.50E-5
3.70E-5
3.80E-5
3.92E-5
4.04E-5
4.16E-5
4.28E-5
4.20E-5
4.32E-5
4.44E-5
4.56E-5
4.68E-5

Fig. 16.3 (**a**) Quantitative assay of PMZ, ENP, and VPM involves three separate TOF/TOF events in a single laser shot. (**b–d**) Zoomed-in regions of the x-axis show the relevant time (mass) range for each drug transition, with the figure insets showing the chemical structures of the three drugs. For each of (**b**) PMZ, (**c**) ENP, and (**d**) VPM, the isotopically labeled internal standard (denoted with an asterisk) and analyte precursor and fragment ions are all clearly visible in the mass spectrum. The colored portions of each structure inset correspond to the fragment ion of each species that was detected. The asterisks in the structures denote the locations of the heavy isotope labels. Spectrum represents an average of ~1000 laser shots (Reprinted with permission from Ref. [8])

References

1. Bucknall M, Fung KYC, Duncan MW (2002) Practical quantitative biomedical applications of MALDI-TOF mass spectrometry. J Am Soc Mass Spectrom 13(9):1015–1027. doi:10.1016/s1044-0305(02)00426-9
2. Duncan MW, Roder H, Hunsucker SW (2008) Quantitative matrix-assisted laser desorption/ionization mass spectrometry. Brief Funct Genomic Proteomic 7(5):355–370
3. Duncan MW, Matanovic G, Cerpapoljak A (1993) Quantitative analysis of low molecular weight compounds of biological interest by matrix-assisted laser desorption/ionization. Rapid Commun Mass Spectrom 7(12):1090–1094. doi:10.1002/rcm.1290071207
4. Hatsis P, Brombacher S, Corr J, Kovarik P, Volmer DA (2003) Quantitative analysis of small pharmaceutical drugs using a high repetition rate laser matrix-assisted laser/desorption ionization source. Rapid Commun Mass Spectrom 17(20):2303–2309. doi:10.1002/rcm.1192
5. Sleno L, Volmer DA (2005) Some fundamental and technical aspects of the quantitative analysis of pharmaceutical drugs by matrix-assisted laser desorption/ionization mass spectrometry. Rapid Commun Mass Spectrom 19(14):1928–1936. doi:10.1002/rcm.2006
6. Reich RF, Cudzilo K, Levisky JA, Yost RA (2010) Quantitative MALDI-MSn analysis of cocaine in the autopsied brain of a human cocaine user employing a wide isolation window and internal standards. J Am Soc Mass Spectrom 21(4):564–571. doi:10.1016/j.jasms.2009.12.014
7. Erickson BK, Jedrychowski MP, McAlister GC, Everley RA, Kunz R, Gygi SP (2015) Evaluating multiplexed quantitative Phosphopeptide analysis on a hybrid Quadrupole mass filter/linear ion trap/Orbitrap mass spectrometer. Anal Chem 87(2):1241–1249. doi:10.1021/ac503934f
8. Prentice BM, Chumbley CW, Hachey BC, Norris JL, Caprioli RM (2016) Multiple time-of-flight/time-of-flight events in a single laser shot for improved matrix-assisted laser desorption/ionization tandem mass spectrometry quantification. Anal Chem 88(19):9780–9788. doi:10.1021/acs.analchem.6b02821
9. Prentice BM, Chumbley CW, Caprioli RM (2017) Absolute quantification of rifampicin by MALDI imaging mass spectrometry using multiple TOF/TOF events in a single laser shot. J Am Soc Mass Spectrom 28(1):136–144. doi:10.1007/s13361-016-1501-2
10. Prentice BM, Chumbley CW, Caprioli RM (2015) High-speed MALDI MS/MS imaging mass spectrometry using continuous raster sampling. J Mass Spectrom 50(4):703–710. doi:10.1002/jms.3579
11. Vestal ML (2010) TOF-TOF with high resolution precursor selection and multiplexed MS-MS. USA patent US 7,838,824 B2, Nov. 23, 2010
12. Biollaz J, Schelling JL, Combes BJD, Brunner DB, Desponds G, Brunner HR, Ulm EH, Hichens M, Gomez HJ (1982) Enalapril maleate and a lysine analog (MK-521) in normal volunteers- relationship between plasma drug levels and the renin-angiotensin system. Br J Clin Pharmacol 14(3):363–368
13. Arafat T, Awad R, Hamad M, Azzam R, Al-Nasan A, Jehanli A, Matalka K (2005) Pharmacokinetics and pharmacodynamics profiles of enalapril maleate in healthy volunteers following determination of enalapril and enalaprilat by two specific enzyme immunoassays. J Clin Pharm Ther 30(4):319–328. doi:10.1111/j.1365-2710.2005.00646.x

Chapter 17
Detection and Monitoring of Subsurface Contamination by Using Geophysical Methods and Data: Case Studies of Contamination from Chemical and Biological (Biogeochemical) Processes

Ali Ismet Kanli

Abstract Monitoring and distribution of bacterial transport, bioclogging processes, imaging to infer biogeochemical processes associated with bioremediation, remediation-induced biogeochemical end-products are detectable using geophysical methods. Biogeochemical processes alter the physical and chemical properties in the subsurface, and efficient monitoring of these changes is important for successful implementation of soil and groundwater remediation. In the study, two different geophysical studies are discussed from Germany and Hungary.

17.1 Introduction

One of the most critical encountered tasks to remediation of polluted environment is determining the 3-D spatial distribution of contaminants which stimulate us to provide accurately modeled subsurface. Although there are several conventional methods for modeling the subsurface, to understand the hydrogeological process and spatial distribution of contamination, these methods can be gathered under two main branches that are invasive (e.g. invasive, semi-invasive, less invasive) and non-invasive techniques. Classical borehole techniques, SPT and CPT are among the invasive or less-invasive techniques whereas electrical and electromagnetic based ERT and GPR, most widely used techniques in imaging the subsurface in the last decade [2], are called as non-invasive geophysical methods. The first study area is given in Fig. 17.1 which is located at Germany.

A.I. Kanli (✉)
Faculty of Engineering, Department of Geophysical Engineering, Istanbul University, Avcilar Campus, Istanbul 34320, Turkey
e-mail: kanli@istanbul.edu.tr

J.H. Banoub, R.M. Caprioli (eds.), *Molecular Technologies for Detection of Chemical and Biological Agents*, NATO Science for Peace and Security Series A: Chemistry and Biology, DOI 10.1007/978-94-024-1113-3_17

Fig. 17.1 Study area from Germany (From Ref. [3])

The determination method of 3D resistivity distribution from CPT petrophysical parameters was tested on the certificated test area of Research Centre Jülich (Fig. 17.1). The test site is located in the Lower Rhine Embayment in Western Germany and has an extension of 200 × 70 m. The test area was first set up in 1993 by the Research Centre Jülich to research and execute some experiments on water flow and solute transport processes.

The 3D geological profile of the area was defined after 73 boreholes, although only four of them reached the depth of 15–20 m. The aquifer material consists mainly of gravelly and sandy sediments deposited by the braided river system of the Rur. The clay and silt content of the aquifer sediments vary between 0.5% and 7.5% and the mean total porosity is 26 ± 7%. The local base of the aquifer in a depth of 11–13 m consists of intermitting thin layers of clay and silt (Fig. 17.2a).

3D distribution of the electrical resistivity in the investigated depth range of 0–15 m is determined (Fig. 17.2b). Comparing the resistivity and the geological sections a good correlation can be observed among the calculated resistivity distribution and the geological structure. The effect of the clay layer F and the thicker intermitting layers can be also marked clearly. The simplified explanation of the subsurface geology corresponded to their resistivity values in the investigated test area of Krauthausen can be given as follows; A and B layers which are characterized by gravel or debris series are about 1–2 m depth and A has 10–30 Ω-m or 50–90 Ω-m resistivity values whereas B has 190–290 Ω-m resistivity values. The layer C is about 2 m depth and it has the range of 110–180 Ω-m resistivity values. The mixture of the D-E series is about 7 m depth and they are in the range of 50–90 Ω-m or 110–170 Ω-m resistivity values. The C, D and E series form the uppermost aquifer which is characterized by sand and gravel. The layer F which is consisted of silt and

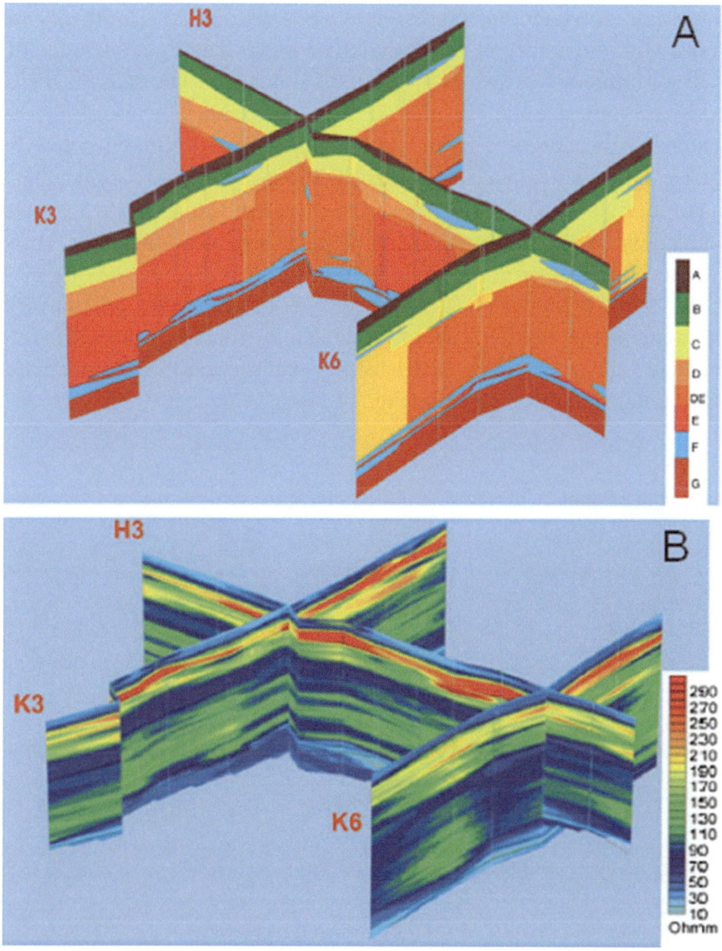

Fig. 17.2 3D geological interpretation (**a**) and 3D electrical resistivity distribution (**b**).of the study area (From Ref. [3])

clay series has 10–30 Ω-m resistivity values. The layer G which is characterized by sandy gravel series has the range of 50–80 Ω-m resistivity values [3].

Molybdenum does not occur naturally as a free metal on Earth, but rather in various oxidation states in minerals. It readily forms hard, stable carbides in alloys, and for this reason most of world production of the element (about 80%) is in making many types of steel alloys, including high strength alloys and super alloys. Most molybdenum compounds have low solubility in water, but the molybdate ion $MoO4^{2-}$ is soluble and forms when molybdenum-containing minerals are in contact with oxygen and water. Industrially, molybdenum compounds are used in high-pressure and high-temperature applications, as pigments and catalysts. About 86% of molybdenum produced is used in metallurgical applications such as alloys, with the rest of molybdenum used as compounds in chemical applications.

A congenital molybdenum cofactor deficiency disease, seen in infants, results in interference with the ability of the body to use molybdenum in enzymes. It causes high levels of sulfite and urate, and neurological damage. The cause is the inability of the body to synthesize molybdenum cofactor, a heterocyclic molecule that binds molybdenum at the active site in all known human enzymes that use molybdenum. Although human toxicity data is unavailable, animal studies have shown that chronic ingestion of more than 10 mg/day of molybdenum can cause diarrhea, growth retardation, infertility, low birth weight and gout; it can also affect the lungs, kidneys and liver [1]. Therefore it is very important to monitoring and detection of the contamination. The second study area is located nearby both agriculture and industrial environment having some problems about molybdenum contaminations. Therefore, we have conducted electromagnetic based geophysical surveys that are adequately used in environmental studies in the area (Fig. 17.3).

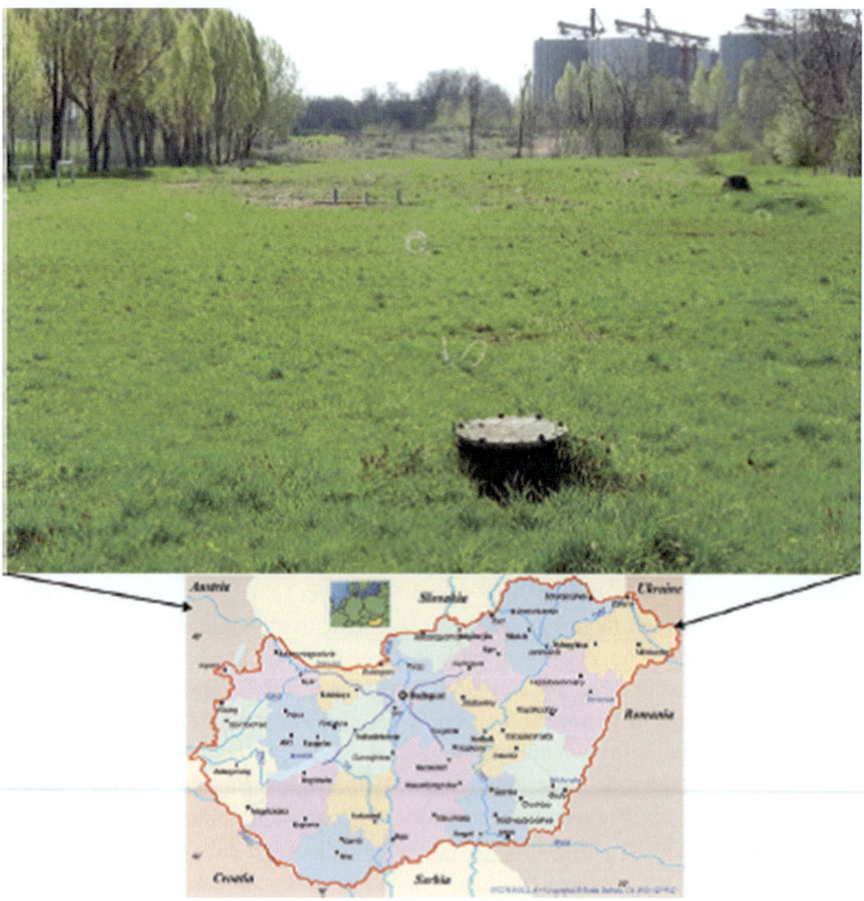

Fig. 17.3 Study area from Hungary (From Ref. [1])

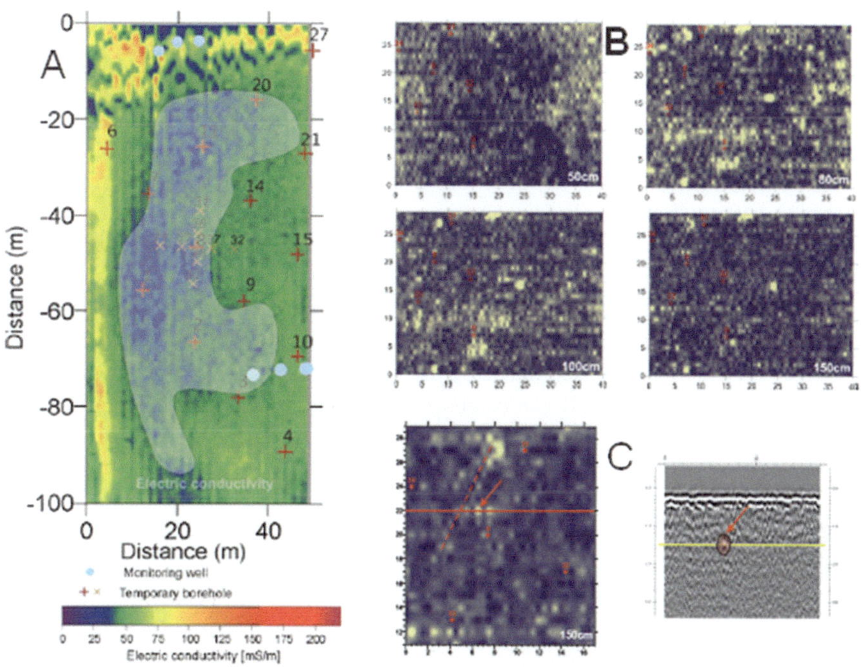

Fig. 17.4 Contour map of the *low* conductivity zone (**a**). Depth slice from 150 cm processed with 450 MHz GPR data is given on the *left*. GPR section from the relatively *high* reflectivity zone is given on the *right* (**b**). Depth slices of processed 450 MHz GPR data measured above the shallow molybdenum contamination (**c**) (From Ref. [1])

In the Fig. 17.3, some monitoring wells located in the survey area can be seen. These wells have been monitored to acquire information on molybdenum contamination with hydrogeological conditions of the investigated area. When the acquired maps are interpreted (Fig. 17.4a), an "L" shaped (reversed) anomaly can be identified at the upper end of the map which has high conductivity (low resistivity) and low susceptibility. On the other hand, when compared to their surroundings, three higher conductivity and lower susceptibility anomalies can be identified about ~2 m long stripes that showing up at different frequencies on the maps at near the well no. 13. Detailed interpretation results also show us that there is a shapeless, low conductivity zone, observed on conductivity maps that cannot be seen on the susceptibility maps. The maximum of this zone is located near the high molybdenum concentration.

In the second stage of measurements, GPR surveys were carried out in the study area. The aim of this investigation was to find a container or other any geological unit that can be a source of contamination. In geo-radar measurements, two different type of antennas were used, 450 MHz and 50 MHz frequency respectively. During the measurements, the attenuation of the waves was very high with thb2e maximum penetration depth was about ~1.5 m. The data acquisition was carried out along profiles with 1 m interval. The data was collected in the area among the wells numbered 34, 31, 32 and 33. The spatial sampling rate was less than 10 cm within the sections. In radar measurements, we get some reflections from the border of the materials with different conductivity properties due to the existence of plastic and metal tubes. We processed the sections concerning reflection strength, and the results are sliced at several depth levels. The maps at different depth levels from 50 cm to 150 cm are given in Fig. 17.4.b.

The important anomaly locations highlighted on the slices were searched in details on the individual records to understand the causes of the reflections. As seen on the 1.5 m depth slice, location of the significant anomaly can be recognized. The depth slice with the recorded GPR data is displayed in Fig. 17.4.c. The reflections, observed on the record, are coloured by transparent red. In the figure, a dashed red line that indicates an anomaly can be observed on all the sections independently [1].

17.2 Conclusions

Biogeophysics, the use of time-lapse geophysical methods, has recently evolved as an approach to remotely monitor biogeochemical transformations. Whether for hydrogeophysical or biogeophysical applications, the main advantage of using geophysical data is that geophysical methods provide spatially extensive information about the subsurface in a minimally invasive manner. Some recent biogeophysical research that illustrates the potential of geophysical methods for providing information about biogeochemical changes associated with natural plume evolution or remediation treatments. Understanding how hydrological and biogeochemical properties change over space and time in response to remedial treatments is hindered by our ability to monitor these processes with sufficient resolution and over field relevant scales.

Acknowledgements I would like to thank Dr. Richard M. Caprioli, Dr. Joseph H. Banoub and Dr. El Mokhtar Essassi for inviting me to the Molecular Technologies for Detection of Chemical and Biological Agents NATO SPS ASI 984915.

References

1. Kanli AI, Neducza B (2015) Electromagnetic measurements for monitoring molybdenum contamination in near-surface survey. Earth Sci Res J 19:107–111
2. Nyari Z, Kanli AI (2007) Imaging of buried 3D objects by using electrical profiling methods with GPR and 3D geoelectrical measurements. J Geophys Eng 4:83–93
3. Nyari Z, Kanli AI, Stickel J, Tillmann A (2010) The use of non-conventional Cpte data in determination of 3-D electrical resistivity distribution. J Appl Geophys 70:255–265

Chapter 18
Prospect for the Detection of the Jihadists Drug Catagon in Waste Water

Emanuela Vitale

Abstract Fenethylline, commonly known by the trademark name "captagon", "biocapton" or "fitton", is one of the most popular illegal drugs used by young communities in the Middle East. It is a theophylline derivate of amphetamine having stimulant effects similar to those of other amphetamine-type derivates. Fenethylline was used as medicament for hyperactivity disorders in children, narcolepsy and depression, but it is also used as a drug of abuse. Unlike other drugs of abuse, the clandestine synthesis of fenethylline is simple, using inexpensive laboratory instrumentation and raw materials legal to obtain. An innovative technique to detect and quantify captagon in waste water is showed, in order to help the police and the other army forces to find terrorists in a smaller district of city.

18.1 Introduction

Fenethylline, also known as amphetaminoethyltheophylline or amfetyline, commerical name Captagon, is a psychostimulant synthetic drug derived by a linkage of amphetamine and theophylline obtained via an alkyle chain. In 1960, the German industry Degussa AG introduced fenethylline as 50 mg tablet in the Federal Republic of Germany as a milder alternative to amphetamine and related compounds for use as a psychostimulant; 3 years later after its production a strict prescription status was requested. Consequently, its disposal was controlled, and the drug was used only under medical supervision [1]. The history of misuses of Captagon was born in 1980 when thugs in Bulgaria (Sofia) commenced importing few quantities of Captagon from West Germany, before starting their own illegal production on the industrial scale. In order to increase its foreign currencies the Bulgarian regime started trafficking captagon into war torn conflicts confirming that blood and money are a perfect marriage.

E. Vitale (✉)
Science of life Department, Magna Graecia University, Catanzaro, Italy
e-mail: emanuela.vit@virgilio.it

J.H. Banoub, R.M. Caprioli (eds.), *Molecular Technologies for Detection of Chemical and Biological Agents*, NATO Science for Peace and Security Series A: Chemistry and Biology, DOI 10.1007/978-94-024-1113-3_18

The production of Fenethylline drug was forbidden in 1986, however, after the fall of Berlin wall in 1989, the distribution routes and production places have survived for many years. When trafficking became “private”, captagon drug made it to the strongest Bulgarian mafia. Strangely enough, this big illegal development commerce permitted Bulgaria to shown excellent economic wealth that allowed its entry in the European Union in 2007.

Captagon have played a major role also in the “Arabian spring” as its previous uses were demonstrated by Ukrainian soldiers as stimulant during war operations in the east part their country. Since then, Captagon production has shifted to Middle East and, from 2011, in particular in Syria, where, nowadays, its production is out of control. After its production in Syria, the trafficking exportation initiate in Lebanon and the primary drug market for fenethylline are countries located in the Arabian Peninsula and also in North Africa. Recently, on 26th October 2015, prince of Saudi Arabia Abdel Mohsen Bin Walid Bin Abdulaziz, and four other people were detained in Beirut on charges of drug trafficking when two tons of captagon tables and cocaine were seized on a private jet by airport security before departure for Saudi capital of Riyadh [2] In November 2015, Agency France Press reported that the Turkish authorities had seized 2 tons of captagon on the Syrian border. The seized tablets (approximately 11 million), that have been produced in Syria, were being shipped to countries in the Arab states of Persian Gulf. It has been reported that 7.3 million captagon tablets were seized “in one raid and were set to be shipped to gulf countries by sea hidden in 1300 oil filters”, while the rest 3.6 million tablets were seized in a depot [3]. These quantities represents a big source of income both from drug party and from war and the amount is around 960 million € (1.3 billion $). In other words, if on the hand captagon has become the “rich drug” because each pill is sold 10–15 $ (between 7 and 15 €), on the other hand the proceeds bankrolls war with weapons to destroy Assad regime [4].

18.2 Chemistry and Toxicology

Fenethylline is amphetamine conjugated with theophylline via an alkyl chain [5–7] (Fig. 18.1). Its systematic IUPAC name is (*R,S*)-1,3-dimethyl-7-[2-(1-phenylpropan-2-ylamino)ethyl]purine −2,6-dione, but it can be also encountered as amphetaminoethyltheophylline and amfetyline. Fenethylline is a crystal solid with the molecular

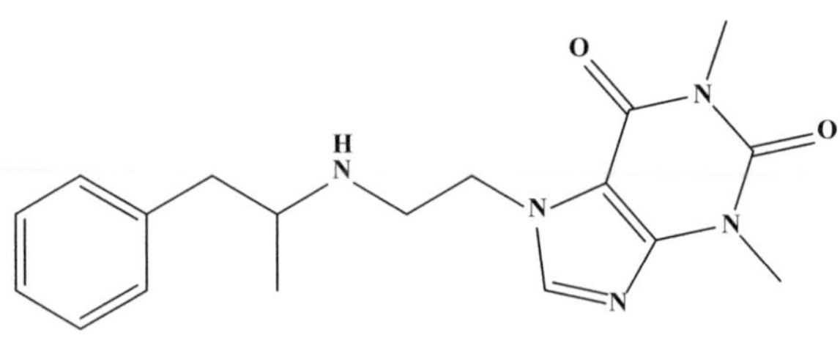

Fig. 18.1 Fenethylline structure

formula $C_{18}H_{23}N_5O_2$ (molecular weight 341.408 Da), and it is usually found as its hydrochloric salt (molecular weight 377.95 Da) [8, 9].

There are many studies on captagon effects discussed in the literature. Various metabolic studies have shown that fenethylline was prodrug for amphetamine and/or theophylline. It should be noted that theophylline is a weak stimulant agent, whereas amphetamine is an active stimulant [9–11], so the final biological effect is the result of the combined action of both these two stimulant agents [12].

Fenethylline stimulates the central nervous system increasing alertness, concentration ability and physical performance, while providing a feeling well-being and appetite suppression, due to its chemical similarity to natural neurotransmitters, like dopamine and adrenaline [13–15]. It also increases productivity and ability to concentrate. Captagon has also numerous side effects, due to the long-term use of it. The most common of them are extreme depression, lethargy, sleep deprivation (insomnia), anxiety, mood swings, feelings of anger and rage, irritably and impatience. The chronic use of captagon could lead to hemorrhagic central retina vein occlusion.

The abuse of captagon has been also correlated with acute myocardial infarction and aggressive behavior. This is the principal reason whereby jihadists take pills before their terrorist attacks. Captagon causes a state of amoral excitement leading extreme aggression and moral irresponsibility.

As regarding fenethylline metabolism, Yoshimura et al. [16] demonstrated that six metabolites are produced. More specifically amphetamine, p-hydroxyamphetamine, acetyltheophylline, 7-aminoethyltheophylline, hydroxyethyl theophylline and carboxymethyl theophylline are detected. They concluded that captagon metabolism proceeds via oxidative cleavage. The last mentioned metabolite disappears earlier and it can be detectable between 24 and 48 h, while amphetamine remains in the urine over the same time interval [16, 17].

18.3 Analytical Measurements

In order to help army and police counter-terrorism units to discover terrorists in a more restricted area of a city, a new ultra-concentration technique was developed trying to search Captagon metabolites and its traces in waste water. We have used as model the ultra-concentration technique developed for cocaine drug, which is one of the most popular and used drug in Calabria (South Italy). This technique involved liquid/liquid extraction from waste water, followed by SPME adsorption of analytes onto fibers; following detection and quatification of analytes by GC-MS. Therefore, we have taken 1 L of waste water and reduced it to 1 μl. Then, we extracted cocaine and its metabolites with SPME technique and, finally, injected in GC-MS. We obtained a LOD and LOQ, respectively, of 1.88 ng and 4.2 ng. After optimization of this method, we tested it on the two principal Captagon metabolites, amphetamine and theophylline, spiking 1 L of waste water. We obtained good results in term of linear correlation coefficient ($R^2 = 0.999$) of LOD and LOQ (Table 18.1).

Table 18.1 LOD and LOQ of Captagon drug

Drug	LOD (ng)	LOQ (ng)	N
Amphetamine	3.0	6.1	12
Theophylline	0.25	0.5	12

The next step will be to test this method on the Captagon analyte itself and on real samples.

18.4 Conclusion

Captagon is the drug of abuse most used in the Arabian Peninsula with Syria being the major manufacturer, consumer and exporter, due to its strategic position at a crossroads in the Middle East. It is likely to remain popular among the young affluent population in the Arabian Peninsula States, as well as among militants and jihadist fighters. In Europe captagon is not common, so the ability to highlight metabolites in waste water could facilitate the identification of abusers and their preventive surveillance.

References

1. Kristen G, Schaefer A, von Schlichtegroll A (1986) Fenethylline: therapeutic use, misuse, and/or abuse. Drug Alcohol Depend 17:259–271
2. Kuwait Times Saudi prince held in record Beirut airport drug test (2015). http://news.kuwaittimes.net/website/saudi-prince-held-in-record-beirut-airport-drug-bust/
3. Turkey seizes 11 million pills of "Syria war drug": reports (2015). http://timessofindia.indiatimes.com/world/middle-east/Turkey-seizes-11-million-pills-of-Syria-war-drug-reports/articleshow/49860478.cms
4. La lettera 43. Guerra in Syria. I soldati drogati di Captagon. http://www.lettera43.it/it/articoli/economia/2013/10/01/guerra-in-siria-i-soldati-drogati-di-captagon/100096/
5. Pubchem Fenethylline https://pubchem.ncbi.nlm.nih.gov/compound/fenethylline#section=2D-Structure
6. EllisonT LL, Bolger JW, Okun R (1970) The metabolic fate of 3H-fenethylline in man. Eur J Pharmacol 13:123–128
7. Nickel B, Ninch G, Peter G, von Schlichtegroll A, Tibes U (1986) Fenethylline: new results on pharmacology, metabolism and kinetics. Drug Alcohol Depend 17:235–257
8. http://lookchem.com/cas_1892-80-4_suppliers.html
9. Wikipedia Fenethylline (2015) https://it.wikipedia.org/wiki/Fenethylline
10. Kreamer T, Maurer HH (2002) Toxicokinetics of amphetamines: metabolism and toxicokinetic data of designer drugs: amphetamine, methamphetamine, and their N-alkyl derivates. Ther Drug Monit 24:277–289
11. Shulgin AT (1975) Drugs of abuse in the future. Clin Toxicol 8:405–456
12. Abdalla MA (2005) Chemical characterization of counterfeit captagon tablets seized in Jordan. Forensic Sci Int 152:185–188

13. Drahl C (2015) What you need to know about captagon, the drug of choice in war-torm Syria. Forbes http:// www.forbes.com/sites/carmendrahl /2015/11/21/what-you-need-to-know-about-captagon-the-drug-of-choice-in-war-torn-syria/#76b431f31697
14. Drug Enforcement Administration (DEA) (2003) Fenethylline and the Middle East: a brief summary DEA-03046
15. Inaba DS, Cohen WE (2000) Uppers, Doeners, all Arounders: physical and mental effects of Psycoactive drugs. CNS Pubblications Inc, Ashland, pp 102–103
16. Yoshimura H, Yoshimitsu T, Yamada H, Koga N, Oquiri K (1983) Metabolic fate of fenethylline in rat adnd man. Xenobiotica 18:929–940
17. Kikura R, Nakahara Y (1997) Hair analysis for drugs of abuse XVI. Disposition of fenethylline and its metabolite into hair and discrimination beween fenethylline use and amphetamine use by hair analysis. J Anal Toxicol 21:291–296

Chapter 19
Monitoring of Cadmium and Lead in the Kidneys of Ovine and Caprine Animals, in Albanian Area

Ilir Pecnikaj, Elona Shahu, and Ederina Ninga

Abstract Exposure of human populations to cadmium (Cd) and Lead (Pb) from food and products thereof may produce effects in organssuch as the kidneys, liver, lungs, cardiovascular, immune and reproductive systems. Since Cd and Pb have been identified as a human carcinogen, biomarkers for early detection of susceptibility to cancer are important for the public health (Fowler, Toxicol Appl Pharmacol 238:294–300, 2009). The ability to evaluate Cadmium and Lead exposure of these elements through monitoring is a step towards understanding their health effects. We have carried out a monitoring program, in 3 years, in different areas of Albania. Our target was the monitoring of the accumulation of Cd and Pb in the kidneys of ovine/caprine. The samples were chosen because their products are commonly used by the population. The determination of these two chemical elements was performed in according with EC No 1881/2006. During the entire monitoring period only two samples have been found to be noncompliant. Both samples resulted to be from south of Albania, (Gjirokastra region).

19.1 Introduction

Heavy metal pollution is known to be a global environmental problem. Many of these metals manifest high affinity for sulfur-containing ligands and strongly bind the latter. Therefore, when such heavy metals enter the cell, they interact with SH-groups, inactivate many enzymes and disturb many metabolic processes [1, 2].

I. Pecnikaj (✉)
Department of Evaluation Chemical – Toxicological and Pharmacological of Drugs, Faculty of Pharmacy, Catholic University "Our Lady of Good Counsel", Rr. D. Hoxha, Tirane, Albania
e-mail: i.pecnikaj@@unitir.edu.al

E. Shahu • E. Ninga
Food Safety and Veterinary Institute of Albania, Rr. A Moisiu 82, Tiranë, Albania

J.H. Banoub, R.M. Caprioli (eds.), *Molecular Technologies for Detection of Chemical and Biological Agents*, NATO Science for Peace and Security Series A: Chemistry and Biology, DOI 10.1007/978-94-024-1113-3_19

The idea that all heavy metals are highly toxic is misleading: this series comprises such elements as copper, zinc, manganese, iron, and other micronutrients essential for all animals and plants. Among non-nutrient heavy metals, Cd and Pb are the most widespread [3]. Most of Pb and Cd contamination results from four human economic activities [4]. Burning liquid and solid fuels, smelting and foundry works, sewage high in Pd and Cd, and soil-applied chemicals, including fertilizers [4–7]. Cd and Pb enter into the food chain mostly through products of plant origin, which often accumulate heavy metals from contaminated soils [8, 9]. When bound on the cell surface and within the cells, Cd and Pb ions interact with the functional groups of proteins, nucleic acids, polysaccharides, etc. and substitute for other metal ions already bound to these functional groups (Fig. 19.1).

Various metabolic disorders arise, and it is usually difficult to tell which disorders are primary and which are of secondary origin. In spite of the number of monographs and reviews on heavy metals published over the last years [4–10] many issues stay insufficiently elucidated, including the distribution of heavy metals, their toxic effects, and the mechanisms of plant responses to an excess of heavy metals.

During the last 50 years in Albania as well in the other countries in the region, there has been a huge industrial evaluation in which the production and the processing of heavy metal has been a big part of it. During those years no environmental requiems were asked for the industry and as well no waste treatment was performed. After several interventions of the government, such as closing the factory production and processing of lead in 1992, the prohibition of the use of petrol and the lead-based colorants in 2000 (DCM Nr.147, dated 21.03.2000 "On the quality of fuels, gasoline and diesel") and duty to control the collection and disposal of batteries in 2003 (Law No. 9010, dated 13.2.2003 "On the management of solid waste"), at the end "Law No. 9537, dated 18.5.2006 "On the management of hazardous wastes", have meant that control of them to be more rigorous and the dispersion in the environment is minimized to the maximum.

In the present study lead and cadmium were analyzed in ovine/caprine kidneys, these trace metals, are included in National Residue Control Plan. Periodic monitoring of cadmium and lead was performed in ovine and caprine samples. The samples were taken over years 2013–2015. Our goal was to discuss the major trends in Cd and Pb distribution in our country, through the control of accumulations in the ovine and caprine kidneys of the Cd and Pb.

19.2 Materials and Methods

Sampling of ovine/caprine kidney were taken by Food Safety Inspectors in accordance with EC No 333/2007. All the samples have been transported to the lab in thermo box within the same day. The samples that could not be handled and analyzed during the arrival day have been stored in deep frozen conditions. Before analytical determination the samples were cleaned fom fat and homogenized using Ultraturrax homogenizer.

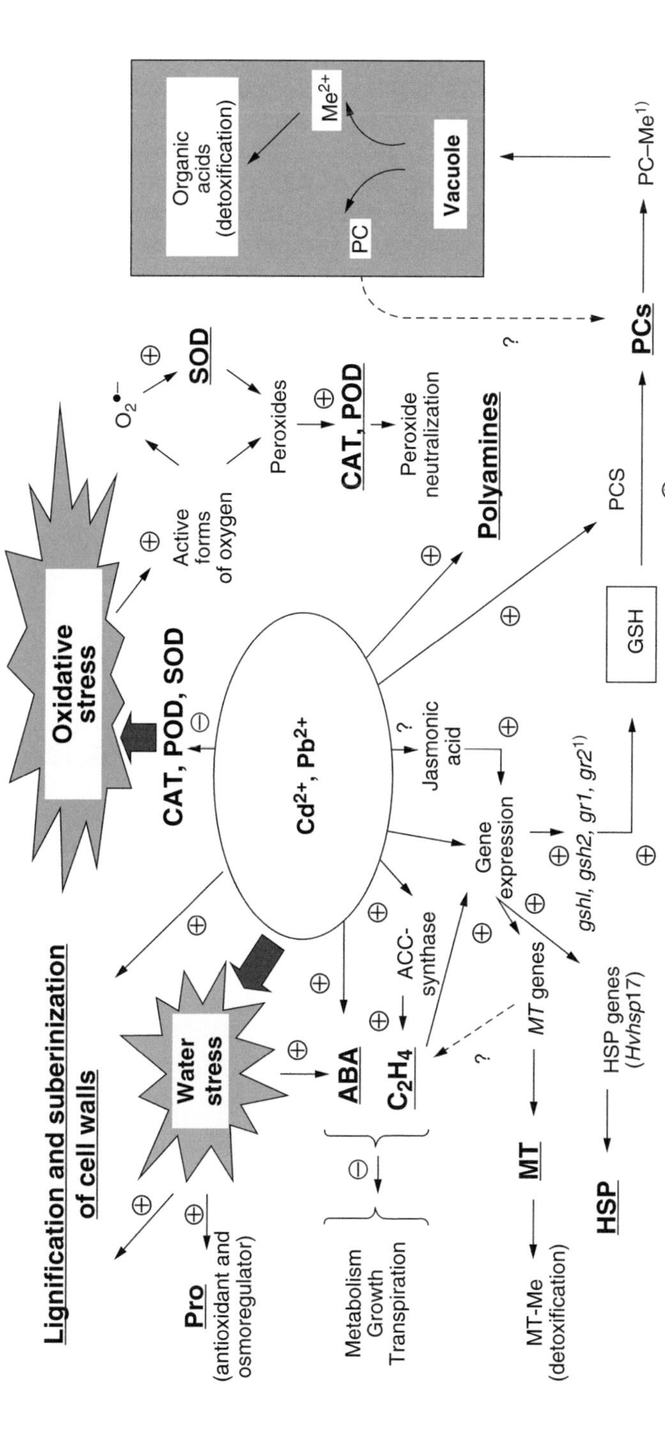

Fig. 19.1 The mechanism of cell response to the toxic Cd and Pb effects. *CAT* Catalase, *GSH* reduced glutathione, *MT* metallothioneins, *MT–Me* metallothionein-metal complex, *PCS* phytochelatin synthase, *PCs* phytochelatins, *PC–Me* phytochelatin–metal complex, *POD* peroxidase, *SOD* superoxidedismutase [11]

The trace metal determination was carried out following standardized methods for the analysis of cadmium and lead such as SSH EN 13804:2003, SSH EN 14084:2003, CE 1881/2006. Two parallels for each sample were prepared according to the following procedure. Samples (1.0 ±0.2 g) were weighed into Teflon digestion vessels; 4.0 ml nitric acid (69.5% m/V) and 1.0 ml hydrogen peroxide (30% m/V) were subsequently added. The sealed containers were placed in a microwave oven (Microwave Digestion BERGHOF) for 17 min and the samples were heated according to the temperature program. After complete digestion the sample solutions were cooled for 20 min to the room temperature. The sample solutions were diluted with ultra-pure water to the total volume of 20 ml and transferred into volumetric flasks, capped and stored at room temperature.

Atomic Absorption Spectrometer (ThermoSolaar GF 95/FS 95), equipped with graphite furnace system, was used to determine the concentration of cadmium and lead in kidney samples. The quantitative determination of Pb and Cd were performed by using a four different point calibration curve for lead and three different point calibration curve for cadmium. For both the element were used matrix modifiers.

Statistical analyses were performed by using IBM SPSS Statistics 20 Software.

19.3 Results and Discussion

Concentrations of the two elements investigated during this study are represented in the Table 19.1. Their values are expressed in mg.kg^{-1}, wet weight, as the mean value of two parallel samples. Sampling was performed in different region of Albania, during 2013-2015, in frame of National Residue Control Plan, on measures to monitor certain substances and residues thereof in live animals and animal products.

Most of the samples, both ovine and caprine, have been shown to be compliant with the Maximum Limit set in Commission Regulation 1881/2006.

Table 19.1 Concentrations of Cd and Pb (mg.kg^{-1}) in the kidneys of ovine and caprine animals

	Caprine		Ovine	
Region	Cd	Pb	Cd	Pb
Berat	0.12–0.83	0.012–0.044	–	0.02–0.141
Diber	0,745	0.094	0.42–0.86	0.007–0.25
Elbasan	–	–	0.56	0.028
Fier	–	–	0.03–0.34	0.016–0.275
Gjirokaster	0.18–0,824	0.006–0.54	0,157–1.3	0.02–0.074
Korce	0.320–0,512	0.0082–0.070	0.086–0.49	0.0035–0.103
Kukes	–	–	0.72	0.059
Lezhe	0.034–0.089	0.025–0.045	0.08–0.523	0.08–0.41
Shkoder	0,048–0.076	0.005–0.057	–	–
Tirane	0.09–0,653	0.009–0.041	0.146–0.48	0.011–0.105
Vlore	0,083–0.483	0.1–0.121	0.15–0.358	0.014–0.043

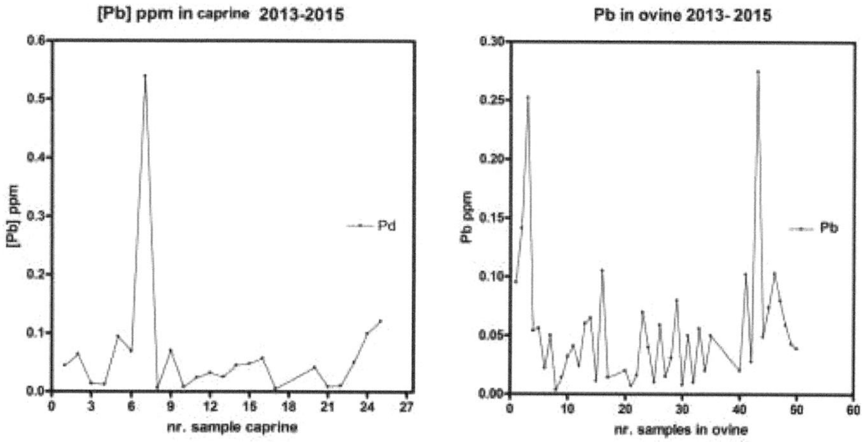

Fig. 19.2 Pb in ppm in caprine and ovine in 3 years (2013, 2014, 2015)

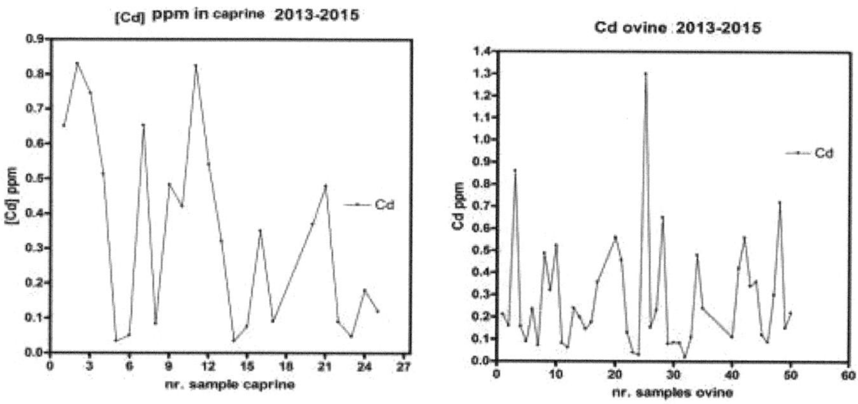

Fig. 19.3 Cd in ppm in caprine and ovine in 3 years (2013, 2014, 2015)

During all the monitoring period only one of caprine samples exceeded the maximum limit of 0.5 mg.kg^{-1} lead in wet weight (the amount of lead in this sample was 0.54 mg.kg^{-1}) (Fig. 19.2).

From the data represented in the Table 19.1, cadmium seem to be more present in all the samples although only one all the samples exceeded the Maximum Limit. It should be mentioned that the ML for this element is relatively high (1 mg.kg^{-1}). The sample which was noncompliant with the EC regulation requirement was from Gjirokastra region and the level of Cd on it was found to be 1.3 mg. Kg^{-1}, Fig. 19.3.

All the date was evaluated by statistic programs using Principal Component Analysis (PCA), Fig. 19.4. As it can be seen from the Fig. 19.4 the values of Cd and Pb in ovine is quite stable. Almost all results are in the same area. Only three of them are outside of the area, where one of these three is over of limits established

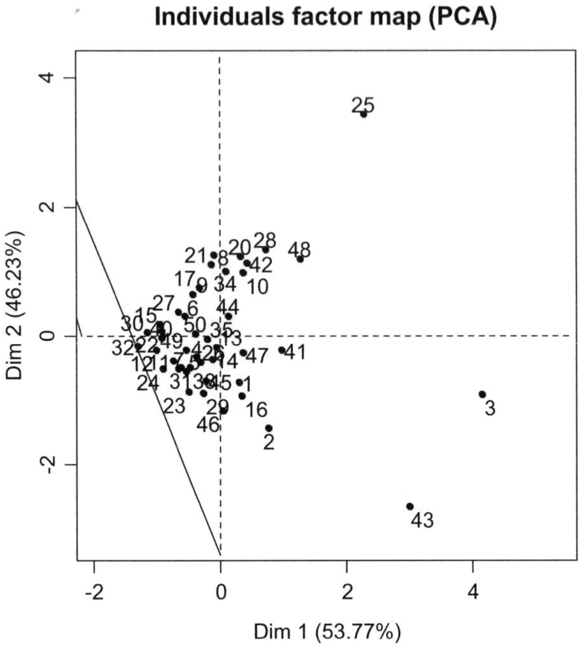

Fig. 19.4 PCA for samples of kidneys ovine

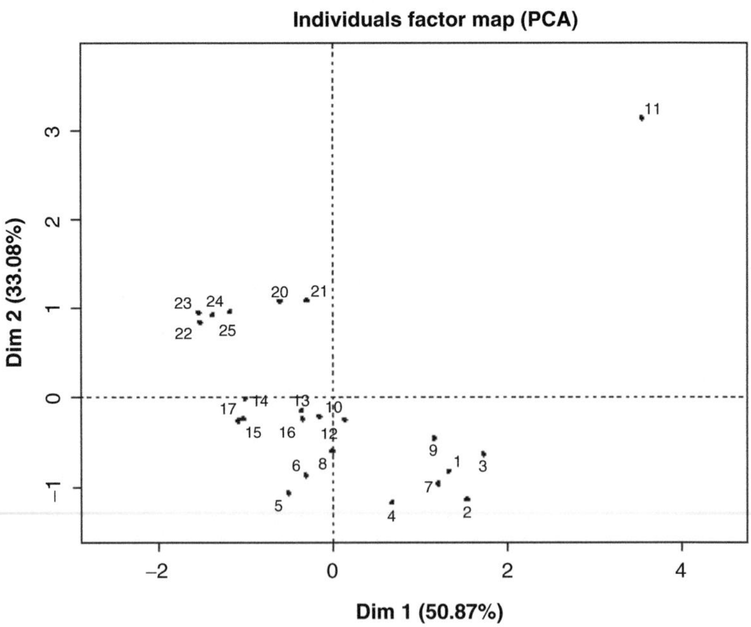

Fig. 19.5 PCA for samples of kidneys caprine

by Law (example 25). While for the caprine, the trend they are decreased (Fig. 19.5). All results are separated almost in years, where their trend is decreasing. Only one of them is over of limits established by Law (example 11).

19.4 Conclusions

During 3 years of monitoring only two samples exceed the maximum limit established by EC No 1881/2006. Both cases were founded to be in south of Albania. One ovine sample for cadmium and the other one was caprine for lead.

These cases were in the same year (2014), but were only two isolated cases. By repeated control in that area, we have not found any more similar cases.

Statistical data analysis have shown that the trend of the rate for Cd and Pb in these years seems to be in marginally decline, regards caprine, and more stable although in lower values in ovine.

References

1. Fowler BA (2009) Monitoring of human populations for early markers of cadmium toxicity: a review. Toxicol Appl Pharmacol 238:294–300
2. EN 13804:2003 “Foodstuffs – Determination of trace elements-Performance criteria, general consideration and sample preparation; EN 14084:2003 “Foodstuffs – Determination of trace elements – Determination of lead, cadmium, zinc, copper and iron by AAS after microwave digestion”. Polish Committee for Standardization Oublications
3. Commission Regulations (EC) No 1881/2006 on setting maximum levels for certain contaminants in foodstuffs. Regulation as amended by the European Parliment and the Council. OJL 284:1–4
4. Sanita di Toppi L, Gabbrielli R (1999) Response to cadmium in higher plants. Environ Exp Bot 41:105–130
5. Baker AJM (1987) Metal tolerance. New Phytol 106:93–111
6. Antosiewicz DM (1992) Adaptation of plants to an environment polluted with heavy metals. Acta Soc Bot Polon 61:281–299
7. Salt DE, Blaylock M, Kumar NPBA, Dushenkov V, Ensley BD, Chet I, Raskin I (1995) Phytoremediation: a novel strategy for the removal of toxic metals from the environment using plants. Biotechnology 13:468–474
8. Alekseeva-Popova, N.V. (1991) Toxic effect of lead on higher plants, Ustoichivost’ k tyazhelymmetallamdikorastushchikhvidov (Tolerance of Plant Species Grown in the Wild to Heavy Metals), Alekseeva-Popova NV, (ed) Leningrad, Lenuprizdat, pp 92–100
9. Bingham, F.T. (1986) Metal ions in biological systems. In: Singel H and Singel A (eds) Concepts on metal ion toxicity. New York, Marcel Dekker. Translated under the title Nekotoryevoprosytoksichnostiionovmetallov, Moscow: Mir, 1993
10. Ernst, W.H.O. (1999) Effects of heavy metals in plants at the cellular and organismic level. In: Schuurmann G and Markert B (eds) Ecotoxicology. Ecological fundamentals, chemical exposure and biological effects. Heidelberg, Wiley and Sons, pp 587–620
11. Seregin IV, Ivanov VB (2001) Physiological aspects of cadmium and lead toxic effects on higher plants. Russ J Plant Physiol 48:523–544

Chapter 20
Quantitative Analysis of Dysregulated Proteome in Methylmalonic Acidemia

Costanzo Michele, Caterino Marianna, Minopoli Giuseppina, Santorelli Lucia, P. Venditti Charles, and Ruoppolo Margherita

Abstract Methylmalonic Acidemias (MMNAs) are heterogenoeus severe autosomal recessive inborn errors of metabolism, MMNAs are caused by deficency of vitamin B12-dependent methylmalonyl-CoA mutase enzyme (MUT). Therefore, we used MUT silencing and a proteomic quantitative approach to define the pathways directly perturbed by this metabolic disease.

20.1 Background-AIM

Methylmalonic Acidemias (MMAs) are heterogeneous and severe autosomal recessive inborn errors of metabolism (IEM) belonging to the rank of organic acidemias (OAs). MMA is caused by deficiency of the vitamin B12-dependent methylmalonyl – CoA mutase enzyme (MUT), which converts L-methylmalonyl – CoA into

C. Michele (✉) • M. Giuseppina • S. Lucia
CEINGE – Biotecnologie Avanzate scarl, Naples, Italy

Department of Molecular Medicine and Biotechnologies, University of Naples "Federico II", Naples, Italy
e-mail: michele.costanzo@hotmail.it

C. Marianna
CEINGE – Biotecnologie Avanzate scarl, Naples, Italy

Discimus RFC, Naples, Italy

P. Venditti Charles
Organic Acid Research Section, Genetics and Molecular Biology Branch, National Institutes of Health (NIH), Bethesda, MD 20892, USA

R. Margherita
Discimus RFC, Naples, Italy

CEINGE – Biotecnologie Avanzate scarl, Naples, Italy

Department of Molecular Medicine and Biotechnologies, University of Naples "Federico II", Naples, Italy

J.H. Banoub, R.M. Caprioli (eds.), *Molecular Technologies for Detection of Chemical and Biological Agents*, NATO Science for Peace and Security Series A: Chemistry and Biology, DOI 10.1007/978-94-024-1113-3_20

succinyl – CoA, a Krebs cycle intermediate. This enzymatic block is responsible for the accumulation of high levels of methylmalonic acid in body fluids.

The disorders are caused by deficiency of MUT apoenzyme or defective metabolism of the enzymatic cofactor, 5′-deoxyadenosylcobalamin (Ado-Cbl). When the defect is caused by mutations in the sequence of MUT gene, the pathology is defined isolated Methylmalonic Acidemia. Mutations in one or more of the genes codifying for cobalamin biosynthesis enzymes lead to *cbl*-related MMA. Two pathologic phenotypes related to MUT apoenzyme defect are known: mut^0 patients show a totally absent enzymatic activity, while mut^- patients only have a partial deficiency and are responsive to vitamin B12 treatment. For mut^0 phenotype patients no specific therapy exists but they can partially recover the metabolic instability through liver or combined liver/kidney transplantations.

The hepatic proteome of mut^0-MMA patients has been previously studied in comparison with the proteome of healthy individuals. Differential proteomics experiments have shown the dysregulation of proteins involved in energetic metabolism, gluconeogenesis and Krebs cycle anaplerosis. The observed dysregulations could be due to a metabolic adaptation of the patients and not directly due to mutase deficiency. To test this hypotesis, an *in vitro* silencing of MUT gene was performed using the RNA-interference strategy and the protein expression profile analysed by a quantitative proteomics approach.

20.2 Methods

The expression of MUT gene was reduced in neuronal SH-SY5Y cell line using a specific siRNA against MUT mRNA. Three replicates were prepared for each silencing condition: 0, 24 and 48 h. In parallel, triplicates were also prepared for each control condition at the same times. The control SH-SY5Y cell line was transfected by using a random sequence (scrambled), which has no significant homology to any sequences in the human genome.

Cellular lysates of three replicates for each condition were gathered and fractionated on a preparative 10% gel by 1D–SDS-PAGE. The gel lanes (relative to scrambled-48 h and siMUT-48 h samples) were cut to create 3 mm slices and each slice was excised from gel. Peptide mixtures were obtained by in situ hydrolysis with trypsin and loaded on a LTQ-Orbitap XL spectrometer.

The proteomic analysis was performed using Proteome Discoverer™ platform for protein identifications. For the quantitative analysis, the abundances of each protein present in the different proteomes were estimated by means of the Spectral Counting (SpC) approach. The normalized spectral abundance factor (NSAF) generated from LC-MS/MS datasets was used.

A statistical parameter, the Rsc (Ratio of Spectral Counts), was used to identify the differentially expressed proteins: Rsc values higher than +2,0 and lower than −2,0 were accepted.

20.3 Results

To evaluate if these dysregulations could be due to a metabolic adaptation of the patients, the expression of MUT protein was decresead in order to study in vitro alterations exclusively due to mutase absence. From Spectral Counts results, 198 up-regulated and 144 down-regulated proteins were identified. Dysregulated proteins were grouped according to Gene Ontology (GO) and Kyoto Encyclopedia of Genes and Genomes (KEGG) databases. Differentially expressed proteins result involved in metabolic processes, cell cycle, apoptosis processes and cellular response to stress. The proteomic dataset suggests that the high levels of methylmalonic acid could determine cell death and, consequently, clinic manifestations affecting the nervous system in MMA patients.

20.4 Conclusions

MUT silencing and the proteomic quantitative approach allow us to define the pathways directly perturbed by the metabolic disease. They represent targets to better understand its molecular basis and for the design of therapies to alleviate clinical manifestations of MMA.

Chapter 21
Sensors for Electrochemical Determination of Various Oxidizable Analytes with a Graphene Oxide (GO) and/or Multi Walled Carbon Nanotubes (MWCNTs) Modified Glassy Carbon Electrode

Vivek Vishal Sharma and Domenica Tonelli

Abstract Glassy carbon electrodes were modified by drop casting carbon nanomaterials, graphene oxide, GO, and multi-walled carbon nanotubes, MWCNTs, alone, mixed together (Composite) or in the form of bi-layers. The reduction of GO was carried out by means of a green approach consisting of 10 voltammetric cycles (CV) from 0 to –1.5 V, vs SCE at the scan rate of 0.05 V s^{-1}, in acetate buffer 0.1 M at pH 4.5. The modified electrode was characterised by SEM, ATR-IR and UV-Vis analysis. Various compounds, like catechol and paracetamol, were investigated by cyclic voltammetry to find out which one responds most effectively to the modified electrode. Chronoamperometry was emplyed to prepare an analytical method for the determination of various analytes like L-cysteine, hydrazine.

21.1 Introduction

In the field of chemically modified electrodes (CMEs), multiwalled carbon nanotubes (MWCNTs) have been widely used due to their unique structural, electronic and physical properties with one of the most important characteristics being their reported ability to promote electron-transfer processes [1]. Very recently graphene (GR), a 2D carbon material comprised of a single sheet of hexagonally packed carbon atoms, has attracted tremendous attention because of its unique nanostructure and extraordinary electro-catalytic properties such as high surface area, excellent conductivity, and high mechanical strength. Based on its properties, GR is considered as an ideal electrode material for electrochemical and biosensing [2]. However, synthesis of graphene is expensive, hence we use graphene oxide which can be

V.V. Sharma (✉) • D. Tonelli
Dipartimento di Chimica Industriale "Toso Montanari", Università di Bologna, Viale del Risorgimento 4, 40136 Bologna, Italy
e-mail: vivekvishal.sharma2@unibo.it

J.H. Banoub, R.M. Caprioli (eds.), *Molecular Technologies for Detection of Chemical and Biological Agents*, NATO Science for Peace and Security Series A: Chemistry and Biology, DOI 10.1007/978-94-024-1113-3_21

easily reduced electrochemically to obtain a material almost equivalent to graphene for use as a modifier.

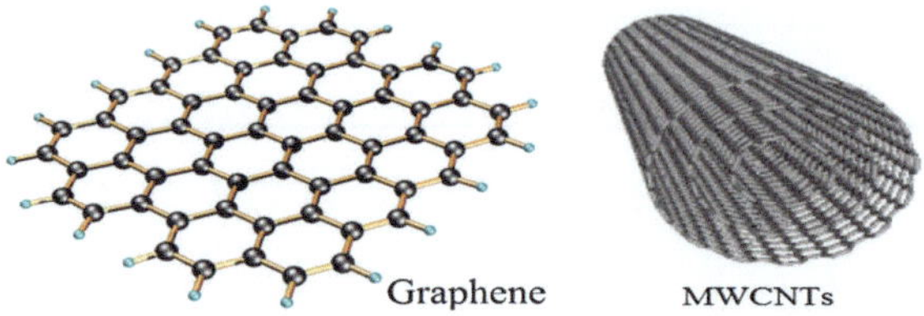

21.2 Preparation of Chemically Modified Electrode (CME)

Prior to surface modification, the glassy carbon surface was gently polished on a fine (4000 grit) wet SiC sandpaper until a mirror finish was obtained. Then it was thoroughly rinsed with DD water. Aqueous solution of graphene oxide (GO) in varying concentrations was prepared from the stock solution of 4 mg/ml. To prepare the MWCNTs solution, 2 mg Of MWCNTs was taken in a vial and to it 2 ml of N,N′-dimethyl formamide (DMF) was added. The vial was placed under ultrasonic treatment for 20 min. From the above 1 mg/ml stock solution of MWCNT in DMF, varying concentrations of the MWCNTs dispersed in DMF was prepared.

For modification of the GC surface, 10 μL of nanocarbon material in the suspension was drop-casted on bare GC and left to dry in an oven at 60 °C for 15 min. After drying, the CME was electroreduced in pH 4 phosphate buffer solution (PBS) to convert the graphene oxide to reduced graphene oxide. In altogether five different configurations of the CME was prepared (Fig. 21.1):

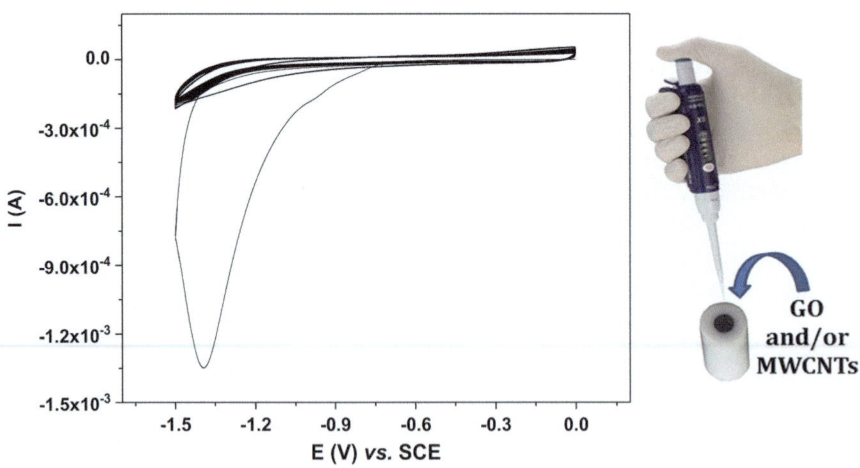

Fig. 21.1 Electroreduction (ER) of the Chemically Modified electrode (CME) carried out in pH 4 Phosphate buffer solution (PBS) to get electroreduced GO (rGO)

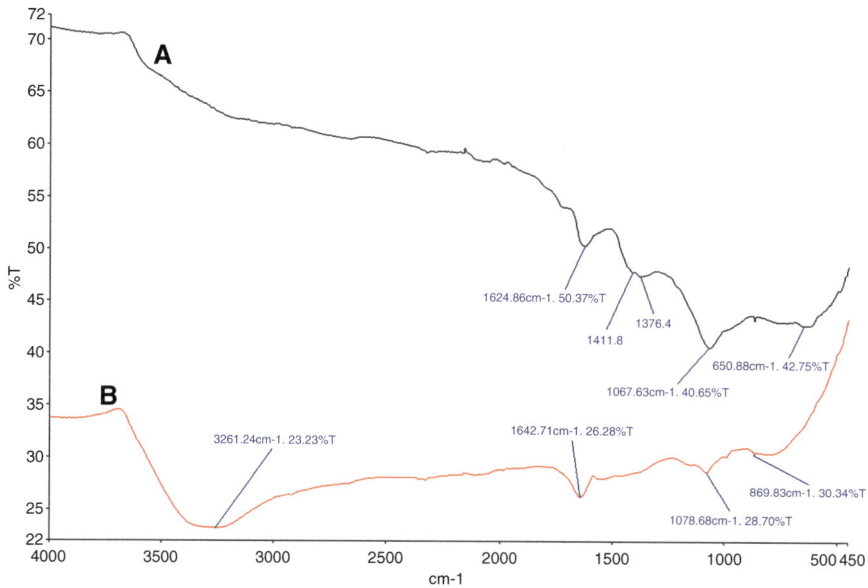

Fig. 21.2 ATR-IR of the Chemically Modified electrode (CME) for the composite configuration (A) before and (B) after electroreduction, carried out in pH 4 PBS

1. Only rGO
2. Only MWCNTs
3. Bilayer MWCNTs/rGO
4. Bilayer rGO/MWCNTs
5. Composite (rGO + MWCNTs)

21.3 Characterisation of CME

Various configurations of the CME was characterised using Attenuated total reflectance infrared spectroscopy (ATR-IR) and scanning electron microscopy (SEM). As shown in Fig. 21.2, for the composite configuration of the CME, after electoreduction in pH 4 PBS, there is a reduction in the peaks for the C-O and C = O stretching as is evident on comparing the spectrum A with that of B. Similar reduction in the peaks of C-O and C = O was observed in the spectra for other configurations [3, 4]. SEM images were recorded for the various configurations of the CME.

21.4 Cyclic Voltammetry Studies

Typical voltammetric responses for various analytes like L-cysteine, paracetamol and catechol towards the CME in 0.1 M PBS solution, pH 4 is shown in Fig. 21.3. Herein, we can observe two couples of symmetrical peaks in case of catechol, an

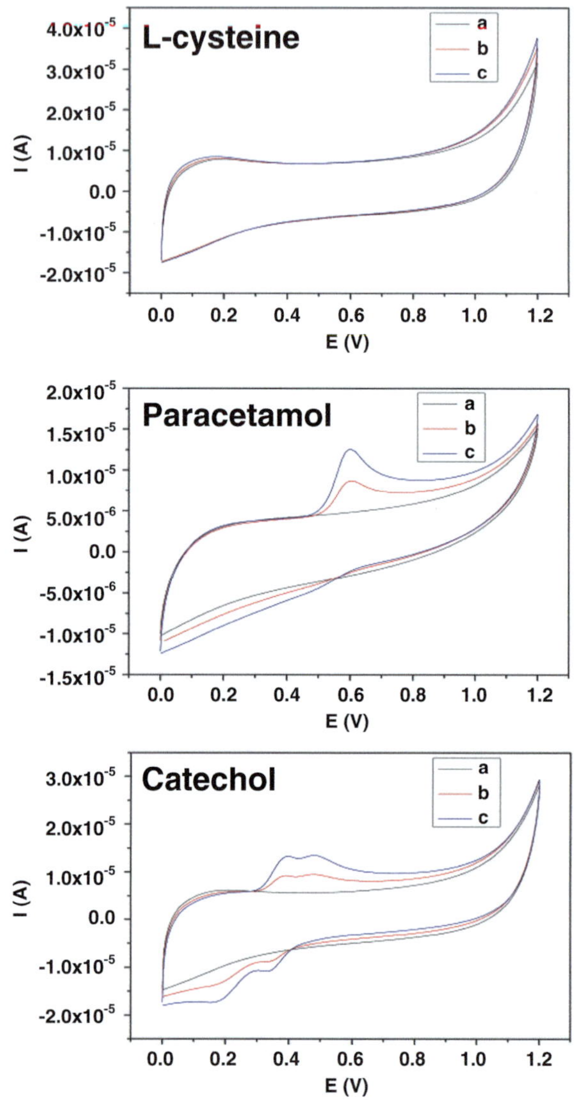

Fig. 21.3 CV of (**a**) Blank Buffer solution (**b**) 0.1 mM and (**c**) 0.2 mM of various analytes in 0.1 M pH 4 PBS, Scan rate = 0.05 V/s, CME prepared by drop-casting of 10 μL a solution containing GO (0.2 mg/ml) and MWCNTs (0.2 mg/ml) dispersed in DMF

asymmetrical anodic peak in case of paracetamol and virtually no response towards L-cysteine. Further experiments have been carried out to find out the relationship between the peak current vs. scan rate and also for the determination of the heterogeneous electron transfer constant. Additional experiments were performed at pH 4, 7 and 9 to ascertain the optimum conditions for the determination of each analyte [5].

21.5 Chronoamperometric Studies

Based on the results obtained from cyclic voltammetric studies, chronoamperometric measurements were performed for various analytes under the optimized conditions to ascertain the limit of detection (LOD), limit of linearity (LOL) and range of detection for each of the analytes. Preliminary results showed that it was possible to detect L-cysteine in the range of 5 μM to 3 mM with the limit of detection as 2.5 μM. Similar results were obtained for other analytes (Fig. 21.4).

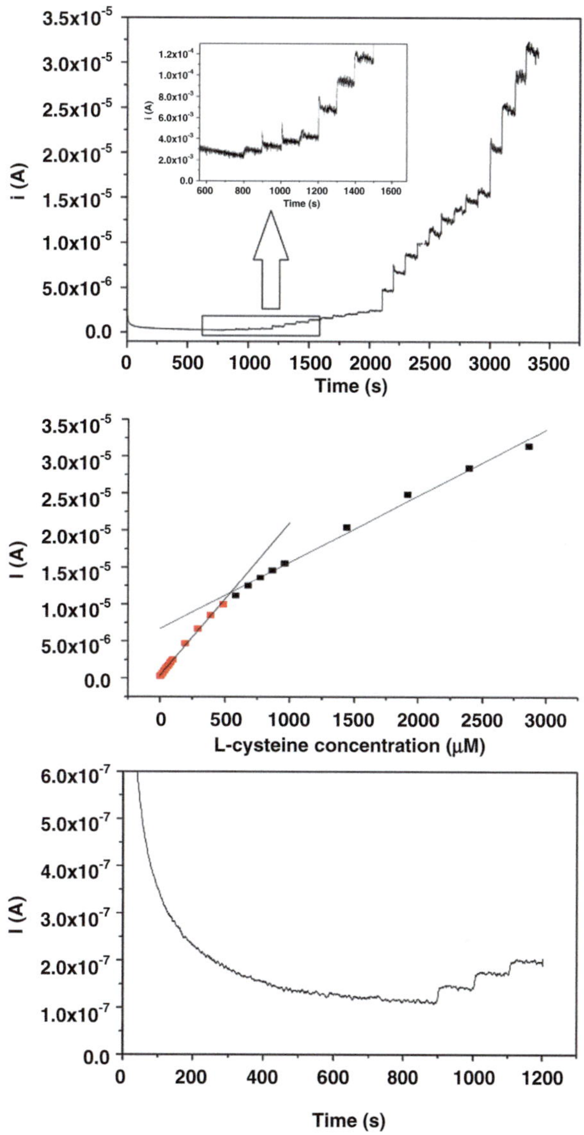

Fig. 21.4 (a)– Chronoamperometric response for various additions (2.5 μM to 3 mM) of L-cysteine to 20 ml of 0.1 M pH 4 PBS, biased at a potential of 0.65 V vs. SCE. (**b**) – Calibration plot of anodic current vs. concentration of L-cysteine obtained from the CA response above. (**c**) – CA response of 100 μM additions of hydrazine to 20 ml of 0.1 M pH 7 PBS, biased at a potential of 0.65 V *vs*. SCE

References

1. Elrouby M (2013) Electrochemical applications of carbon nanotube. J Nano Adv Mat 1(1):23–38
2. Guo HL, Wang XF, Qian QY, Wang FB, Xia XH (2009) A green approach to the synthesis of graphene nanosheets. ACS Nano 3(9):2653–2659
3. Li D, Muller MB, Gilje S, Kaner RB, Wallace GG (2008) Processable aqueous dispersions of graphene nanosheets. Nat Nanotechnol 3(2008):101–105
4. Zhou M, Zhai Y, Dong S (2009) Anal Chem 81(14):5603–5613
5. Alshahrani LA, Li X, Luo H, Yang L, Wang M, Yan S, Liu P, Yang Y, Li Q (2014) The simultaneous electrochemical detection of catechol and hydroquinone with [cu(Sal-β-Ala)(3,5-DMPz)2] /SWCNTs /GCE. Sensors 14:22274–22284

MIX
Papier aus verantwortungsvollen Quellen
Paper from responsible sources
FSC® C105338

If you have any concerns about our products,
you can contact us on
ProductSafety@springernature.com

In case Publisher is established outside the EU,
the EU authorized representative is:
Springer Nature Customer Service Center GmbH
Europaplatz 3, 69115 Heidelberg, Germany

Printed by Libri Plureos GmbH
in Hamburg, Germany